CLIMATE CHANGE *and* FORESTS

CLIMATE CHANGE *and* FORESTS

Emerging Policy and Market Opportunities

Charlotte Streck, Robert O'Sullivan,
Toby Janson-Smith and Richard Tarasofsky
Editors

Concept Publishing Company, New Delhi-110 059

CLIMATE CHANGE and FORESTS
Emerging Policy and Market Opportunities

Edited by
Charlotte Streck, Robert O'Sullivan
Toby Janson-Smith and *Richard Tarasofsky*

Published in agreement with The Brookings Institution Press, Washington
for publication and sale only in the Indian Subcontinent
(India, Pakistan, Bangladesh, Nepal, Maldives, Bhutan & Sri Lanka)

First Published in India 2009

ISBN 13: 978-81-8069-653-4
ISBN 10: 81-8069-653-7

Published and Printed by

Concept
Since 1974

Concept Publishing Company
A/15-16, Commercial Block, Mohan Garden
New Delhi-110059 (India)
Phones : 25351460, 25351794, *Fax* : 091-11-25357109

Editorial Office:
H-13, Bali Nagar, New Delhi-110 015, India.

Email : publishing@conceptpub.com,
Website: www.conceptpub.com

Contents

PART FOUR
OUTLOOK: AVOIDED DEFORESTATION AND THE POST-KYOTO AGENDA

PART FIVE
NATIONAL SYSTEMS AND VOLUNTARY CARBON OFFSETS

Foreword

DAVID FREESTONE

I am delighted and honored to have been invited to write a foreword for this excellent and timely work. It is particularly timely because in December 2007, in a historic decision, the parties to the 1992 UN Framework Convention on Climate Change (UNFCCC), meeting in Bali, Indonesia, decided to include the issue of avoided deforestation—or "reducing emissions from deforestation and forest degradation" (REDD), as it is known in UNFCCC argot—in the Bali Action Plan. This plan is the so-called road map for negotiations that aim to develop by 2009 a legal instrument to replace the 1997 Kyoto Protocol when it expires in 2012. The Kyoto Protocol to the UNFCCC requires its developed country parties to make reductions in their emissions of greenhouse gases by an average of some 5.2 percent from 1990 levels throughout its five-year commitment period, 2008–2012. The Bali road map is of particular importance in that the UNFCCC parties agree to consider "measurable, reportable and verifiable nationally appropriate mitigation actions" for *all* parties, including developing country parties, although developed country parties also agree to consider "commitments, . . . including quantified emission limitation and reduction objectives." For present purposes, even more significant is the provision of the action plan that commits the parties to consider "policy approaches and positive incentives on issues relating to reducing emissions from deforestation and forest degradation in developing countries; and the role of conservation, sustainable management of forests and enhancement of forest carbon stocks in developing countries."

This decision represents a major breakthrough in the UNFCCC negotiations. In addressing the issues of climate change, it is worth recalling that the text of the 1992 convention puts the important role of sinks, such as forests, in absorbing carbon on a par with the need for the reduction of greenhouse gas (GHG) emissions. For example, article 4(1)(d) requires the parties to "address stabilization of climate by sources and use of sinks," and article 3(2) requires that parties "should take precautionary measures . . . [and] lack of scientific certainty should not be used as a reason for postponing such measures . . . which [should] . . . cover all relevant sources, sinks and reservoirs [of GHGs]." Nevertheless, the issue of sinks was highly controversial in the negotiation of the Kyoto Protocol and thereafter in the protracted process leading to the development of the "guidelines, policies and rules" for the implementation of the protocol contained in the now-famous Marrakech Accords—agreed at the seventh session of the Conference of the Parties (COP) to the UNFCCC in Marrakech in November 2001. The Marrakech Accords set out the basic regulatory framework for the protocol and its so-called flexibility mechanisms, including the Clean Development Mechanism (CDM). Under the CDM, industrialized countries can invest in projects in developing countries that reduce emissions of GHGs and then use the "certified" emission reductions produced by those projects toward their own reduction targets. Despite the fact that agriculture, forests, and other land uses (AFOLU) account for some 20 percent of the total amount of carbon that exists on the planet, it was decided at Marrakech that only reforestation and afforestation projects would be eligible for consideration under the CDM—and indeed to date only one such project has been approved by the CDM Executive Board. In the same vein, the European Emissions Trading Scheme (ETS), which began operations among the EU countries in 2005, does not count sinks at all.

There are a number of reasons for this—some political but others methodological. However, as we move into the beginning of the Kyoto commitment period and attention is focused on the post-2012 regime, it is important that sinks, and particularly forest sinks, are firmly back on the agenda. In 2005, at the eleventh session of the COP, Papua New Guinea and Costa Rica—with support from a number of important forested countries—put forward a formal proposal that considers the crediting of benefits from avoiding further deforestation. This is a vitally important issue. The 2006 *Stern Review on the Economics of Climate Change,* commissioned from a team led by economist Sir Nicholas Stern by the then U.K. chancellor of the exchequer, Gordon Brown, identified avoided deforestation as the cheapest option for mitigating increases in emissions of greenhouse gases.[1] It is not a free option, nor is it even particularly cheap. Sophisticated mon-

1. See www.hm-treasury.gov.uk/independent_reviews/stern_review_economics_climate_change/sternreview_summary.cfm.

itoring mechanisms, often involving satellite surveillance, need to be put into place and effective compensation systems devised to encourage governments and their nationals to stop cutting down trees.

Nevertheless, the fourth assessment of the UN Intergovernmental Panel on Climate Change (IPCC) has shown that the current situation and prognosis for dangerous climate change is already far worse than had previously been envisaged. The Stern Review highlights the fact that if serious action is not taken within the next fifteen to twenty years, then the costs of coping with climate change could be in excess of 20 percent of total global income annually. The World Bank has estimated total global income currently at some U.S.$35 trillion a year, rising by 2050 to perhaps U.S.$350 trillion, when the global population is estimated to be some 9 billion, with major relocations of population in developing countries and huge demand for new infrastructure and power sources.[2] Twenty percent of U.S.$350 trillion in 2050 is some U.S.$70 trillion a year. This is an enormous sum of money that puts the current need for serious investments in new technology and innovative approaches to tackling both mitigation and adaptation into proper perspective.

It is also clear that the public sector by itself is unlikely to be able to mobilize the scale of resources necessary. The success of the emerging carbon market has shown how the private sector can respond. The World Bank annual *State and Trends of the Carbon Market* has shown huge growth in carbon trading.[3] In 2006 total trades topped U.S.$30 billion. Although the market is dominated by the European ETS, with some U.S.$24 billion in trades, the CDM, mobilizing resources for developing countries, reached nearly U.S.$5 billion. The system is a long way from perfect, but it does demonstrate vividly the scale of resources that can be mobilized, mostly from the private sector. By contrast, Charlotte Streck and colleagues pointw out that the public sector response—the Global Environment Facility (which includes five other focal areas besides climate change within its mandate)—mobilized U.S.$3.2 billion in 2006 for a four-year replenishment.

Forests are no quick fix for these issues. There is no quick fix. This scale of challenge will require a wide range of approaches and technologies. However, agriculture, forests, and other land uses are an important part of the mix of mitigation efforts that will be necessary to achieve the reduction in carbon emissions that we will need. Avoided deforestation, or REDD, in particular carries with it benefits that are not simply carbon-centric. Loss of forests worldwide has put strains on the lifestyles of local communities and indigenous peoples, the conservation of biological diversity, and a wide range of ecosystem services. In particular, loss of forests

2. K. Hamilton and I. Johnson, *Responsible Growth to 2050* (Washington: World Bank, 2004).

3. K. Capoor and P. Ambrosi, *State and Trends of the Carbon Market 2007* (Washington: World Bank, 2007).

has caused problems of water conservation and drainage, of air quality (through haze and dust), and of erosion and loss of topsoil. It has caused land slips and siltation problems, which in turn affect the viability of other marginal land.

It is against this background that the contributors to this book reexamine some of the key issues necessary for an informed view of the realistic role that forests (and AFOLU) should play in a post-Kyoto regime and the role that carbon trading might play in support of this agenda.

The book is divided into five parts. After an excellent introductory section that sets out the basic issues of the forestry and climate change agenda (chapter 1, by the editors) and the use of market mechanisms for forest conservation (Portela, Wendland, and Pennypacker), the contributors to part two look at the basic issues raised by forests under the 1992 UNFCCC and the Kyoto Protocol. They cover the history and background of the forest issue in the UNFCCC process (Trines) and why it is so controversial (Ebeling). Fehse considers the important synergies between the Rio Conventions, and Scholz and Jung look at the fate of forest projects under the Kyoto mechanisms—not an encouraging story. Schlamadinger and colleagues then look at the related issue of bioenergy and at the way bioenergy projects have been considered, especially in relation to their status as "renewables."

Part three is a thoughtful section on methodological lessons learned. Its authors offer "state of the art" observations about the issues faced by forest carbon projects. After a detailed discussion of the way forest projects have been considered in the CDM process (Locatelli, Pedroni, and Salinas), chapters are devoted to permanence (Lecocq and Couture), the quantification of sequestration (Pearson, Walker, and Brown), experiences with attempting to do this in Chile (Hervé and Claro), and the contractual aspects of such projects (Miller, Wilder, and Knight).

The core of the book is part four, in which contributors look at the post-Kyoto agenda and the ways in which avoided deforestation might be operationalized in a new post-Kyoto regime. Here the book really pushes the agenda importantly. After a look at ways of creating incentives to reduce emissions from deforestation (O'Sullivan), there is a highly technical but important and accessible assessment of accounting for such activities (Mollicone and others). After consideration of Latin American perspectives on deforestation (Estrada Porrua and García-Guerrero) and compensation possibilities for deforestation (Schwartzman and Moutinho), the section ends with an excellent discussion of ways in which national and project approaches to avoided deforestation could be combined in a complementary way—the "nested approach" (Streck and others).

Part five looks at the experience of national systems in Australasia (Gould, Miller, and Wilder) and North America (Kelly and others) in providing incentives for forestry projects at various levels of government. Authors from the United States (Hamilton, Bayon, and Hawn) and Australia (Meizlish and Brand) discuss the methodologies adopted for schemes offered in the important voluntary sector of the carbon market. Throughout the book, contributors offer useful case studies describing lessons learned, which reinforce some of the more theoretical pieces.

In 2005 Charlotte Streck and I edited a volume based on a series of expert workshops convened by the World Bank.[4] The book was designed to share more widely the considerable experience and expertise that the World Bank and a wide range of partners had developed in the field of carbon finance in the first few years of working on the pioneering Prototype Carbon Fund (PCF). Established by the bank in 2000, well in advance of the coming into force of the Kyoto Protocol, with some U.S.$180 million in contributions from public and private sector participants, the PCF had a strong "learning by doing" agenda. Many of its operations and the instruments it developed were truly first of a kind.

In the few years since then the growth in the carbon market has been astonishing. From 2000, when the PCF was virtually all that was available, the market has grown to be worth U.S.$10 billion in 2005 and more than U.S.$30 billion in 2006. What is equally impressive, albeit understandable with such a high level of investment, is that in this relatively short time the carbon market and its participants have developed a very high degree of sophistication. This book draws extensively upon the practical expertise developed in the fashioning of carbon projects in other sectors and the general methodological lessons learned in presenting projects for review by project sponsors as well as under the Kyoto mechanisms. It also reflects the considerable research and thought that has already gone into trying to make forest projects meet CDM criteria and into pushing the envelope beyond the artificial straitjacket put upon such projects by their restriction to reforestation and afforestation activities.

The editors are to be commended for having assembled an extremely impressive and highly qualified group of authors at the cutting edge of their subject. The result is a book that is valuable for the insights if offers to those interested or involved in current forest sector projects. However, it is more than this—it draws on the widest spread of thinking to offer ways forward in the area of avoided deforestation, or REDD—the sector regarded as the "lowest cost option" in the Stern Review. Now that the Bali Action Plan has recognized the importance of reducing emissions from deforestation and forest degradation in developing countries, it is clear that any new post-Kyoto regime must consider forests and land use much more centrally in its approach. This book will be invaluable in that process.

4. D. Freestone and C. Streck, eds., *Legal Aspects of Implementing the Kyoto Protocol Mechanisms: Making Kyoto Work* (Oxford University Press, 2005).

PART ONE

Introduction

1

Climate Change and Forestry: An Introduction

CHARLOTTE STRECK, ROBERT O'SULLIVAN,
TOBY JANSON-SMITH, AND RICHARD TARASOFSKY

Climate change is one of the most significant global challenges of our time, and addressing it requires the urgent formulation of comprehensive and effective policy responses. A changing climate affects nearly every sector of the world's economy and is intricately intertwined with other major environmental threats such as population growth, desertification and land degradation, air and water pollution, loss of biodiversity, and deforestation. To date, most of the international attention directed toward combating climate change has been strikingly insufficient and focused primarily on the industrial and energy sectors. The agriculture, forestry, and other land use sector—AFOLU in current climate policy jargon—has so far been treated as an unwelcome distraction from tackling industrial and energy-related emissions, rather than being seen as an integral part of the climate change problem for which we must develop comprehensive solutions.[1]

The resulting bias has led international climate negotiators to disregard the major role forests and agricultural systems play in climate change. In the context of the Kyoto Protocol, widespread controversies and a lack of knowledge made negotiators agree to too little too late.[2] This result is not withstanding the recognition in 1997 that, with the adoption of the Kyoto Protocol, any attempt to stabilize atmospheric greenhouse gas (GHG) concentrations will have to bring land-use-related emissions and removals into the equation. According to a 2006 study led by former World Bank chief economist Nicholas Stern, the costs of reducing the effects of climate change can be significantly lowered if reduced deforestation and reforestation options are used effectively: "Curbing deforestation is a highly cost-effective way of reducing greenhouse gas emissions and has the potential to offer significant reductions fairly quickly. It also helps preserve biodiversity and protect soil and water quality. Encouraging new forests and enhancing the potential of soils to store carbon offer further opportunities to reverse emissions from land use change."[3]

The idea for this book was triggered by the conviction that an effective post-Kyoto agreement must include a comprehensive system that allows for the accounting of land-use-related emissions and removals and establishes incentives to reduce emissions from deforestation. With a view to the forthcoming debate, we thought the time was ripe to compile existing knowledge, expertise, and experience and make it available in one volume. At the same time we sought to produce a practical reference manual outlining the history of AFOLU in international climate change negotiations, identifying key lessons learned from implementing the various policy frameworks and from actual forestry project experience to date, and drawing on all this to propose solutions for how best to move forward. This book has benefited from contributions and input by the leading forestry and climate change experts in government, international organizations, academe, civil society, and the private sector.

In this chapter we provide a short overview of the role of forestry and agriculture in current climate policies—a recurring theme throughout the book. Like the other contributors, we focus our review on potential approaches to incorporating carbon sequestration and emission avoidance into emerging climate policy frameworks, rather than addressing the scientific debate that has surrounded the topic of climate change and forestry, which has already been written about at length.

Forestry and Climate Change

Forests are the world's most important terrestrial storehouses of carbon, and they play an important role in controlling its climate. The world's remaining forest ecosystems store an estimated 638 gigatonnes (Gt) of carbon, 283 Gt of which are in the forest biomass alone.[4] This is a significant amount of carbon—approximately 50 percent more than all the carbon in the atmosphere. Forest ecosystems are sensitive to climatic change. Over long periods of time plants have adapted to local climatic, atmospheric, and soil conditions, and this, combined with temperature and rainfall patterns, is what characterizes an ecosystem. A change in these variables can dramatically affect species viability. Stress caused by a change in the conditions of an ecosystem may also increase its vulnerability to pests and fires. Thus, massive areas of forests could be lost from these climate-induced threats, which in turn could further accelerate climate change in a vicious positive feedback loop.

On the other hand, land-based activities represent one of the most significant untapped opportunities for mitigating climate change:

—Simply leaving mature forests intact will lock up significant amounts of carbon that might otherwise be released into the atmosphere. Land-use changes, predominately deforestation, currently contribute about one-fifth of global carbon emissions (see chapter 15). Deforestation is the greatest source of GHG emissions in many developing countries, including Brazil and Indonesia, the world's biggest GHG emitters after the United States and China. Reducing emissions from deforestation may be one of the most cost-effective tools for reducing GHG emissions globally and could give people the time needed to mobilize the resources and develop the technology for "decarbonizing" the world's energy and industrial production.

—Sustainably managed forests can produce wood and other biomass that is a renewable, carbon-neutral alternative to fossil fuels and other construction materials. In this way sustainable forest management can help to reduce energy-related emissions (chapter 7).

—Forest ecosystems contain the majority (approximately 60 percent) of the carbon stored in terrestrial ecosystems and have the potential to absorb about 10 percent of global carbon emissions projected for the first half of this century into their biomass, soils, and associated products and, in principle, to store them in perpetuity.[5]

Forestry in Climate Negotiations

Both the UN Framework Convention on Climate Change (UNFCCC) and the Kyoto Protocol acknowledge the role that forests play in global climate (chapter 3).[6] Whereas the UNFCCC tends to refer to the reduction of GHG emissions and the increase of atmospheric GHG removals by sinks as parallel and equally important elements in any climate strategy, the Kyoto Protocol focuses on creating a framework for reducing industry- and energy-related emissions. The defining element of the Kyoto Protocol is a system of GHG emission targets with which all ratifying industrialized nations must comply. Reflecting the protocol's focus on energy and industrial emissions, the targets of individual countries are calculated without taking into account forestry- and land-use-related emissions. During the negotiations that led to the adoption of the Kyoto Protocol, controversy arose over whether parties should be allowed to offset emissions produced in other sectors with removals generated by biological sequestration or whether the effort to combat climate change should be concentrated on the reduction of emissions from the use of fossil fuels (chapter 4).

Those arguing against the accounting and use of forestry offsets were concerned that carbon offsets might be negated in cases where human action or natural events such as wildfires reversed the carbon benefits. If a tree is felled, stored carbon is released and the temporary climate benefit reversed—that is, the benefit is "non-permanent." The existence of this permance risk distinguishes emission removals generated by the forestry section from emission reductions generated by the industrial and energy sectors. The issue has, therefore, been a core concern about credits from activities that rely on sequestration of carbon in trees or soils.

Eventually negotiators decided in Kyoto that "direct human-induced" changes in GHG emissions and removals by sinks since 1990 could be used to meet a portion of the parties' emission commitments. Furthermore, articles 6 and 12, which define the project-based mechanisms called Joint Implementation (JI) and the Clean Development Mechanism (CDM), refer directly, in the case of JI, or at least indirectly, in the case of the CDM, to carbon sinks. AFOLU under the CDM is limited, however, to afforestation and reforestation projects, which are granted credits that can be used only for a limited period of time to comply with Kyoto commitments (chapter 6). The regulatory limitations of forestry under the CDM have subsequently severely hampered the development of this sector (chapter 8).

Nevertheless, the experience with crediting carbon from afforestation and reforestation projects has helped to create knowledge and overcome the scientific uncertainties that, among other things, stood in the way of an early agreement on expanded consideration of the forestry sector (chapter 10). In parallel, countries gained experience in authorizing relevant projects, and lawyers engaged in defining legislative and contractual frameworks (chapters 11, 12). Yet the limitations of the Kyoto Protocol can only be described as deeply unsatisfactory, because they have led to a situation in which there is an incentive to restore and protect forest systems in industrialized countries (chapters 3, 4) but no incentive to reduce emissions from deforestation in developing countries—the most important source of emissions from the land-use sector.

Negotiations toward a post-Kyoto agreement started in the context of the UNFCCC and Kyoto Protocol annual meetings in December 2005. On this occasion Papua New Guinea and Costa Rica put forward a submission to consider whether and how incentives to reduce tropical deforestation could be included in the future climate regime. This submission created a great deal of interest and earned significant support from developing and industrialized countries alike. This kicked off discussions on ways to address emissions from deforestation in developing countries. Since then a number of ideas and policy approaches on how to expand the carbon market to create incentives for forest conservation have been proposed and are being discussed as part of a post-Kyoto agreement (chapters 13–17). There is some hope that progress will be made in formulating an incentive framework that might grant financial awards for reductions in deforestation even ahead of final discussions on a more comprehensive post-Kyoto framework. The Bali round of UNFCCC negotiations held in December 2007 produced encouraging results. Demonstration projects that reduce emissions from deforestation and degradation (REDD) will be formally encouraged and recognized, and REDD in developing countries will be included in the Bali Action Plan, which is the two-year process to negotiate a post-Kyoto agreement.

Forestry and the Carbon Market

Many of the benefits provided by forests are currently considered part of the global commons and are freely available for everybody. Forests purify air and water, stabilize soil, support biodiversity, produce pharmaceutical substances, and act as carbon storehouses—all of which humans treat as unlimited and free services. Typically, no legal rights and consequently no monetary value are assigned to these services. The value of a forest is usually defined solely in terms of things that can be owned and readily traded—the timber in the trees and the land on which the forest grows. This means that those who control or have access to forests often have greater incentives to clear them and turn them to economically productive uses than to conserve them.

These services provided by forests need to be appropriately priced if people are to make decisions about forests that are based on their true value. Schemes that envisage payments for "ecosystem services" try to address this market failure by creating financial incentives to conserve, protect, and restore forests (chapter 5). Assigning value to emission reductions or removals (carbon storage) by creating tradable carbon credits is one of the most developed and promising approaches for tapping the forestry sector in the fight against climate change, and it is therefore a key topic in this book.

The carbon market relies on emission trading and the transfer of carbon credits. The CDM and JI allow countries to invest in emission-reducing projects or programs in countries where abatement costs for emission reductions are lower than in their own economies. In return for their payments, investors or carbon purchasers receive a right to the carbon credits generated by the project. These carbon credits can be used to meet compliance obligations under international and national regulatory regimes. The carbon market created under the Kyoto Protocol and a number of regional and national emission-trading schemes is worth billions of dollars each year (chapters 6, 8, 18, 19).

Because the Kyoto Protocol does not address forest conservation—that is, the prevention of deforestation—in developing countries, these countries are restricted in their opportunities to benefit from the CDM. Most of the offsets generated by AFOLU projects are currently traded in the so-called voluntary market, where the rules are typically more flexible and accommodating of such projects (chapters 20, 21). Companies invest in voluntary offsets for marketing purposes, to meet voluntary corporate social responsibility objectives, or to get ahead of emerging regulations. Increasing numbers of individuals are now joining them in wishing to offset their carbon "footprints."

A New International Framework

Despite a common understanding that the AFOLU sector is far too important, both as a sink and as a source, to be marginalized again, differences remain regarding when, to what extent, and how land-use-related sinks and emissions should be integrated into a post-Kyoto regime. Regardless of the design details, it is important that any post-Kyoto agreement provide the right framework and incentives for the following:

—Rewarding decreased deforestation; sustainable forest, land, and wetland management; forest restoration; and the sustainable use of biomass

—Establishing a reliable accounting system that includes the flux of biological carbon

—Promoting sustainable development and an inclusive climate policy

—Capturing synergies between the Convention on Biological Diversity, the Convention to Combat Desertification, and the Millennium Development Challenge goals

Far from having embraced the full complexity of the issue, negotiators have engaged in the initiative taken by Papua New Guinea and Costa Rica to narrowly focus on defining an instrument to reduce GHG emissions from deforestation in developing countries. Until now, the broader question of how to effectively integrate AFOLU emissions, sequestration, and emission reductions into a post-Kyoto regime has been sidelined.

At their annual summit held in Heiligendamm, Germany, in June 2007, the Group of Eight leading industrialized countries (G8) expressed their commitment to taking a leadership role in future efforts to reduce GHG emissions. The summit's declaration stressed, however, that developing countries had to contribute to a global effort to reduce emissions. In subsequent statements the European Union made it clear that it saw a commitment by developing countries to reduce emissions from deforestation as a promising example of how developing countries could demonstrate their commitment to mitigating climate change. The EU envisaged a system of national targets under which emission reductions would be rewarded with carbon credits. Although forestry per se still ranks low in the priority list of most EU member countries, developing countries' willingness to consider sectoral targets provides a welcome opportunity to negotiate industrial targets with major emitting, developing nations, which remains the priority of most EU negotiators.

When it comes to the debate over REDD, a number of options are being proposed, including both market- and non-market-based approaches. Market-based approaches rely on the carbon market and aim to create incentives for avoiding further deforestation. In most cases they include the awarding of tradable carbon credits once a country or project has generated a proven climate benefit by reducing GHG emissions below the business-as-usual scenario.

A number of proposals are associated with baselines developed at the national level whereby countries take on voluntary commitments to reduce their national deforestation rates in the form of reduction targets vis-à-vis the national baselines (chapter 16). In order to account for challenges developing countries face in establishing national-scale systems, it has been proposed to combine national approaches with the authorization of a project or subnational approach. Those arguing in favor of including project-based activities refer to the required level of resource mobilization, which goes beyond what public funds could make available and thus must tap private capital (chapter 17). Proponents of other approaches, although less likely to prevail, argue that an efficient system has to move away from a baseline-and-credit approach toward a cap-and-trade approach, which will allow developing countries to access financing on the basis of a binding conservation commitment.

Another question—whether credits generated by activities or programs that reduce emissions from deforestation should be fully fungible with other carbon markets—gets back to the old debate of whether GHG reductions achieved in forestry should be fully fungible with industrialized-country reduction commitments. Taking into account the volume of reductions that may be achieved through REDD activities, and the need to avoid flooding the market, carbon credit fungibility must be matched by strict emission limitations in industrialized countries. Mandating a tighter overall cap than would be possible without REDD crediting could create a win-win situation for the environment and the global economy. Another option would be to create a separate or parallel market for REDD credits, although the economic viability of such an approach is questionable.

Conclusion

Omitting deforestation in developing countries from a post-Kyoto agreement would leave out a major source of carbon emissions, which could undermine many of the gains made through fossil fuel reductions. Many of the insecurities that made an international agreement at Kyoto impossible have been addressed in the last decade, including the development of robust measurement and monitoring (that is, accounting) protocols and various means for addressing the permanence problem.

Forest and biodiversity conservation are intrinsically linked to the mitigation of climate change and humans' adaptation to such change. If we lose our forests, we lose our biggest sink of terrestrial carbon and a system that regulates and influences local and regional climate patterns and extreme weather events. It is therefore necessary that a post-Kyoto regime include a comprehensive carbon accounting mechanism with the necessary incentives for conserving our forests, especially in the tropics, where they are most threatened and can play a vital role in supporting sustainable livelihoods for the world's poor.

Certainly from a development perspective, AFOLU carbon projects represent one of the few means by which many of the world's poorest people, including most Africans, will be able to meaningfully participate in and benefit from the global carbon market. For the first time these people have the promise of being able to sustainably capture an ecosystem service value associated with their land, instead of being forced to liquidate forest resources just to survive.

Keeping in mind the broader context, we hope readers find the chapters in this book both interesting and useful for understanding the increasingly complex climate change negotiations under way today. We further hope that a deeper understanding of the interlinkages between climate policy and forestry will help negotiators define a robust and enduring international framework for reducing GHG emissions from all sources while providing the right incentives for the conservation and sustainable use of the earth's most precious natural resources.

Notes

1. International climate change experts have traditionally referred to "land use, land-use change, and forestry," or LULUCF, as the prevailing term for this sector. However, the most recent Intergovernmental Panel on Climate Change (IPCC) guidelines refer to AFOLU, a more consistent and complete term by which to describe this sector. See IPCC, *2006 IPCC Guidelines for National Greenhouse Gas Inventories,* vol. 4, *Agriculture, Forestry, and Other Land Uses,* prepared by the National Greenhouse Gas Inventories Programme, H. S. Eggleston and others, eds. (Institute for Global Environmental Strategies, Japan).

2. FCCC/CP/1997/L.7/Add.1 Decision 1/CP.3, "Adoption of the Kyoto Protocol to the United Nations Framework Convention on Climate Change," Annex, reprinted in 37 ILM 22 (1998), entered into force February 16, 2005 (hereinafter "Kyoto Protocol").

3. N. Stern and others, *Stern Review on the Economics of Climate Change* (London, 2006), p. 537.

4. This is the amount of carbon stored in the world's biomass, deadwood, litter, and soil. See Food and Agriculture Organization (FAO), "Global Forest Resources Assessment 2005: Progress towards Sustainable Forest Management," Forestry Paper 147 (Rome: FAO, 2006).

5. IPCC, *Land Use, Land-Use Change, and Forestry: A Special Report of the IPCC* (Cambridge University Press, 2000).

6. UN Doc Distr General A/AC.237/18(Part II)/Add.1, May 15, 1992 (hereinafter "the UNFCCC"). The UNFCCC entered into force on March 21, 1994, and currently has near-universal membership, with 191 countries having ratified it.

2

The Idea of Market-Based Mechanisms for Forest Conservation and Climate Change

ROSIMEIRY PORTELA, KELLY J. WENDLAND, AND LAURA LEDWITH PENNYPACKER

The world has approximately 4 billion hectares of forests, roughly 30 percent of them primary forests.[1] Their provision of goods and services plays an important role in the overall health of the planet and is of fundamental importance to human economy and welfare. These goods and services—collectively called ecosystem goods and services, or simply ecosystem services—include, among other things, food and timber, the formation of soils, the regulation of climate and hydrological processes, and the spiritual, aesthetic, and recreational opportunities associated with people's enjoyment of nature.[2]

Regulating greenhouse gases is one of the most significant ecosystem services provided by forests today. It is estimated that forests store more than 280 gigatonnes (Gt) of carbon in biomass.[3] Conserving this biomass prevents the emission of carbon dioxide (CO_2) and other greenhouse gases (GHGs) that cause climate change. When forests are destroyed, heat-capturing gases and aerosols are released into the atmosphere, aggravating climate change effects.[4] This will most certainly have ecological and physiological effects on forests, possibly altering the lengths of growing seasons, biomass production, and species competition and distribution and ultimately causing major shifts in biomes.[5] The combined effects

The views expressed in this chapter are those of the authors and do not necessarily represent the views of Conservation International, with which the authors are affiliated.

of climate change and habitat fragmentation pose the most important human-induced threat to biodiversity and ecosystems.[6] One effect may be the major loss of cover of tropical forests as a result of drying conditions.[7]

Despite the numerous benefits that forests offer humans, decisions to convert forests to other land uses are often based on market incentives that do not incorporate the value of ecosystem services. There are several reasons for this. In some cases people lack an understanding of the functioning of large-scale natural systems and the effects that human actions can have on them. In other cases losses of forests are associated with the fact that conventional market systems undervalue ecosystem services in everyday decisionmaking.[8] In order to correct for this market failure and change undesirable behaviors, policy instruments or incentives are required.

In this chapter we first review the policy instruments that have been used to achieve forest conservation, looking particularly at the role of market-based mechanisms. We then discuss the opportunities that climate change and carbon markets present for forest conservation. The current climate regulation regime, through voluntary and regulatory markets, particularly the Kyoto Protocol's Clean Development Mechanism (CDM), is already providing opportunities to harness market forces toward the conservation and restoration of forests. The effects of these markets on forests, however, are still limited. The implications for forest conservation become more interesting when one starts to consider several potential market opportunities for forest carbon. These new market opportunities could dramatically shift forest conservation market practices and ensure the provision of many other ecosystem goods and services, including globally important biodiversity. These new approaches offer a holistic approach to human-induced climate changes and would support ecological sustainability, human welfare, and the world economy.

Policy Mechanisms for Forest Conservation

Although forest loss has been occurring for many centuries, the last few decades have shown alarming rates of deforestation, with about 13 million hectares of forest converted to other land uses every year. This cumulative loss of forest biomass represents a significant decline in total carbon stocks, which decreased by about 1.1 Gt annually between 1990 and 2005. South America and Africa had the largest net forest losses from 2000 to 2005, with averages of 4.3 and 4.0 million hectares of forest cleared in those years, respectively.[9] Decisionmaking generally prioritizes private, direct benefits from forest clearing, even though the loss threatens the existence of forests and deprives society of biodiversity and other important ecological benefits. This problem stems from the failure of policymakers and market institutions to translate demand for forest goods and services into income for local landowners and to account for the social costs incurred by the clearing of forests for private benefits.

Market Failure in Forest Conservation

In a competitive economy, markets control the way the resources available to society, influenced by preferences and available technologies, combine to produce goods and services that meet human needs. For many situations the market economy works well and provides the socially optimal level of goods and services. Yet for many other situations, such as the consumption of natural resources, market systems have led to excessive and unsustainable extraction. The reason many natural resources are not traded efficiently in market systems is that they do not meet two crucial conditions necessary for a market outcome to be efficient: the good or service should be private rather than public, and there should be no difference between the private and social costs of producing the good or service.[10]

Forests and the ecosystem services they provide fail to meet these criteria. Along with producing a variety of marketable products, forests provide many unpriced ecosystem services that benefit human well-being at local, national, and global scales. Private benefits of forest resources are normally obtained from the economic returns of cleared forests, such as timber, crops, and pasture. Social benefits, on the other hand, are normally associated with the functional properties of intact forests, such as the regulation of regional and global climate patterns, the provision of clean water, and the stabilization of soil—each of which contributes to overall public well-being. Whereas the private benefits of forest exploitation are valued through the market, social benefits are not valued in the market without some type of policy intervention.[11] Because landowners do not face the full costs of their actions in terms of foregone ecosystem services, they often use resources unsustainably, leading to inefficient allocation of forest resources and the services they provide for greater human well-being.[12] Private landowners behave in these ways because the marginal private benefits of their cleared land (for example, the monetary return from crop cultivation) are not balanced against the marginal costs to society of forest loss (for example, losses of important forest services).

Market systems also fail to provide incentives for conserving forest ecosystem services because most of these services are (quasi) public goods: they are nonrival and nonexcludable, meaning that it is impossible to preclude someone's using a good (nonexclusiveness), and use by one leaves no less for others (nonrivalry). This leads to what is known in economics as the "free rider" problem. That is, if one landowner maintains forest cover to ensure clean water or clean air, then everyone else gets a free ride at the landowner's expense. Thus the incentive for any one person to conserve forests is low.

Policy Mechanisms to Correct Market Failure

Given the pervasiveness of market failure and the consequent depletion of public environmental goods and services, different policy tools have been used to alter landowners' behaviors for about the past five decades.[13] Historically, some of the most common initiatives to protect and manage forests and the services they provide have been regulatory, or command-and-control, instruments. Although these tools are still common, a shift has taken place toward more participatory mechanisms and, recently, toward the use of economic incentives and market-based instruments.

At the national and local level, command-and-control mechanisms such as the establishment of protected areas are common, and although they are effective in many situations, their usefulness is often limited by lack of sufficient information about what to protect, lack of funds, and vulnerability to political pressures.[14] At the international level, many environmental treaties, such as the UN Convention on Biological Diversity and the UN Convention to Combat Desertification, are in essence command-and-control regulatory regimes. And although they represent important global efforts to protect natural resources, their ability to do so is often limited by their voluntary nature (no enforcement of compliance), their lack of binding obligations (with the exception of the Convention on International Trade in Endangered Species, or CITES), and the absence of sustainable funding mechanisms.[15]

Other approaches to trying to change people's behavior from deforestation to forest conservation have relied on education and local participation. Environmental education is a component of many conservation projects, with the hope that more information about the benefits of ecosystems will lead to greater interest in conservation. A recent approach has been to attempt to give local communities a stake in resource management by establishing integrated conservation and development projects and community-based conservation programs. These participatory approaches strive to create private interest in protecting public goods but have faced many roadblocks in developing countries.[16] As a general rule, both kinds of initiatives fail to address the fundamental issues of market failure. Thus their effectiveness at protecting ecosystems is limited unless they are implemented along with other policy instruments.

In contrast to nonmarket approaches to forest conservation, market-based mechanisms encourage a particular behavior by changing the incentives for individual agents. These instruments are often described as "harnessing market forces" because they encourage individuals to take actions that meet both their private interests and policy goals.[17] Market-based mechanisms differ from command-and-control approaches in that they allow more flexibility in the way policy targets are met. This reduces the amount of information policymakers need and reduces overall costs. If designed well, market-based mechanisms can even encourage participants to exceed policy targets and can spur innovation. Market-based mechanisms for pollution control have achieved notable success in developed countries.[18] The use of market mechanisms for forest conservation in developing countries, although still limited, is growing and has the potential to lead to more cost-efficient and effective conservation projects. In many situations the role of market-based mechanisms will be to complement rather than substitute for nonmarket instruments, and many market-based mechanisms can be used concurrently to meet policy goals. A review of several market-based interventions found that many of the most successful relied on a mix of regulatory and market-based tools.[19] Table 2-1 highlights the types of market-based instruments being used to meet forest conservation objectives in developing countries and provides brief descriptions of the main characteristics of the tool, its advantages, some examples, and its potential limitations.[20]

Climate and the Market: Opportunities for Forest Conservation

The role of forests in climate regulation is important for both human welfare and the economy. However, because the social benefits of conserving forests outweigh the private benefits, public policy interventions will be necessary if forest conservation is to play a significant role in carbon sequestration or emission reductions. Some of the most cost-effective and efficient interventions are likely to be market-based mechanisms. Well-designed and well-implemented market-based mechanisms for forest carbon sequestration can direct significant technical and financial resources toward climate change mitigation—at a relatively low cost compared with the cost of shifting to cleaner energy technologies.[21] Unfortunately, despite some progress, the forest carbon emissions market has experienced limited development. This is due in part to (1) poor understanding of the extent to which forests contribute (both positively and negatively) to climate change and GHG emissions; (2) burdensome rules and constraints in regulatory markets that limit potential demand; and (3) technical and methodological concerns surrounding the verification of forest carbon emission reductions. In what follows we provide a brief overview of current and potential market-based opportunities for forest conservation in the context of global climate abatement.

Current Market Opportunities

Current carbon markets are complex and evolving rapidly. Both regulatory and voluntary carbon markets have developed in recent years as a result of international, national, and regional policies, as well as from increased public awareness of climate change. Regulatory markets have been developed to help meet emissions targets as mandated by international and national regulatory authorities.

Table 2-1. *Market-Based Mechanisms for Forest Conservation*

Type of mechanism	*Key characteristics*	*Advantages over other market-based instruments*	*Example(s)*	*Potential limitations*
Economic incentives	Direct incentives provided to landowners to make forest conservation economically competitive with other land uses.	A more direct method for achieving conservation objectives than some of the alternatives.	*Land acquisition:* Land is bought outright from the landowner for conservation purposes. *Conservation easements and leases:* Voluntary agreements in which land is leased from the landowner in perpetuity or for some predetermined length of time, conditional on conservation objectives. *Direct payments:* Cash or in-kind payments are made directly to landowners and communities for meeting conservation objectives.[b]	Can be expensive to implement.[a]
Market enhancement	Environmentally friendly production or marketing strategies, or both, are used to make conservation activities profitable.	Can attract new funding sources to meet conservation objectives.	*Certification or eco-labeling:* Encourages people to make better-informed decisions about their purchases or consumption habits (for example, certified timber). *Marketing of many nontimber forest products:* Marketing strategies used to appeal to environmentally and socially conscious consumers (for example, shade-grown coffee).	Scaling up production systems can have adverse effects on conservation objectives.[c]

			Sustainable timber management: Harvesting strategies for timber that protect the value of other ecosystem services and biodiversity.	Not proven to be economically competitive against more conventional logging practices.[d]
			Bioprospecting permits: The sale of permits for rights to search for and use genetic materials in a forest.	Profitability of bioprospecting for local people has been shown to be small and inefficient.[e]
			Eco-tourism: A sustainable form of natural-resource-based tourism that is intended to contribute to the conservation of such areas.[f]	Does not work in all areas, such as those that lack spectacular scenery or charismatic species.[g]
Market creation	Expansion of the market system by extension of the scope of what is considered marketable or by establishing new markets.[h]	Can attract new funding sources to meet conservation objectives.	*Markets for watershed services:* Downstream water users pay upstream landholders for adopting practices that secure the provision of clean, ample water.[i] For example, in Ecuador payments are being implemented to protect upland forests that provide downstream watershed services.[j]	Generating demand for new markets can be a challenge.[k]
			Carbon markets: Landholders are paid for avoiding deforestation or for planting trees. For example, in Madagascar the government is receiving international funding to store carbon in standing forests and to sequester carbon by planting trees.[l]	

(continued)

Table 2-1. *Market-Based Mechanisms for Forest Conservation* (Continued)

Type of mechanism	*Key characteristics*	*Advantages over other market-based instruments*	*Example(s)*	*Potential limitations*
Taxes and subsidies	A tax is used to raise an individual's cost of performing an action to equal the cost borne by society.[m] A subsidy is used to decrease an individual's costs to a point where the conservation activity is competitive with alternative economic opportunities.[n]	Project costs can be better anticipated using these instruments.[o]	*Private Natural Patrimony Reserves (RPPN)*: A Brazilian system in which landowners can designate private property as conservation areas to protect biodiversity in perpetuity, in exchange for rural property tax exemptions, preference in the state concession of rural credits, and cooperation with private and public entities in the protection, management, and handling of the RPPN.[p]	Often, uncertainty about the level at which to set the tax or subsidy in order to achieve the desired conservation objectives.[q]
Tradable permits (or quotas)	An allowable level of an activity (for example, deforestation) is set and allocated among individuals. Individuals who reduce their activity below this level can sell excess permits or use them to offset activities in other areas.[r]	Requires less external regulation and control than some of the other approaches once the total number of permits is set.[s]	*Tradable development rights (TDRs) or habitat preservation credits:* Allows landowners to transfer rights to develop on one piece of land to another landowner. For example, several Brazilian states are implementing regulations that allow for cross-property compensation and trading to meet the country's "legal forest reserve" obligation.[t] A more established form of TDR is the United States' wetlands mitigation banking system, in which landholders can sell wetland credits to developers.[u]	Determining the appropriate unit of measurement for permits involving habitat conservation can be complicated.[v]

a. S. Polasky, C. Costello, and A. Solow, "The Economics of Biodiversity," in *Handbook of Environmental Economics*, vol. 3, edited by K.-G. Maler and J. R. Vincent (New York: Elsevier, 2005).

b. P. J. Ferraro, and R. D. Simpson, "Cost-Effective Conservation: A Review of What Works to Preserve Biodiversity," *Resources* 143 (2001): 17–20.

c. N. M. Belcher, M. Ruiz-Perez, and R. Achdiawan, "Global Patterns and Trends in the Use and Management of Commercial NTFPs: Implications for Livelihoods and Conservation," *World Development* 33 (2005): 1435–52; Polasky, Costello, and Solow, "The Economics of Biodiversity."

d. D. W. Pearce, F. Putz, and J. K. Vanclay, "Sustainable Forestry in the Tropics: Panacea or Folly?" in *The Earthscan Reader in Forestry and Development*, edited by J. A. Sayer (London: Earthscan, 2005), pp. 280–304; D. Rice, R. Gullison, and J. Reid, "Can Sustainable Management Save Tropical Forests?" *Scientific American* 276 (1997): 34–39.

e. D. Simpson, R. Sedjo, and J. Reid, "Valuing Biodiversity for Use in Pharmaceutical Research," *Journal of Political Economy* 104, no. 1 (1996): 163–85.

f. D. A. Fennell, *Ecotourism: An Introduction* (New York: Routledge, 1999).

g. S. Wunder, "Ecotourism and Economic Incentives: An Empirical Approach," *Ecological Economics* 32 (2000): 465–79.

h. G. Heal, *Nature and the Marketplace: Capturing the Value of Ecosystem Services* (Washington: Island Press, 2000).

i. S. Wunder, "Payments for Environmental Services: Some Nuts and Bolts," CIFOR Occasional Paper 42, 2005.

j. See www.cifor.cgiar.org/pes/_ref/projects/ecuador.htm.

k. Wunder, "Payments for Environmental Services."

l. One example is the Ankeneny-Zahamena-Mantadia Biodiversity Conservation Corridor and Restoration Project in Madagascar. See http://carbonfinance.org/Router.cfm?Page=BioCF&FID=9708&ItemID=9708&ft=Projects&ProjID=9638.

m. In land conservation programs there are few examples of taxes being used in the conventional sense to raise the costs of a land-use activity. Instead, taxes are frequently used by giving tax breaks or tax exemptions in exchange for some conservation activity. This is referred to as tax differentiation. R. Stavins, "Experience with Market-Based Environmental Policy Incentives," in *Handbook of Environmental Economics*, vol. 1, edited by Maler and Vincent, pp. 355–435.

n. For some environmental goods and services, the removal of perverse subsidies can also play a role in correcting market failures. See N. Myers, "Lifting the Veil on Perverse Subsidies," *Nature* 392, no. 6674 (1998): 327–28. However, although removal of perverse subsidies is important for correcting market failure, it is considered insufficient by itself to create large changes in forest conservation outcomes. See Polasky, Costello, and Solow, "The Economics of Biodiversity."

o. Heal, *Nature and the Marketplace.*

p. See www.ibama.gov.br/siucweb/rppn/.

q. Heal, *Nature and the Marketplace.*

r. Stavins, "Experience with Market-Based Environmental Policy Incentives."

s. Ibid.

t. For a description of this program, see K. M. Chomitz, T. S. Thomas, and A. S. P. Brandao, "The Economic and Environmental Impact of Trade in Forest Reserve Obligations: A Simulation Analysis of Options for Dealing with Habitat Heterogeneity," *Revista Economia e Sociologia Rural* 43, no. 4 (2005): 657–82.

u. See www.epa.gov/owow/wetlands/facts/fact16.html.

v. This is because the size of the trading area can affect cost-effectiveness, and different land units may not contain the same number of conservation objectives. See Chomitz, Thomas, and Brandao, "The Economic and Environmental Impact of Trade in Forest Reserve Obligations."

Voluntary carbon markets offer alternatives for companies, governments, organizations, and individuals operating outside of regulatory mandates to reduce carbon emissions.

Opportunities for forest conservation in carbon markets rest in the project-based sector, in which a project must demonstrate that it reduces GHG emissions relative to a business-as-usual scenario. Opportunities for forest carbon projects currently exist in forest conservation, restoration, and sustainable forest management. The most common forest carbon emission projects are afforestation and reforestation (AR) projects, followed by agroforestry and avoided deforestation. Forest carbon projects are often designed and implemented according to the rules and restrictions of the regulatory and voluntary markets, as well as according to opportunity and transaction costs.

Forest carbon projects have the potential to generate benefits beyond carbon sequestration, especially strong community livelihood and biodiversity benefits. These additional benefits should allow such projects to generate higher prices for carbon sequestration by forests, accurately reflecting the bundling of forest ecosystem services. Unfortunately, opportunities for this are currently limited in the regulatory market, and voluntary carbon projects generally fetch lower prices than their regulatory counterparts.

THE REGULATORY CARBON MARKET. The regulatory carbon market is dominated by the Kyoto Protocol of the United Nations Framework Convention on Climate Change (UNFCCC). The Kyoto Protocol has created a framework in which market-based forestry activities can help mitigate climate change. Activities are allowed under both of the protocol's flexible mechanisms, the CDM and the Joint Implementation (JI) framework. The Kyoto Protocol market framework has influenced the creation of subsequent regulatory markets, including the European Union's Emissions Trading Scheme (EU ETS), Norway's Emissions Trading system, and the United States' Regional Greenhouse Gas Initiative (RGGI).

As shown in table 2-2, the development of forest carbon projects within the regulatory markets is currently constrained. The CDM holds the greatest—albeit small—opportunity for forest conservation. CDM forestry carbon projects (frequently referred to as LULUCF projects, for "land use, land-use change, and forestry") earn credits for carbon sequestration only through AR, and project development is considerably restricted. For example, AR projects are limited to 1 percent of base-year emissions of industrialized countries and are temporary, requiring replacement with permanent non-LULUCF credits regardless of the state of the forest. Moreover, project development and reporting is especially complex and burdensome for LULUCF projects within the CDM, and thus few projects have been approved.[22]

No other forest carbon activities, such as avoided deforestation, are allowed under the CDM or at any meaningful scale within the other regulatory markets.

Table 2-2. *Forestry Project Types and Allowances in Regulatory and Voluntary Markets*

Market	*Eligible forestry options*	*Potential for incorporation of other forest benefits*[a]
Regulatory		
Kyoto Protocol CDM	Afforestation and reforestation[b]	Low-medium
EU Emissions Trading Scheme	All forestry excluded until at least 2008	Low
Emerging U.S. regulatory markets (RGGI, Calif., federal)	U.S.-based forest conservation and restoration	Medium
New South Wales Abatement Scheme, Australia	Australian forest restoration only	Medium
Voluntary		
Voluntary retail carbon market	Tropical forest conservation and restoration	High
Chicago Climate Exchange	Tropical forest restoration and forest conservation	High

a. Authors' compilation.

b. AR projects restricted to 1 percent of base-year emissions of Kyoto industrialized countries.

Because of this constraint, opportunities for forest conservation are severely limited, and the potential to capture additional forest ecosystem values is lost.

The Voluntary Carbon Market. The voluntary and retail carbon market consists of companies, governments, organizations, and individuals seeking to reduce the climate effects of their operations by voluntarily investing in projects that credibly reduce GHG emissions. Several market frameworks exist for voluntary projects, including the Chicago Climate Exchange (CCX)—the world's first voluntary, legally binding, rules-based system to reduce and trade GHG emissions—and the retail market. Voluntary retail carbon offsets, including forestry carbon offsets, are often marketed through intermediaries who sell shares to consumers at desired quantities. There are an estimated thirty to forty retail carbon offset providers worldwide, most of them based in Europe, the United States, and Australia.[23]

A large majority of these voluntary markets offer forestry carbon offsets with multiple benefits. Table 2-2 shows the potential of two of them for including additional benefits such as sustainable development, biodiversity, and other ecosystem services. Standards are even being developed in the retail markets to certify that carbon projects provide these multiple benefits. One example is the Climate Community and Biodiversity Alliance (CCBA) standards.[24]

Despite these opportunities, the voluntary carbon market is still in its early stages of development. Typically, projects are small scale, in part because of limited (although growing) demand and the prevalence of smaller-scale forestry activities.[25] Opportunities for forest conservation are expanding, however, with the inclusion of most, if not all, forest carbon project types. And although most voluntary carbon markets impose standards and verification guidelines, considerable flexibility exists to invest in small-scale projects with multiple co-benefits, because compliance with stringent CDM rules and paperwork is not required.

Regulatory versus Voluntary Markets: Market Share, Value, and Volume. Forest carbon projects account for a very small share, in both volume and value, of all emission reduction projects in the regulatory and voluntary carbon markets. The voluntary markets see a larger number of forest carbon projects,[26] but as shown in table 2-3, the regulatory carbon markets are orders of magnitude larger than the voluntary markets. Thus, given the constraints the regulatory markets impose on forest projects and the lack of opportunities for multiple-benefits projects, considerable forest conservation opportunities are being lost.

The estimates in table 2-3 also show the significant differences in market value between regulatory and voluntary carbon projects: traditionally, regulatory projects receive higher prices per unit than their voluntary counterparts. In addition, both regulatory and voluntary forest carbon projects receive a fraction of the price that many other carbon markets realize.[27] As mentioned earlier, this is due to regulatory complexities and the restrictions and burdens placed on forest carbon projects, which lead to limited demand within the market.

Potential Market Opportunities

Given the limited effects current forest markets have on both forest conservation and GHG emissions, recent initiatives in the regulatory sector have begun to reevaluate the role of forest conservation and protection. New opportunities in the forest carbon project sector include California's climate regulatory regime and Australia's proposed cap-and-trade system. These initiatives have the potential to affect forest conservation significantly if international forest conservation opportunities are included in them at a meaningful scale. However, what holds the greatest promise for cost-effective climate abatement and forest conservation is allowance-based emissions credits for developing countries that make commitments to reduce deforestation. We briefly highlight the mechanisms proposed to implement such an approach and discuss their potential to reduce emissions from deforestation.

The UNFCCC has begun to explore policy instruments that could provide an incentive to "reduce emissions from deforestation and degradation" (REDD) in developing countries.[28] Tropical forests play a crucial role in regulating the global carbon cycle; they take up carbon from the atmosphere and accumulate it in their biomass. When forests are destroyed, the carbon they store is released into the air,

Table 2-3. *Market Values and Volumes of Regulatory and Voluntary Carbon Markets, 2005 and 2006*

Market	2005		2006	
	Market value (millions of \$)	*Volume (Mt CO_2)*[a]	*Market value (millions of \$)*	*Volume (Mt CO_2)*
Regulatory				
Kyoto Protocol CDM	2,417	341	5,477	508
EU Emissions Trading Scheme	7,908	321	24,357	1,101
New South Wales Abatement Scheme, Australia	59	6	225	20
Voluntary				
Voluntary retail carbon market	44	6	100	>10
Chicago Climate Exchange	3	1	38	10

Source: K. Capoor and P. Ambrosi, "State and Trends of the Carbon Market 2007" (Washington: World Bank, 2007) (http://carbonfinance.org/docs/StateoftheCarbonMarket2007.pdf).

a. EU Emissions Trading Scheme, Australia, and Chicago Climate Exchange volume data include allowance-based transactions published from various trading platforms as well as volume known to be exchanged over the counter. Kyoto Protocol CDM and voluntary retail market volume data include project-based transactions supplied by direct interviews, major carbon-industry publications, and Natsource. Transactions include only signed contracts—that is, those resulting in Emissions Reductions Purchase Agreements (ERPAs).

contributing to global warming. Current rates of tropical deforestation, which is occurring primarily in developing countries, account for about 18 percent of total GHG emissions worldwide.[29] In Brazil, which in past years has had the highest deforestation rates in South America and the second highest in the world, carbon emissions from deforestation account for more than twice the emissions from fossil fuel use.[30] Forest losses lead to additional effects beyond climate—effects on biodiversity and human livelihoods. For example, biologists estimate that 1,800 species *populations* go extinct every hour from forest destruction, or 16 million populations a year.[31]

Given these trends, it is imperative that the regulatory sector, particularly the UNFCCC, properly address the effects of deforestation on climate change. Inclusion of REDD as a viable mitigation strategy in the next global climate agreement (post-2012) would give developing countries significant financial and technical capacity for forest conservation and the protection of forest ecosystem services. Carbon financing, by helping to align the preferences of private individuals with public needs, would help minimize the current market failures that allow for the destruction of tropical forests worldwide.

Developing policy mechanisms that provide incentives for forest protection in a regulatory scheme, however, is complex. Significant risk and uncertainty exist in generating credible and verifiable emission reductions and in awarding such emission reductions for forest protection at the national level, not just restoration at the project level. "Additionality," "leakage," and "permanence" are among the common concerns expressed.[32] Developing reference scenarios and accurate measurements of carbon forest stocks, emissions, and credits from forest conservation (carbon inventory, monitoring, and verification) poses technical and scientific challenges for most forest carbon projects, and particularly for a REDD policy. In addition, designing a framework that includes all current *and potential* tropical forest emitters is essential if meaningful emission reductions are to occur at the global scale.[33]

The scientific community believes that the challenges associated with estimating and monitoring deforestation emissions are surmountable in a cost-effective manner.[34] Presently, several alternative approaches are being proposed for implementing a UNFCCC REDD policy; two of these are dealt with in more detail in subsequent chapters of this book. One is the compensated reduction approach, in which a country receives post-facto compensation when it reduces deforestation below a national baseline of average historical deforestation. The country is then obligated to use these funds in activities that further contribute to the goals of the UNFCCC and Kyoto.[35] The second approach is the carbon stock approach, which advocates a cap-and-trade system for regulating forest carbon.[36] In this system, carbon stock credits (called carbon stock units) are allocated to a country on the basis of existing carbon pools stored in tropical forests at a certain reference date. A certain percentage of these credits would have to be held constant within the country, and forest areas would be placed under protection. A quota of the credits would be available for trading among countries to allow for deforestation. More recently, an approach called the "nested" approach has been proposed, which seeks to combine the national baseline approach, outlined in the compensated reduction approach, with a project-based approach that allows flexibility in forest carbon crediting for countries unable to measure or control national deforestation changes.[37]

Market-Based Mechanisms for Forest Conservation and Carbon: Additional Considerations

For a market-based mechanism to operate efficiently, certain institutional and regulatory systems must first be in place. What many critics of market-based systems do not acknowledge is that these same systems are necessary to implement any policy instrument. We highlight some of the general conditions needed for market-based mechanisms aimed at forest conservation and carbon outcomes and touch on the issue of equity within market systems. Among the general conditions needed are the following:

—*Property rights:* Property rights are central to the idea of "privatizing" the forest resource so that an exchange can be made between the supplier of the good or service and those who demand it.[38] The type of property right institution appropriate for a particular good or service will vary depending on the nature of the resource and the characteristics of the user group.

—*Legal framework:* A legal framework is necessary to establish who is responsible and liable for different aspects of market transactions. How the legal framework is set up determines which party, the buyer or the seller, bears the brunt of transaction costs, ultimately affecting the viability of the market system.[39]

—*Regulatory framework:* A regulatory framework defines the conditions under which the market will operate. It can be used to identify how potential buyers and sellers will be brought together, how market exchanges will be carried out, and appropriate market prices. In general, the better established a market is, the more rules and regulations it has.[40]

—*Monitoring and enforcement:* Some type of monitoring and enforcement system is necessary to ensure that sellers adhere to the rules and conditions of transactions. For many forest services this requires satellite imagery and statistical capacity. As with command-and-control mechanisms, lack of monitoring and enforcement can severely compromise the effectiveness of market-based instruments.[41]

An additional consideration is equity. A number of researchers have recently questioned whether market solutions in developing countries will effectively benefit rural populations who live near forests and other landscapes that provide ecosystem services.[42] The difficulties of ensuring equitable benefits in market transactions stem from the fact that many of the poorest lack property rights, and concern exists that as new economic opportunities are introduced for forest resources, the poor may be unable to retain access to or control over them. Another concern is that transaction costs will exclude the poorest of the poor from participating in emerging opportunities. Some project-based evidence shows that these factors can be an issue and that regulatory measures may be required to ensure equitable access to markets.[43]

Concluding Remarks

The current rate of forest loss is unsustainable given the role that ecosystem services play for human economy and welfare. Capturing the value of these services in market economies can be challenging, but a number of innovative market-based mechanisms hold promise for bringing ecosystem services into market transactions in a more efficient and cost-effective way than has been achieved by alternative policy instruments. Market-based systems are already being used to create incentives to conserve and restore forests that regulate climate-changing GHGs. However, the effects of these measures on forest conservation and the provision of other ecosystem goods and services are limited by a complex and forest-constrained regulatory system and a voluntary market that fetches low carbon prices. For real effects on forest conservation to occur, the potential market opportunities highlighted in this chapter would need to be adopted in a post-Kyoto climate regime.

In an optimal world, the multiple benefits that forests provide, both now and in the future, would be incorporated into every forest conservation decision. As the world's nations move forward with climate negotiations, they have a great opportunity to strengthen the linkages between forest carbon market initiatives and market incentives for other forest ecological services, a strategy similar to that being explored by the Joint Liaison Group for the Rio Conventions.[44] Until this happens, the world can expect to continue to lose forests and their associated services in ways that harm ecological sustainability, human welfare, and the economy.

Notes

1. Food and Agriculture Organization of the United Nations (FAO), "Global Forest Resources Assessment 2005: Progress towards Sustainable Forest Management," Forestry Paper 147 (Rome: FAO, 2006).

2. World Resources Institute, "Ecosystem and Human Well-being: Biodiversity Synthesis," Millennium Ecosystem Assessment (Washington, 2005).

3. It is estimated that forest ecosystems as a whole contain 638 Gt of carbon (to a soil depth of 30 cm), roughly half of it being biomass and deadwood combined and half being soils and litter combined. FAO, "Global Forest Resources Assessment 2005."

4. Land-use change and forestry account for an estimated 18 percent of all global GHG emissions. World Resources Institute, "Navigating the Numbers: Greenhouse Gas Data and International Climate Policy" (Washington, 2005).

5. Study of the dynamics and composition of central Amazonian forests over the past two decades has shown accelerated productivity and dominance of faster-growing canopy and emergent trees, as well as a decline in slower-growing trees, most likely a result of rising atmospheric CO_2 concentrations. See W. F. Laurance and others, "Pervasive Alteration of Tree Communities in Undisturbed Amazonian Forest," *Nature* 428 (2004): 171–75.

6. L. Hannah, T. E. Lovejoy, and S. H. Schneider, "Biodiversity and Climate Change in Context," in *Climate Change and Biodiversity,* edited by T. E. Lovejoy and L. Hannah (Yale University Press, 2005), pp. 3–14.

7. Dynamic global vegetation models (DGVMs) used to simulate the response of global vegetation to a climate scenario produced with the HadCM2 general circulation model (GCM) have shown dramatic reduction in rainfall in southwestern Africa and the Amazon. See R. A. Betts and H. H. Shugart, "Dynamic Ecosystem and Earth System Models," in *Climate Change and Biodiversity,* edited by Lovejoy and Hannah, pp. 232–51.

8. C. S. Holling, "The Resilience of Terrestrial Ecosystems: Local Surprise and Global Change," in *Sustainable Development of the Biosphere,* edited by W. C. Clark and R. E. Munn (Cambridge University Press, 1996), pp. 107–32; K. M. Chomitz, E. Brenes, and L. Costantino, "Financing Environmental Services: The Costa Rican Experience and Its Implications," *Science of the Total Environment* 240 (1999): 157–69.

9. Preceding figures are from FAO, "Global Forest Resources Assessment 2005."

10. G. Heal, *Nature and the Marketplace: Capturing the Value of Ecosystem Services* (Washington: Island Press, 2000).

11. S. Pagiola, J. Bishop, and N. Landell-Mills, eds., *Selling Forest Environmental Services: Market-Based Mechanisms for Conservation and Development* (London: Earthscan, 2002).

12. E. B. Barbier, "Valuing Environmental Functions: Tropical Wetlands," *Land Economics* 70, no. 2 (1994): 155–73.

13. A. Schneider and H. Ingram, "Behavioral Assumptions of Policy Tools," *Journal of Politics* 52, no. 2 (1990): 510–29.

14. S. Polasky, C. Costello, and A. Solow, "The Economics of Biodiversity," in *Handbook of Environmental Economics,* vol. 3, edited by K.-G. Maler and J. R. Vincent (New York: Elsevier, 2005), pp. 1517–60.

15. Parties to CITES are required to monitor trade in endangered species and to report their trade records, which are compiled to inform the public about the global volume of trade in listed species. The treaty is enforced through reports on alleged infractions. See R. Bonnie, M. Carey, and A. Petsonk, "Protecting Terrestrial Ecosystems and the Climate through a Global Carbon Market," *Philosophical Transactions of the Royal Society of London,* Series A, 360 (2002): 1853–73.

16. M. Wells and K. Brandon, "People and Parks: Linking Protected Area Management with Local Communities," World Bank, Washington, 1992.

17. R. Stavins, "Experience with Market-Based Environmental Policy Incentives," in *Handbook of Environmental Economics,* vol. 1, edited by K.-G. Maler and J. R. Vincent (Amsterdam: Elsevier, 2001), pp. 355–435.

18. One example is the tradable permit system used to reduce sulfur dioxide emissions in the United States. For more information see Heal, *Nature and the Marketplace.*

19. J. Freeman and C. D. Kolstad, "Prescriptive Regulations versus Market-Based Incentives," in *Moving to Markets in Environmental Regulation: Lessons from Twenty Years of Experience,* edited by J. Freeman and C. D. Kolstad (Oxford University Press, 2007), pp. 3–17.

20. Because all market-based instruments are considered to be advantageous over traditional approaches for the reasons already given, we focus on the relative advantages of one type of market-based tool over another.

21. N. Stern and others, *Stern Review on the Economics of Climate Change* (London, 2006) (www.hm-treasury.gov.uk/media/8AC/F7/Executive_Summary.pdf); K. M. Chomitz and others, "At Loggerheads? Agricultural Expansion, Poverty," Policy Research Report (Washington: World Bank, 2006).

22. For additional information on UNFCCC CDM LULUCF project rules and requirements, see http://unfccc.int/kyoto_protocol/mechanisms/clean_development_mechanism/items/2718.php.

23. N. Taiyab, "Exploring the Market for Voluntary Carbon Offsets," International Institute for Environment and Development, London, 2006.

24. See www.climatestandards.org/index.html.

25. E. Harris, "Working Paper on the Voluntary Carbon Market: Current and Future Market Status, and Implications for Development Benefits," paper presented at the International Institute for Environment and Development, London, October 26, 2006.

26. Ibid. Evaluating 40 percent of the voluntary market, Harris estimated that LULUCF projects dominated the voluntary carbon market with 56 percent of market share.

27. See K. Capoor and P. Ambrosi, "State and Trends of the Carbon Market 2007" (Washington: World Bank, 2007).

28. See UNFCCC, "Reducing Emissions from Deforestation in Developing Countries: Approaches to Stimulate Action," 2005/CP/L2, December 6, 2005.

29. World Resources Institute (WRI), "Navigating the Numbers: Greenhouse Gas Data and International Climate Policy" (Washington: WRI, 2005). There is some uncertainty in the numbers. According to Intergovernmental Panel on Climate Change (IPCC) estimates, global land-use change and forestry emissions in the 1990s averaged 1.6 Gt of carbon per year ± 0.8 Gt. Recent estimates of global net emissions from land-use change in the tropics point to 1.1 ± 0.3 Gt of carbon per year, a much lower rate of deforestation and carbon emission than previously thought. See F. Achard and others, "Improved Estimates of Net Carbon Emissions from Land Cover Change in the Tropics for the 1990s," *Global Biogeochemical Cycles* 18, no. 2 (2004): GB2008, doi:10.1029/2003GB002142 (www.geo.ucl.ac.be/LUCC/lucc.html).

30. FAO, "Global Forest Resources Assessment 2005." Estimates based on calculations of emissions from deforestation of the Brazilian Amazon in R. A. Houghton and others, "Annual Fluxes of Carbon from Deforestation and Regrowth in the Brazilian Amazon," *Nature* 403 (2000): 301–04, and on calculations for top GHG emitting countries in WRI, "Navigating the Numbers."

31. J. Hughes, G. Daily, and P. Ehrlich, "Population Diversity: Its Extent and Extinction," *Science* 278 (Oct. 24, 1997): 689–92; G. Ceballos and P. Ehrlich, "Mammal Population Losses and the Extinction Crisis," *Science* 296 (May 3, 2002): 904–07.

32. "Additionality" means that carbon credits obtained from forest projects must be additional to the business-as-usual scenario. "Leakage" is shorthand for a project's not leading to losses outside the project area. "Permanence" means that measures are taken to prevent the carbon gain from eventually being lost as a result of major disturbances.

33. This requires not only that incentives be given to deforestation reductions but also that countries with large forest reserves and low deforestation rates be included in market design in order to prevent perverse incentives or negative market externalities that might lead to deforestation's becoming "profitable" for these countries. See G. A. B. da Fonseca and others, "No Forest Left Behind," *PLoS Biology* 5, no. 8 (2007): 216.

34. Bonnie, Carey, and Petsonk, "Protecting Terrestrial Ecosystems"; see also UNFCCC/SBSTA/2006/10, pp. 7–8.

35. M. Santilli and others, "Tropical Deforestation and the Kyoto Protocol," *Climatic Change* 71 (2006): 267–76; Instituto de Pesquisa Ambiental da Amazônia (Amazon Institute for Environmental Research), "Reduction of GHG Emissions from Deforestation in Developing Countries," 2006 (http://unfccc.int/resource/docs/2006/smsn/ngo/007.pdf); Environmental Defense, "Reducing Emissions from Deforestation in Developing Countries: Approaches to Stimulate Action," 2006 (http://unfccc.int/resource/docs/2006/smsn/ngo/009.pdf); C. Streck and S. M. Scholz, "The Role of Forests in Global Climate Change: Whence We Come and Where We Go," *International Affairs* 82, no. 5 (2006): 861–79.

36. S. Prior, C. Streck, and R. O. Sullivan, "A Carbon Stock Approach to Creating a Positive Incentive to Reduce Emissions from Deforestation and Forest Degradation," Centre for International Sustainable Development Law, 2006 (http://unfccc.int/resource/docs/2007/smsn/ngo/001.pdf).

37. L. Pedroni and C. Streck, "Mobilizing Public and Private Resources for the Protection of Tropical Rainforests: The Need to Create Incentives for Immediate Investments

in the Reduction of Emissions from Deforestation within the International Climate Change Regime," Climate Focus and CATIE, 2007 (www.climatefocus.com/newspubs/downloads/publications/REDD_Policy_Brief.pdf).

38. R. Coase, "The Problem of Social Cost," *Journal of Law and Economics* 3 (1960): 1–44.

39. Organization for Economic Cooperation and Development (OECD), *Handbook of Market Creation for Biodiversity: Issues in Implementation* (Paris: OECD, 2004).

40. Ibid.

41. Chomitz and others, "At Loggerheads?"; Freeman and Kolstad, "Prescriptive Regulations versus Market-Based Incentives."

42. N. Landell-Mills and I. Porras, "Silver Bullet or Fool's Gold? A Global Review of Markets for Forest Environmental Services and Their Impact on the Poor," International Institute for Environment and Development, London, 2002; M. Grieg-Gran, I. Porras, and S. Wunder, "How Can Market Mechanisms for Forest Environmental Services Help the Poor? Preliminary Lessons from Latin America," *World Development* 33, no. 9 (2005): 1511–27; S. Wunder, "Payments for Environmental Services: Some Nuts and Bolts," Occasional Paper 42 (Jakarta, Indonesia: Center for International Forestry Research, 2005).

43. Grieg-Gran, Porras, and Wunder, "How Can Market Mechanisms for Forest Environmental Services Help the Poor?"

44. The Joint Liaison Group was established in 2001 to improve the coordination and synergy of activities among the Rio Conventions: the UNFCCC, the United Nations Convention to Combat Desertification, and the Convention on Biological Diversity. See H. Hoffmann, "The Joint Liaison Group between the Rio Conventions: An Initiative to Encourage Cooperation, Coordination and Synergies," *Work in Progress: A Review of Research Activities of the United Nations University* 17, no. 1 (2003): 23–25.

PART TWO

The International Arena

3

History and Context of LULUCF in the Climate Regime

EVELINE TRINES

The subject of land use, land-use change, and forestry (LULUCF) was introduced in the run-up to the third Conference of the Parties (COP 3) in 1997 in Kyoto, Japan, at a very late stage in the negotiations that resulted in the adoption of the Kyoto Protocol. The discussion was hedged with a serious lack of understanding of the subject and a shortage of reliable estimates of existing and potential emissions and removals in the forest sector. This resulted in confusion and opposing positions, and as a result the decisions reached in Kyoto were vague and internally inconsistent. A period of intensive study and debate followed, which led to the creative interpretation of rules, modalities, and guidelines to operationalize what was thought to have been agreed upon in Kyoto. An onerous and complex set of rules was finally adopted at the seventh COP in Marrakech (2001) and at the ninth COP in Milan (2003).[1] Once the Kyoto Protocol entered into force, all these decisions were formally accepted by the COP, which also served as the first Meeting of the Parties (MOP) to the Kyoto Protocol in Montreal in 2005.

Since Kyoto, significant insights have emerged regarding the role "sinks" play in the global carbon budget and the potential role of forests in mitigating climate change and contributing to sustainable development, particularly in developing countries. These insights have gradually created a new context in which the role of forest sinks under a future climate regime could be shaped to effectively contribute to the mitigation of climate change and to a more sustainable use of natural

Table 3-1. *Total Potential Reductions for all GHGs for the Year 2030*[a]

Price per t CO_2e (U.S.$)	*Reduction in Mt CO_2e*
0–20	1,900–2,100
0–50	2,400–2,600
0–100	3,100–3,300

a. Total biophysical potentials for 2030 are 5,500–6,000 Mt CO_2e. The biophysical potential is the same as the technical potential and is the potential that could be achieved on the land available if there were no economic or other barriers.

resources. In this chapter I reflect on the negotiations context at the time of the Kyoto conferences and offer some new insights and a number of possible "lessons learned."

The Significance of the Sector

The role of LULUCF in the global carbon cycle is significant. Since the industrial revolution approximately 270 gigatonnes of carbon (Gt C) have been emitted as CO_2 into the atmosphere through fossil fuel burning and cement production, and about 136 Gt C as a result of land-use change, predominantly in forest ecosystems.[2] Tropical deforestation accounts for one quarter of global carbon emissions.

On the other hand, significant potential exists for economic greenhouse gas (GHG) mitigation in the agricultural sector.[3] Table 3-1 sets out a range of potential GHG reductions for the year 2030 at different prices for a tonne of CO_2 equivalent (CO_2e).[4] The overall economic mitigation potential in the forestry sector can be estimated in the range of 2,000–4,000 megatonnes (Mt) CO_2 per year against all prices.[5] Considering these numbers, the overall economic mitigation potential in agriculture and forestry can be regarded as significant in comparison with total global CO_2 emissions (excluding agriculture and forestry) of 20,000–25,000 Mt CO_2 per year. The LULUCF sector can be considered part of the "problem" but also potentially part of the solution. Despite relative low costs and many co-benefits of this sector, its potential is barely tapped, owing to a number of types of barriers. A large proportion of this mitigation potential lies in developing countries, or those with economies in transition. For instance, 80 percent of the total global agricultural mitigation potential is found in developing countries.[6]

The Origin and Nature of the LULUCF Agreement

In Kyoto in 1997, parties to the United Nations Framework Convention on Climate Change (UNFCCC) agreed to "Quantified Emission Limitations or Reduction Commitments" (QELRCs) for industrialized countries and countries with economies in transition that would reduce total net greenhouse gas emissions below 1990 levels by 5.2 percent.[7] The countries are listed in Annex I of the convention (they are the so-called Annex I Parties), and the QELRCs are included in Annex B of the Kyoto Protocol. These targets apply to the first commitment period (CP1) under the Kyoto Protocol, the five years from 2008 through 2012.

The Kyoto Protocol established three flexibility mechanisms that can be used by Annex I Parties to meet their QELRCs: Joint Implementation (JI) between industrialized countries and countries with economies in transition (Annex I Parties), emissions trading (ET), and the Clean Development Mechanism (CDM).[8] Under these mechanisms, private and public entities can generate net emission reductions that can be counted against the QELRCs. Emission credits can also be generated through LULUCF projects. Whereas JI does not limit eligible LULUCF activities to particular sectors, the CDM allows for only a small percentage of afforestation and reforestation (AR) project activities under stringent conditions in the first commitment period. These conditions were agreed to only in Milan during COP 9 in 2003, six years after the establishment of the Kyoto Protocol.

Annex A of the Kyoto Protocol lists the sectors and gases that must be used by Annex I Parties to calculate QELRCs. The list includes agriculture but excludes land-use change and forestry (LUCF).[9] In practice this means that to the extent that a country uses net emission reductions achieved through LUCF activities to achieve compliance, it does not need to achieve emission reductions in other sectors. In this way LUCF credits can offset emission reductions in other sectors; therefore they are sometimes referred to as "offsets." If LUCF had been an integral part of Annex A, it would have formed part of the QELRCs rather than a potential offset. Some viewed this offset capacity of LUCF credits as an unacceptable way of "watering down" the Kyoto agreement.

The only LUCF activities that have to be taken into account by Annex I Parties to reach their QELRCs in the first commitment period are afforestation, reforestation, and deforestation. In addition, Annex I countries may elect to include cropland and grazing land management, revegetation, and forest management during CP1.[10] Forest management may be used only to a limited extent: country-specific caps have been agreed upon to limit the offset capacity of forest management. Annex I Parties who want to use any of these additional activities under article 3.4 must decide to do so in advance of the commitment period and cannot change this decision afterward. Credits from AR project activities under the CDM may also contribute to achieving emission reduction targets in Annex I countries. They are limited, however, to 1 percent of the base-year emissions of the Annex I country times five (for the number of years of the commitment period). Figure 3-1 sets out how commitments to reduce emissions in non-LUCF sectors can be reduced by the LUCF sector plus AR credits from the CDM during the first commitment period. It must be emphasized that under the Kyoto Protocol, reducing emissions from

Figure 3-1. *How Commitments to Reduce Emissions from Non-LUCF Sectors Can Be Reduced by LUCF Net Emissions and Removals*[a]

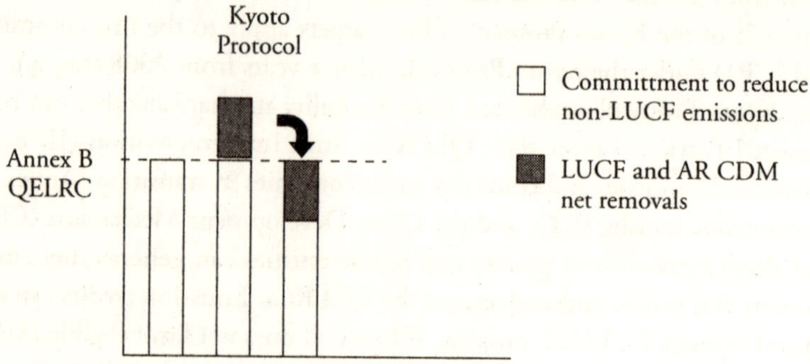

Source: Modified from E. Trines and others, "Policy Options for the Integration of Agriculture and Forestry in a Post-2012 Climate Regime," unpublished paper, 2007.

a. Includes CDM AR project activities in the first commitment period. Because QELRCs were set before modalities for LUCF were decided, any effort to reduce emissions in that sector or to sequester carbon does not lead to greater mitigation of climate change but instead removes the necessity to make similar efforts in non-LUCF sectors. Hence, LUCF credits became "offsets," offsetting non-LUCF emissions.

tropical deforestation in developing countries cannot be credited in CP1. Thus a large mitigation potential in non–Annex I countries has been ignored.

The Evolution of the Current Agreement

One of the principal reasons behind many of the rules governing LULUCF is that QELRCs were set before agreement was reached on LULUCF (leading to the situation shown in figure 3-1). The proponents of using LULUCF activities to mitigate climate change (the United States, Australia, Canada, Japan, New Zealand, a group of Latin American countries excluding Brazil, and some nongovernmental organizations [NGOs] such as the Union of Concerned Scientists, Environmental Defense, and the Nature Conservancy) and the opponents of doing so (the EU, Brazil, the Alliance of Small Island States, and some other NGOs such as the World Wildlife Fund, Greenpeace, and Friends of the Earth) consequently tried, respectively, to increase and decrease the share of net removals that could fall under the agreement. As a result of this debate, the design of the rules to operationalize the implementation of the Kyoto Protocol regarding LULUCF became the instrument by which the parties for and against including LULUCF tried to achieve their objectives. If LULUCF had been included in Annex A in the first place and therefore had been an integral part of the QELRCs, this could have been avoided. Because this was not the case, the degree to which LULUCF could

Figure 3-2. *Potential Effects of Including LULUCF as an Integral Part of Annex A of the Kyoto Protocol*[a]

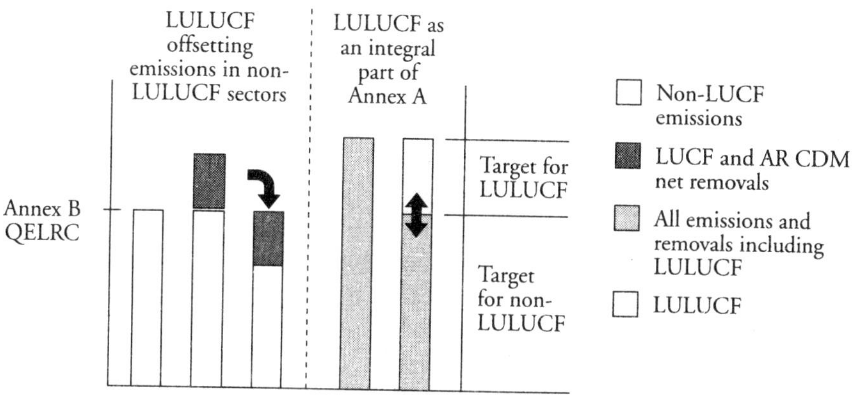

a. Left, current situation under the Kyoto Protocol: LULUCF offsetting emissions in non-LULUCF sectors. Right, alternative option: LULUCF as an integral part of Annex A.

contribute to mitigation commitments became increasingly a subject of debate. Unfortunately, reliable country data were lacking at the time, and it was impossible to make reliable projections of what part of that mitigation potential could be realized by individual countries during the commitment period if the use of LULUCF options were unlimited. In this information vacuum, opponents tried to keep the scale of admissible LULUCF activities even further restrained in order to limit the extent to which LULUCF activities could assist parties in meeting their emission reduction targets.

Figure 3-2 illustrates that net removals in the LULUCF sector could have raised commitments to cut emissions overall if the sector had been included as an integral part of Annex A of the Kyoto Protocol. If this were the case, then the LULUCF sector would not be able to offset net reductions in non-LULUCF sectors.

Uncertainty related to measuring and monitoring carbon sequestration, as well as baseline determination and projections of trends, was another concern. Some sources of uncertainty were subsequently addressed by the "Good Practice Guidance on LULUCF" issued by the Intergovernmental Panel on Climate Change (IPCC) in 2003, and some were addressed by the new IPCC Inventory Guidelines (2006),[11] but these modalities did not exist at the time of the negotiations.

The issue of "nonpermanence" added to the complexity of the discussions, because carbon sequestration can be reversed: net emission removals by forests and in agriculture are not permanent. Nonpermanence is unproblematic in Annex I countries, because they continuously measure and monitor sinks. This means that any future release of carbon into the atmosphere will be accounted for at the appropriate point. However, in respect of CDM projects, nonpermanence is a large problem, because the obligation to monitor the existence of the carbon stocks expires at the end of the project's lifetime, and therefore carbon releases ultimately may not be accounted for.

The most contentious issues surrounding LULUCF were resolved by the seventh Conference of the Parties in Marrakech in 2001, except for CDM AR project activities, which took Parties until COP 9 to resolve in Milan in 2003.[12] The additional issues related to CDM AR project activities were the following:

—The "permanence" issue was resolved by creating temporary crediting. Credits for sinks under the CDM will be issued only for a fixed period of time, either five years (temporary Certified Emission Reductions [tCERs]) or the length of the project (long-term Certified Emission Reductions [lCERs]). Different modalities apply to tCERs and lCERs, transferring the liability for the reversal of sequestration from the project in the case of tCERs to the user or buyer of the credits in the case of lCERs.[13]

—"Additionality" and "baseline" were defined as the effects of undertaking an activity in comparison with the business-as-usual (baseline) scenario, to ensure that any issued credits represent additional removals.

—"Leakage" was defined as an activity undertaken at one location whose carbon effect might be negated by that same activity or another activity undertaken elsewhere. Credits generated by the activity in such cases do not represent real overall emission reductions.

—Issues relating to national sovereignty were discussed. Some countries were afraid that situations might arise in which a country would threaten to deforest if no funds were made available for conservation. By contrast, a risk was perceived that foreign entities would buy land in other countries with the objective of taking that land out of production in order to conserve the vegetation (carbon stocks) on it. This phenomenon was sometimes referred to as "hostage taking," which could interfere with a country's national development goals.

—Social and socioeconomic issues arose. Emissions and removals in the LULUCF sector are most often related to land use and not to a "point source" as is the case in other sectors. For instance, the local population might lose full access to all services provided by the land as carbon sequestration becomes the dominant one, ruling out other uses such as logging and food production. This could lead to situations in which the AR activity might not be durable because the local population did not support the use of land as a forest.

A suite of additional rules, modalities, and guidelines was developed to deal with these issues, in addition to a cap that was placed on the use of CDM AR.[14]

Current Situation

Where does that leave us now? For a start, it is commonly agreed that the rules, modalities, and guidelines for LULUCF are complex and onerous. In addition, the scale at which LULUCF activities are undertaken and the extent to which they will help Parties meet their targets during the commitment period are limited. One of the reasons for this is that after the withdrawal of the United States from the Kyoto Protocol, ratification by Russia, Japan, and Canada was essential for the agreement to enter into force.[15] As a result, generous caps on the crediting of GHG emission reductions and removals from forest management were established for these countries, but at the same time the most important *demandeur* of emission reductions, the United States, left the scene. The rules for CDM AR project activities, on the other hand, were set only after years of negotiations. This left the marketplace in doubt for a long time about whether or not this was an interesting investment opportunity. Finally the EU, which by now is probably the most important *demandeur* for credits, decided to exclude LULUCF from its Emissions Trading Scheme (EU ETS), putting up yet another barrier to investments in this sector and in developing countries.[16]

All these factors together have distorted a healthy balance between supply and demand, rendering the market alone unable to sort out these imperfections. For instance, even though the cap on the use of credits from CDM AR projects was set at a very low level, not even 1 percent of the allowed 1 percent of base-year emissions of Annex I parties times five will be realized during the first commitment period, judging from the list of projects that are currently in the CDM pipeline to be officially registered.[17] This is mainly due to a lack of investor interest resulting from, among other things, the slow decisionmaking of the parties regarding rules governing LULUCF and the complexity of the rules, which leads to high transaction costs.

This situation can be expected to continue for the foreseeable future during CP1. A significant reason for this is that the EU, after the United States pulled out of the treaty, became the largest buyer of credits but excludes LULUCF from its EU-wide ETS. On top of this, the review of the EU ETS that could lead to amendments for its second phase (2008–2012) was not initiated in time by the European Commission. The review and associated text proposals for amendments should have been completed in June 2006. The de facto result of this late timing meant that even if such a proposal had come forward, and even if it lifted the ban on LULUCF credits, it would have been too late for the market to respond in time to generate major investments in the sector that would lead to significant additional net LULUCF emission reductions in the period between 2008 and 2012.[18]

Looking Ahead

Where do we go from here? At present only a small proportion of the mitigation potential in agriculture and forestry is being tapped in all regions of the world. In non–Annex I countries this is due mainly to barriers outside the influence sphere of the climate change regime, such as poverty and hunger,[19] but also to a lack of political will of Annex I countries to date to put LULUCF on equal footing with other sectors. Consequently, most mitigation that does occur in non–Annex I countries in the LULUCF sector is a by-product of other policies.

The reluctance noted in some developing countries to embrace the mitigation options in agriculture and forestry may also be related to fears that economic growth will be hindered if land-use change is halted. Therefore, solutions are needed that provide economic opportunities as well as carbon sequestration. But in general, as long as the barriers persist, no significant mitigation will be achieved, even if good policy options are available.[20]

There are positive signs as well. Interest is growing among nations and policymakers in including the reduction of emissions from deforestation (and possibly forest degradation) in developing countries as a mitigation option after 2012, when the first commitment period of the Kyoto Protocol expires. Some parties are proposing that past deforestation rates in non–Annex I countries be used as the baselines against which future rates are compared, so that reductions in the rate of deforestation can be rewarded. The difficulty again lies in designing effective, environmentally sound, and equitable accounting mechanisms for turning the principle into an operational system.[21] One serious point of contention seems to be whether the reward for reducing emissions from deforestation should come from a dedicated fund or be generated through the market by issuing tradable carbon credits.

Future climate change regimes may be able to help reduce barriers by promoting the integration of climate change considerations into general development processes in non–Annex I countries. This might include reward systems for mitigating climate change in the LULUCF sector, providing income to local populations by paying for carbon benefits.

A further question is the extent to which the costs of policies and measures that lead to reductions in emissions from deforestation and forest degradation can be covered by revenues arising from international sales of, or funding related to, carbon offsets.[22] This is an open question when one considers the enormous profits being made today by large corporations engaged in logging and other carbon-intensive land uses (for example, cattle ranching, agricultural production of crops such as soybeans). Their profit margins greatly outstrip the prices currently being paid for carbon credits.

A final consideration, and probably the most important one, is policymakers' recognition that if nations want to achieve the ultimate objective of the UNFCCC articulated in its article 2—stabilization of greenhouse gas concentrations in the atmosphere at a level that would prevent dangerous anthropogenic interference with the climate system—then cuts in emissions will have to be deeper in future commitment periods. If that is to happen in a cost-efficient and effective manner, then emissions from agriculture and forestry must be included in a future climate regime. And because a large proportion of the global mitigation potential in those sectors lies in non–Annex I countries or economies in transition, the greater participation of developing countries in a future climate change mitigation regime is of critical importance.[23]

Notes

1. Rules for afforestation and reforestation project activities under the Clean Development Mechanism in the first commitment period of the Kyoto Protocol, draft decision CP/-9 of ninth session of the COP of the UNFCCC in 2003, adopted at the first session of the COP/MOP of the Kyoto Protocol in 2005 as decision 5/CMP.1.

2. R. Watson and others, eds., *Land Use, Land-Use Change and Forestry,* Special Report of the Intergovernmental Panel on Climate Change (IPCC) (Cambridge University Press, 2000). One tonne of C equals 3.67 tonnes of CO_2.

3. The economic potential means the potential that could be achieved on the land available at a specified price paid for carbon dioxide equivalents.

4. The ranges represent the differences among scenarios A1b, A2, B1, and B2 in the IPCC's Special Report on Emission Scenarios. P. Smith and others, "Greenhouse Gas Mitigation in Agriculture," *Philosophical Transactions of the Royal Society* (in press).

5. E. Trines and others, *Integrating Agriculture, Forestry and Other Land Use in Future Climate Regimes: Methodological Issues and Policy Options,* Climate Change Scientific Assessment and Policy Analysis report 500102002 (Netherlands Environmental Protection Agency, 2006); P. C. Benítez and others, "Global Potential for Carbon Sequestration: Geographical Distribution, Country Risk, and Policy Implications," *Ecological Economics* 60, no. 3 (2007): 572–83; B. Strengers, J. Van Minne, and B. Eickhout, "The Role of Carbon Plantations in Mitigating Climate Change: Potentials and Costs," *Climatic Change* (in press).

6. Trines and others, *Integrating Agriculture.*

7. Article 3, Kyoto Protocol.

8. Respectively, articles 6, 17, and 12, Kyoto Protocol.

9. As a result of the 1996 revised IPCC Greenhouse Gas Inventory Guidelines, net CO_2 emissions or removals from agriculture are also inadmissible. Non-CO_2 emissions from agriculture are to be accounted for under "Agriculture," but CO_2 from soils (in both agriculture and forestry) are accounted for under LUCF. In Mauritius in 2006 the IPCC accepted new inventory guidelines that take a more comprehensive approach to emissions and removals from land-based activities by combining agriculture, forestry, and other land use (AFOLU) into one volume of guidelines. However, for the first commitment period the 1996 guidelines apply.

10. Articles 3.3 and 3.4, Kyoto Protocol.

11. Available at www.ipcc.ch/.

12. The Milan decisions were adopted at the first Conference of the Parties to the UNFCCC serving as the Meeting of the Parties to the Kyoto Protocol in Montreal in 2005 in FCCC/KP/CMP/2005/8/Add.1 as Decision 5/CMP.1, "Modalities and procedures for afforestation and reforestation project activities under the clean development mechanism in the first commitment period of the Kyoto Protocol."

13. Ibid., Annex, section J, "Issuance of tCERs and lCERs," and section K, "Addressing non-permanence of afforestation and reforestation projects under the CDM."

14. FCCC/CP/2001/13/Add.2, Decision 17/CP.7, "Modalities and procedures for a clean development mechanism as defined in Article 12 of the Kyoto Protocol," paragraph 7.

15. This is because of the double trigger that had to be met for the Kyoto Protocol to enter into force. See article 25 of the protocol, which states: "This Protocol shall enter into force on the ninetieth day after the date on which not less than 55 Parties to the Convention, incorporating Parties included in Annex I which accounted in total for at least 55 per cent of the total carbon dioxide emissions for 1990 of the Parties included in Annex I, have deposited their instruments of ratification, acceptance, approval or accession."

16. "Directive 2003/87/EC of the European Parliament and of the Council of 13 October 2003 establishing a scheme for greenhouse gas emission allowance trading within the Community and amending Council Directive 96/61/EC (as amended by Directive 2004/101/EC of the European Parliament and of the Council of 27 October 2004)," article 11a, paragraph 3, subsection b.

17. As of July 2007 only one CDM AR project has been registered. See http://cdm.unfccc.int/Projects/index.html.

18. E. Trines and A. Tas, "The Integration of LULUCF in the EU's Emissions Trading Scheme in a Post-2012 Regime to Mitigate Climate Change," report for the Swedish Environmental Protection Agency, 2006.

19. Trines and others, *Integrating Agriculture.*

20. Ibid.

21. M. Skutsch and others, "Clearing the Way for Reducing Emissions from Tropical Deforestation," *Environmental Science and Policy* 10, no. 4 (2006): 322–34.

22. This is independent of the question of how such international finance is arranged. It is assumed here simply that a country will in some way receive financing for carbon savings from outside purchasers or donors and will use at least part of these financial resources to fund domestic policies and measures. E. Trines and others, "Policy Options for the Integration of Agriculture and Forestry in a Post-2012 Climate Regime," unpublished paper, 2007.

23. Ibid.; Trines and others, *Integrating Agriculture.*; Skutsch and others, "Clearing the Way."

4

Risks and Criticisms of Forestry-Based Climate Change Mitigation and Carbon Trading

JOHANNES EBELING

During the discussions leading up to the adoption of the Kyoto Protocol, the inclusion of forestry-related emissions in the emerging treaty turned out to be one of the most contentious points and lay at the center of many heated and emotional debates. The attention of several nongovernmental organizations (NGOs) remains focused on "carbon forestry," and the debate is flaring up again in the context of rapidly expanding voluntary carbon markets and the increasing popularity of "carbon offsets." Proposals aimed at the years after 2012, when the current Kyoto commitment period ends, have also put a spotlight on the land use, land-use change, and forestry (LULUCF) sector as policymakers consider options to include emission reductions from avoiding deforestation in future climate change mitigation regimes.

The concerns of some countries and much of the environmental NGO community in the past have led to restrictive regulations for forestry activities eligible under the Kyoto Protocol's Clean Development Mechanism (CDM). In comparison with energy and industrial CDM projects, in which numerous creative ways to reduce emissions are eligible, forestry projects under the CDM are limited in the following ways: (1) they are confined to afforestation and reforestation (AR) activities; (2) neither emission reductions from forest conservation nor carbon removals from improved forest management are currently eligible under the CDM; (3) CDM forestry projects are awarded "temporary" credits rather than "regular," permanent carbon credits; and (4) forestry credits can be used only within narrow limits by the parties to the Kyoto Protocol. Partly as a result of complicated requirements and the delayed agreement on these requirements, only one forestry project had gained the approval of the CDM Executive Board as of February 2008, versus more than 900 registered projects overall.[1]

Why has the forestry sector faced so many criticisms, and are they justified? In this chapter I discuss the most pertinent issues in the debate and point out their importance in the context of discussions about a future international agreement on climate change mitigation. These issues pertain to the alleged diversion of climate mitigation efforts through forestry, the risk of nonpermanence of carbon stored in the biosphere, negative carbon "leakage" from forestry projects, perverse incentives, the risk of creating "hot air"—credits that are unmatched by additional emission reductions—and potential negative environmental and social effects of forestry projects. Most of these factors are relevant in the context of both "carbon sinks" (the sequestration of atmospheric CO_2 in tree-planting projects) and carbon sources (CO_2 emissions from deforestation). Where applicable, I highlight differences that exist between those two categories of carbon forestry projects.

Diversion of Mitigation Efforts and Market Flooding

One of the main objections to allowing countries to meet their emission reduction targets through forestry activities—both domestically and via the flexible mechanisms—was based on fundamental differences in opinion regarding the desired ways in which climate change should be mitigated. From a purely pragmatic point of view, CO_2 is CO_2, and limiting its atmospheric concentrations through, for example, energy efficiency measures is effectively the same as reducing emissions from the burning of forests or removing them by planting trees.

In contrast, many environmentalists held and continue to hold the view that forestry activities do nothing to address the root cause of climate change, namely, the combustion of fossil fuels.[2] Conserving forests and creating forest plantations might therefore divert attention away from the necessary restructuring of the world's fossil-fuel-based economies and, by delaying such efforts, create a negative legacy for the future. Importantly, the suggestion to include carbon sinks and avoided deforestation in the CDM was made when greenhouse gas (GHG) emission targets were already set by the Kyoto parties. This led to fears that forestry activities would simply "offset" other GHG mitigation efforts, creating no net reduction in overall emissions. This concern was summarized by the pointed NGO slogan "Don't sink Kyoto!"[3]

Depictions of vast quantities of cheap forestry credits flooding the carbon markets and depressing the price of tradable emission permits led to similar concerns. Cheap credits, although commercially desirable, would decrease incentives to invest in energy-related emission abatement. In this line of reasoning, the large potential of reforestation and avoided deforestation to mitigate climate change is the cause for concern. If planting or protecting forests were significantly cheaper, on average, than reducing emissions from other sources, then a large number of low-cost forestry credits could be created. Given a temporarily fixed demand for these credits, the additional supply could quickly saturate ("flood") the global markets, arresting prices at relatively low levels and effectively reducing investments in more expensive renewable energy and energy efficiency measures. Of course no one knows whether the necessary dramatic reforestation efforts or reductions in deforestation rates could be achieved at a competitively lower price, yet the sheer speculation that this might happen has been used to argue that large-scale forestry activities would effectively halt many other, more costly abatement efforts and that this would be undesirable. Again, this concern arises in the context of limited demand for carbon credits caused by a political decision to aim at relatively meager emission reductions over a short time span.[4]

Including an additional sector with a wide range of low-cost to high-cost abatement options, such as forestry, would lower the aggregate marginal and total costs of climate change mitigation. This is especially true if (and it is a big if) reforestation and reduction of emissions from deforestation prove to be considerably cheaper than the main existing mitigation alternatives, as is alleged by critics and hoped for by supporters. From a positive viewpoint, a lower overall cost curve for emission abatement would give the world community two options: reaching a given mitigation target at a lower cost and realizing a more ambitious target for the same total cost. However, the benefit of reaching a more ambitious mitigation target for the same cost entails what can be perceived as a disadvantage, namely, that some of the overall investment now flows into reforestation and forest conservation rather than other sectors. It is largely a normative, political decision whether the added climate benefit justifies a certain redirection of investment flows. If countries strive toward greater emission reductions, an additional mitigation option will provide more opportunities to achieve this. In addition, a stricter emission target would likely undo any crowding out.

How many carbon credits from reforestation and avoided deforestation could realistically be generated is a difficult question. The actual costs of reducing deforestation in particular are unknown (beyond relatively small pilot projects), and only rough modeling estimates exist.[5] It is clear, however, that significant hurdles to reducing deforestation would have to be overcome in the real world. Many governments have struggled for several decades to reduce illegal land-use conversions and illegal logging and are frequently hampered in their attempts by a lack of institutional capacity and rampant corruption. Some of the countries with the highest deforestation rates also score very low in governance indicators, a problem that is particularly severe in forest frontier areas and that may make it difficult for these countries to participate in the carbon market.[6]

This is not to say that forestry does not represent a large untapped mitigation potential, but a flood of cheap carbon credits from this source appears questionable at the very least. Furthermore, the relative scale of tradable credits has to be put into perspective against projected increases in fossil-fuel-based GHG emissions in developing countries. Some developing countries are already among the major emitters of GHGs today, and projections see China and others as surpassing the current top industrialized-country emitters before long. According to at least one study, China has already topped U.S. emissions in 2006.[7] If current trends persist, emissions from present-day developing countries will more than triple by 2050, accounting for more than two-thirds of vastly increased emissions by then (figure 4-1).[8] These will present enormous mid-term abatement challenges in the industrial and energy sectors of those countries.

Some stakeholders will probably remain concerned about "risks" of market flooding through forestry-based mitigation. Fortunately, there are relatively easy ways to counter the theoretical possibility that forestry credits would distort international carbon markets. In the framework of the Kyoto Protocol, negotiators responded to the challenge by setting a cap for credits from AR. Something similar could be applied to all forestry credits and could be revised once more experience has been gathered. The Kyoto Protocol sets the cap for CDM AR at a maximum of 1 percent of an industrialized country's base-year emissions for each year of the protocol's first commitment period.[9] Experiences to date show that the 1 percent limit is in fact far from being reached, and this might repeat itself in the context of avoided deforestation. Ultimately, however, setting more ambitious emission reduction targets would be a far better option and would avoid market flooding while utilizing the full potential of the forestry and other sectors and their lower aggregate marginal abatement costs.

In this context it should be noted that the extremely cheap reductions in hydrofluorocarbons (HFCs) and nitrous oxide (N_2O) emissions are in fact eligible for crediting under the current CDM regulations. This has led to a situation in which one-third of the current CDM mitigation potential (and 70 percent of credits issued to date) is achieved through projects reducing emissions of these two gases.[10] It is easy to imagine the scale of crowded-out investment through these activities, which do not significantly reduce reliance on fossil fuels or provide the arguably immense co-benefits associated with conservation and reforestation activities.

Permanence

Permanence of emission reductions was one of the main controversies during the design of the Kyoto regime's rules. The concern is that when emissions are reduced by implementing a fuel switch project or an energy efficiency measure in the field of fossil-fuel consumption, it will have a permanent effect. For example,

Figure 4-1. *Projected Greenhouse Gas Emissions under a Business-as-Usual Scenario, 1995–2050*

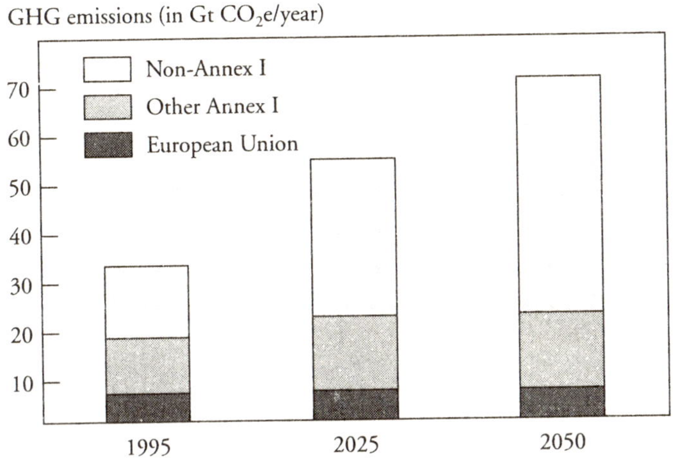

Source: Author's elaboration using data from European Commission, "Greenhouse Gas Reduction Pathways in the UNFCCC Process up to 2025," study commissioned by DG Environment, 2003 (CPI baseline scenario).

if an installation producing electricity from solar energy goes out of service after several years and the old oil-fired power station comes back online, the emission reductions that have been achieved will not become undone and there will be—permanently—less CO_2 in the atmosphere.[11] In contrast, planting a forest as a carbon store carries with it a reversal risk. If the newly created sink burns or is logged, then the sequestered CO_2 will be released back into the atmosphere and there will be no net emission reduction in the end.

Offsetting fossil-fuel-based emissions through forest plantations, critics claim, creates a liability for the future rather than representing a "sustainable" climate mitigation strategy. Offsetting the emissions from fossil-fuel burning by building up a carbon sink in the form of a forest plantation allows the initial emitting activity to continue. In case of destruction of the carbon sink through human activities or nonanthropogenic causes (for example, droughts, fires, pests, storms), the temporarily stored forest carbon would be added to atmospheric GHG concentrations, and no emissions would have been "offset." "You can't sequester the lithosphere in the biosphere" was a conclusion drawn by some NGOs.

The negotiators of the Kyoto Protocol acknowledged the nonpermanence risk of forest carbon and dealt with it in a rather complicated way. AR forestry projects cannot gain permanent certified emission reductions (CERs) but instead may receive temporary credits. The basic idea behind these credits is to limit their

Figure 4-2. *Permanence of Emission Reductions from Forest Conservation*

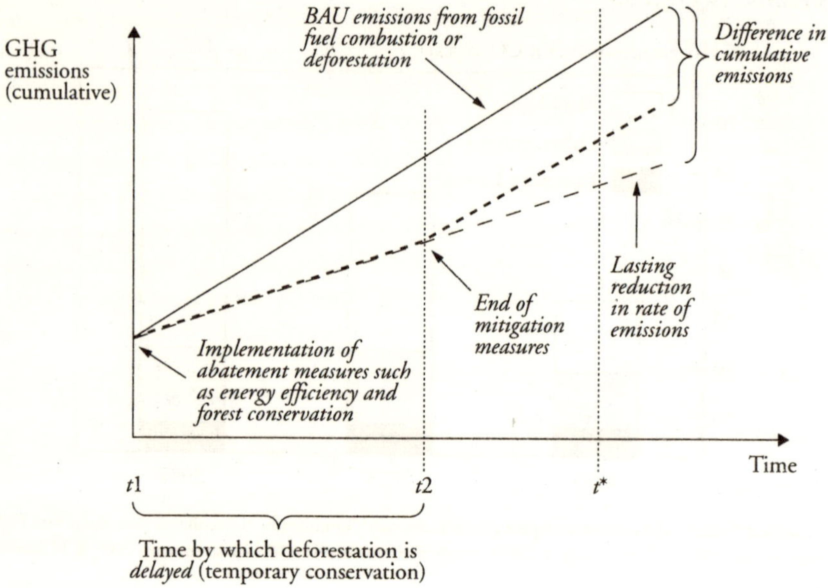

validity to a certain time span. At expiry, the temporary credits have to be replaced by a permanent credit. Each issued forestry CER thus has an attached replacement liability, which can be fulfilled either by demonstrating that the underlying sequestered carbon stock still exists at the time of reverification or by retiring a nonforestry carbon credit (see chapter 6).

One crucial difference emerges when comparing the nonpermanence issue of forest plantations with avoided deforestation. Conserving forests decreases a source of emissions rather than creating a sink for them. In this way, emissions from deforestation are not inherently different from emissions created through the combustion of fossil fuels, nor are there fundamental differences between the respective mitigation measures. Protecting carbon stocks in existing forests now can mean that the stored CO_2 is emitted later, thereby merely delaying emissions from a defined source—but the same argument can be made for fossil fuels. I look at this line of reasoning by considering figure 4-2.

The diagram represents the cumulative effect of GHG emissions from different sources in a country over time. Each year, additional emissions increase the concentration of CO_2 and other gases in the atmosphere. The solid line represents the business-as-usual (BAU) scenario of unabated emissions, whereas the dotted line represents an abatement scenario in which annual emissions are reduced. This leads to a slower rate of change, but cumulative emissions still increase over time because CO_2 and other GHGs are "stock pollutants" with a long atmospheric residence time.

An end of the mitigation measures is conceivable for both energy-based measures and forest conservation: a solar electricity installation could be retired and replaced by a fossil-fuel plant, just as land-use regulations restricting the conversion of forest to agriculture could be reversed. The result in each case is a reversal back to historical emission rates as in the BAU scenario. This development, however, does not correspond to a reversal of the achieved emission reductions. At any time after the end of mitigation measures, the atmospheric concentration of GHGs is lower than would otherwise have been the case. Although the same barrel of oil or the same tree might be burned just a few years later than in the BAU scenario, the time by which this is delayed corresponds directly to permanent benefits for the global climate (as long as there is no "overshoot" relative to BAU emission rates).

An additional benefit of temporary forest protection measures exists if they coincide with national-level "forest transitions"—trends of decreasing deforestation rates over time due to shifts in labor markets and demands for environmental services from forests associated with economic development.[12] In countries with decreasing deforestation rates, forests that are conserved through "temporary" conservation measures could face lower conversion pressure once these measures terminate and remain permanently conserved. Concerns remain, however, about the possibility of large deforestation "rebounds" if countries abandoned RED (reducing emissions from deforestation) policies. Mechanisms such as banking a percentage of RED credits as an assurance for future deforestation increases could accommodate this concern.[13]

In summary, problems of nonpermanence exist for carbon sinks in the form of forest plantations but not necessarily in the case of reducing emissions from deforestation. In relation to the former, they are effectively dealt with through the creation of a special type of nonpermanent carbon credits in the framework of the Kyoto Protocol. With regard to the latter, emission reductions do occur even in the case of temporary emission mitigation efforts and are fundamentally the same as reduction activities in nonforestry sectors.

Carbon Leakage

Reforestation or forest conservation in one place can, in principle, cause deforestation elsewhere. Such effects of climate mitigation activities outside a project's boundaries have been labeled "leakage." Leakage can be subdivided into two basic categories: leakage between locations, or "activity shifting," and indirect "market leakage." It can occur at different scales as domestic and international leakage.[14] In principle, leakage not only exists in a negative form but could also have positive climatic effects, such as by promoting good practices of forest management outside the project area.

Activity shifting is the simple case of displacement of agricultural or logging activities from one place to another because of the implementation of a carbon project in an area. For example, farmers or cattle ranchers could be forced to leave an area because of the establishment of a timber plantation. Similarly, a logging company or farmers at the agricultural frontier might have to cease their activities because of a forest conservation project. In all these cases it can easily be imagined that other remaining forests are targeted in the search for agricultural land or timber and that carbon emissions increase outside the carbon project boundaries when trees are felled.[15] Leakage at the level of a small-scale conservation project can be considerable and can offset up to or even more than 100 percent of carbon benefits if it is not controlled—for example, when farmers simply move on to clear an adjacent forest plot.

Concerns over project-level leakage greatly hampered the inclusion of forestry activities under the CDM and were an important factor in the exclusion of forest conservation from the mechanism. Although land-use modeling has made advances since those days, leakage remains difficult to estimate if forestry competes with other land uses. Therefore, most CDM forestry activities to date have been carried out on marginal, degraded land with few other productive uses, and the first CDM AR methodologies that gained approval allowed for only such cases. In general, whenever risks of leakage cannot be quantified, CDM project developers have to assume a worst-case scenario and deduct the effect of a hypothetical clearing of primary forest from the project benefit.

Market leakage works through the market law of supply and demand as opposed to causing specific actors to increase emissions in another place. In many cases forest conservation, in addition to stopping the clearing of forests, prevents the production of timber and crops on the protected land. Consequently, markets for these products tighten and timber and crop prices may rise.[16] In the most positive scenario, such price increases encourage intensified forestry and agricultural production in other areas without creating an additional risk of deforestation. In a less favorable scenario, particularly when land-use regulations are poorly enforced, higher prices provide an additional incentive to clear forests for timber or agriculture elsewhere, thereby reducing the net benefits of the climate mitigation project. Reforestation may similarly impede agricultural production on plantation areas, potentially reducing its climate benefits through market leakage.

The mechanism could work in a similar way for nonforestry CDM projects. The addition of a renewable energy plant to an electricity grid in most cases increases the supply of energy, thereby decreasing prices of electricity and of fossil fuels used for its production. Given variable consumption rates ("elastic demand," in the language of economics), electricity and fuel use will increase in response to lower prices, and likewise there will be less incentive to increase energy efficiency elsewhere. Because market leakage is difficult to calculate, it is not taken into account in either forestry or energy projects when net emission reductions are calculated.

When we consider climate benefits from reforestation or avoiding deforestation at the national level, leakage becomes irrelevant in terms of carbon accounting. Any subnational offsite effects would be reflected within the same system, as in the emission inventories established by current Annex I countries. This is one of the main advantages of the proposals to measure reduced emissions from deforestation against national deforestation baselines rather than at the level of individual conservation projects.

Even under national-level forestry crediting schemes, however, a theoretical potential for *international leakage* exists. If only some countries participated in a regime for reducing deforestation, then global markets might shift supply and demand patterns for timber or agricultural commodities across national borders. During the design of the Kyoto architecture, similar concerns were voiced regarding "inter-annex leakage" from industrialized (Annex I) to developing (non–Annex I) countries.[17] The argument is that emission reduction regulations increase the costs of industrial production in Annex I countries. This would increase the competitiveness of nations without such legislation and shift production patterns accordingly, through either actual relocation of factories or relative changes in production output within industries or branches of the same firm. Such leakage would be difficult to quantify, and the debate is reminiscent of the one about "pollution havens"—countries that allegedly try to attract industries by lowering environmental standards.[18]

Nevertheless, fears of massive international leakage in the avoided deforestation context seem exaggerated. Reducing deforestation does not mean a complete halt of forest conversion. Countries trying to lower deforestation rates would first target conservation options with low opportunity costs, such as marginally productive land areas. Under most circumstances, improving land-use planning and agricultural productivity—for example, regarding cattle farming—would lower deforestation pressures sufficiently. Thus it is likely that highly profitable ventures such as soy agriculture, palm oil production, and logging for valuable timber would not be strongly affected. These ventures frequently involve internationally mobile actors, and international leakage might indeed occur if countries were to pursue very ambitious avoided-deforestation targets. Eventually, only full participation by all tropical forest countries in an avoided-deforestation regime would completely eliminate risks of international leakage, a fact that also applies to the broader climate regime under Kyoto and the UNFCCC.

Perverse Incentives and Hot Air

The negotiators of the Kyoto Protocol struggled with ways to avoid creating perverse incentives that might increase rather than reduce emissions. If the reference point for emission reductions were to be set in the future, it might create an incentive for countries to increase their baseline emissions up to that point so that emission "reductions" could be undertaken more profitably by starting from a higher level. In the LULUCF field, such perverse incentives might have meant increasing deforestation rates during the baseline period or up to the base year. After it was clear that deforestation avoidance would be ineligible under the CDM, it was feared that landowners might cut forests and later claim carbon credits for reforesting the same land. In addition to creating no climate benefits, this would entail disastrous biodiversity effects.

The Kyoto parties countered the danger of perverse incentives for Annex I countries by establishing 1990 as the historical base year for calculating emission reductions. For CDM forestry activities, it was decided to allow reforestation only on land that could be demonstrated already to have been deforested in 1990.[19] Such an approach increases the environmental integrity and effectiveness of the emission reduction credits created. Policymakers should evaluate, however, whether the 1990 cutoff date is in fact too restrictive, because many credible AR projects could take place on land that was deforested after that year but long enough ago not to create perverse incentives. To give an indication of the magnitude of this potential, roughly 180 million hectares of land were deforested between 1990 and 2005.[20] Some environmental NGOs, on the other hand, vigorously reject any suggestion to amend the 1990 reference year—for example, by setting a later date or a "rolling" reference year—arguing that this would send a signal to landowners that the CDM rewards deforestation.[21]

Another risk to the environmental integrity of carbon credits from all emission sources is the creation of "hot air." Hot air is generated when mitigation actions seemingly lead to a reduction in GHG emissions that on closer examination would have occurred anyway under the BAU scenario. This can take place, for example, because of the breakdown of industries or the ordinary modernization of installations—that is, cases in which emission reductions are not the consequence of specific mitigation activities.[22]

In the case of setting reference values for emissions from deforestation, a risk of "tropical hot air" could be created by incorrectly assuming deforestation rates to be static when in fact they are improving. A number of observations, as well as some economic models, indeed point toward decreasing emissions from deforestation during the next decades as countries pass through "forest transitions" in the course of economic development.[23] For individual countries, however, the large number of factors driving deforestation, such as prices for agricultural commodities, population growth, and land-use policies, leads to considerable inter-annual variation in deforestation and make it difficult to predict and quantify future emissions. This is not fundamentally different from emissions in other (non-LULUCF) sectors or in Annex I countries.

Environmental and Social Effects of Carbon Forestry

While arguing against the inclusion of avoided deforestation in the CDM, which arguably would have preserved biodiversity-rich forests, environmental NGOs painted gloomy pictures of vast monocultures of fast-growing, non-native tree species in the developing world. Such industrial "carbon plantations," established for the benefit of rich countries, would lock up potential agricultural soils for decades or centuries in "Kyoto lands," thereby exacerbating land scarcity and worsening the plight of poor rural communities, possibly even driving them off their land.[24] Some angry voices went so far as to call carbon forestry a thinly veiled form of expropriation and neocolonialism.[25] Such portrayals round out the view of carbon forestry projects as being in fundamental opposition to sustainable development and thus incompatible with the goals of the Clean Development Mechanism. That is, not only are their climate benefits portrayed as being uncertain at best, but they are also alleged to have detrimental local environmental effects that impede rural development.

How much substance do such claims have? I look first at the argument on a conceptual level and consider only reforestation and afforestation, because forest conservation clearly delivers enormous noncarbon co-benefits. Flexible mechanisms such as the CDM and carbon trading between Kyoto parties were designed primarily to lower the costs of climate change mitigation for industrialized countries while providing a way to engage countries without emission reduction targets. One can certainly oppose carbon markets on ideological grounds and criticize the "commoditization of the atmosphere" that might result from assigning property rights to pollution and extending the neoliberal capitalistic system in general.[26] Yet the ability of free markets to find the lowest cost options for producing a good and to channel investments toward the most efficient uses remains unchallenged. Creating monetary incentives for GHG emission reductions by attaching a price to carbon thus addresses one of the largest market failures in the history of humankind. If carbon prices become a sufficiently important economic variable, then commercial actors can be expected to maximize the emission reduction or sequestration potential of their investments.

Creating carbon markets is the first major international attempt to value, in monetary terms, the ecosystem services provided by forests. No tangible market exists to date for the multitude of other services forest ecosystems provide, such as watershed protection, biodiversity, soil conservation, and rural livelihood provision.[27] This could, in principle, create an imbalance in which only the maximization of a project's carbon benefits is encouraged while other environmental or social effects do not enter into the equation or might even be perceived as hampering the former objective. From a purely carbon-focused point of view, fast-growing tree species such as eucalyptus and pine are preferable because they sequester CO_2 more quickly. Similarly, large plantations create economies of scale and reduce relative project transaction costs. In contrast, mixing a large number of species and using slower-growing species with higher biodiversity benefits increases project costs, at least in the short term, while generating no additional carbon finance.[28]

On the development side, a similar logic exists. Despite the dual purpose of the CDM, prominently stated in the Kyoto Protocol, of assisting countries in achieving sustainable development while mitigating climate change, carbon markets contain no design elements that would directly reward sustainable development benefits of CDM projects. Projects within and outside of the forestry sector may indeed promote sustainable development as an intended or unintended by-product, but investors normally cannot directly monetize these benefits.

Parties to the Kyoto Protocol and NGOs discussed potential environmental and sustainable development criteria for the CDM at great length when the rules were being defined.[29] Many civil society actors favored stringent standards for project approval, a demand rejected by developing countries on grounds of sovereignty and opposition to the notion that sustainable development would be defined by northern NGOs. Finally it was decided to include a general requirement that CDM projects "contribute to sustainable development in the host country" but to leave the definition of what this meant up to the respective country's government. The Designated National Authority (DNA) of a non–Annex I country has to assess the effects of potential projects and grant a letter of approval before project developers can apply for registration under the CDM. Some DNAs have since come up with elaborate screening criteria to evaluate environmental and social effects. In addition, stakeholder consultations have to be carried out during CDM project development, which typically involves meetings with local communities and NGOs during which they can voice concerns and suggest design modifications.

What do experiences to date tell us about these concerns? Unfortunately, no recent systematic assessments exist regarding the environmental and social effects of carbon forestry projects. It is clear that many AR projects rely on fast-growing exotic species in order to be commercially viable. This does not mean, however, that the plantations have a negative biodiversity effect, particularly if they have been established on degraded, biodiversity-poor land. The realistic alternative in most cases, the baseline scenario, is continued degradation or fallowing of land on which a CDM project might provide real climatic, employment, and soil conservation benefits. Moreover, many AR projects contain components of assisted natural regeneration and forest conservation, and responsible consultants and project developers have played a key role in introducing these. The establishment of plantations can also have conservation benefits in cases where deforestation or degradation is driven by fuelwood collection.

Only limited experience exists to date regarding the development benefits of CDM forestry. The small number of objective research efforts has yielded mixed results but indicates a clear potential for improving rural livelihoods and diversifying incomes.[30] For example, the establishment of plantations is usually labor intensive, and if forest plantations are managed for timber, then the regular cycle of thinning, harvesting, and replanting can sustain a significant number of jobs for decades. Secondary supply and processing industries can further enhance this potential. Activities intended to reduce carbon leakage, such as agricultural intensification and the development of nonfarm employment opportunities, can deliver additional development benefits.

It is important not to mingle criticisms directed at noncarbon commercial forestry plantations in general with criticisms directed at CDM projects. If environmental and social conditions and risks are properly assessed—something the CDM process clearly provides for—then negative effects can be contained and co-benefits realized. Moreover, in practice few CDM AR projects are purely carbon-market driven, because they often require significant co-financing. Using these other funding sources and seeking out additional ones give developers options to enhance the noncarbon benefits of their projects (see chapter 5). Most project developers pay close attention to potential negative project effects, not least because they, their investors, and the future buyers of forestry carbon credits can ill afford negative public exposure as they face growing demands for corporate social responsibility. Indeed, investors and project developers increasingly apply additional standards such as certification according to the guidelines developed by the Climate Community and Biodiversity Alliance (CCBA) to safeguard and enhance the environmental and social benefits of carbon forestry activities.

Concluding Remarks

It is probably safe to say that no other sector under the CDM contains as many climate safeguards as forestry and is as thoroughly screened for its environmental and social effects. Although many claims by critics of carbon forestry were certainly exaggerated, they may well have succeeded in addressing real problems and driving effective risk-mitigating mechanisms such as expiring carbon credits for AR activities. A great part of the uneasy feeling some NGO stakeholders and market participants may retain regarding forestry projects stems from early phases of the Kyoto negotiations and relates to issues that have long been addressed and for which solutions and safeguards have been devised.

Overall, AR projects are much more restrictively handled than other CDM activities because of their contentious negotiation history. For example, forestry developers need to account scrupulously for project emissions and leakage, and they must calculate items such as vehicle emissions from transporting workers to the forests and carbon leakage in wood posts used for fencing sites, which do not have to be considered for other types of projects. Clearly, such examples illustrate an overshoot in efforts to ensure climate benefits, and now should be the time to address them in a constructive way.

Carbon forestry is no panacea for climate change and undoubtedly needs to be one component of a comprehensive, long-term strategy in which forest restoration and conservation, as well as economic restructuring, are integral approaches, not mutually exclusive trade-offs. Nonetheless, the forestry and land-use sector represents opportunities too significant to be ignored. Sequestering carbon in forestry projects is one of the few ways to actually remove CO_2 that has already been emitted into the atmosphere. Reducing emissions from deforestation addresses the largest source of emissions from developing countries and around 20 percent of the global total.[31] Forestry projects were very much the focus of the CDM and its precursors during the early years and in the eyes of many still represent the classic carbon offset. Using the wealth of experience gathered during the last decade, forestry has the potential to contribute on a much greater scale to tackling climate change while providing significant additional benefits for local development and the environment.

Notes

1. United Nations Framework Convention on Climate Change (UNFCCC), CDM statistics, 2008 (http://cdm.unfccc.int/Statistics/index.html).

2. This is despite the fact that approximately 40 percent of the historical emissions of CO_2 over the last 200 years are from changes in land use and land management, most of which has been deforestation. See Millennium Ecosystem Assessment, *Ecosystems and Human Well-Being: Synthesis* (Washington: Island Press, 2005).

3. See press releases and statements of a number of NGOs: "Don't Sink the Climate Treaty (www.birminghamfoe.org.uk/press/Brummies_Build_Bonn_Boat_2001_07_20.htm); "Sinking the Planet" (www.foe.co.uk/campaigns/climate/news/bonn_talks_july_2001/19_july.html); www.sinkswatch.org/.

4. The Kyoto Protocol mandates that Annex I Parties reduce GHG emissions to an average of 5 percent below 1990 emissions levels by 2012.

5. J. Sathaye and others, "GHG Mitigation Potential, Costs, and Benefits in Global Forests: A Dynamic Partial Equilibrium Approach," Lawrence Berkeley National Laboratory, University of California, Berkeley, 2005; M. Grieg-Gran, "The Cost of Avoiding Deforestation," International Institute for Environment and Development, London, 2006.

6. J. Ebeling and M. Yasue, "Generating Carbon Finance through Avoided Deforestation and Its Potential to Create Climatic, Conservation, and Human Development Ben-

efits," *Philosophical Transactions of the Royal Society B* 363 (1498) (2008): 1917–24; I. Noble, "Some Notes on Avoided Deforestation in the UNFCCC," World Bank, Washington, 2006.

7. T. Sugiyama and D. S. Liu, "Must Developing Countries Commit Quantified Targets? Time Flexibility and Equity in Climate Change Mitigation," *Energy Policy* 32 (2004): 697–704; Netherlands Environmental Assessment Agency, "China Now No. 1 in CO_2 Emissions; USA in Second Position," press release, June 19, 2007 (www.mnp.nl/en/dossiers/Climatechange/index.html).

8. European Commission, "Greenhouse Gas Reduction Pathways in the UNFCCC Process up to 2025," study commissioned by DG Environment, 2003.

9. UNFCCC/CP/2001/13/Add.2, Decision 17/CP.7, "Modalities and procedures for a clean development mechanism as defined in Article 12 of the Kyoto Protocol," paragraph 7.

10. UNEP Risø Centre, "CDM Pipeline Overview," January 2008 (www.cdm pipeline.org/).

11. K. M. Chomitz, "Baseline, Leakage and Measurement Issues: How Do Forestry and Energy Projects Compare?" *Climate Policy* 2 (2002): 35–49.

12. T. K. Rudel and others, "Forest Transitions: Towards a Global Understanding of Land Use Change," *Global Environmental Change* 15 (2005): 23–31.

13. Ebeling and Yasue, "Generating Carbon Finance through Avoided Deforestation."

14. L. Aukland, P. Moura Costa, and S. Brown, "A Conceptual Framework and Its Application for Addressing Leakage: The Case of Avoided Deforestation," *Climate Policy* 3 (2003): 123–36; E. Niesten and others, "Designing a Carbon Market that Protects Forests in Developing Countries," *Philosophical Transactions of the Royal Society*, Series A, 360 (2002): 1875–88.

15. Aukland, Moura Costa, and Brown, "A Conceptual Framework"; R. Schwarze, J. Niles, and J. Olander, "Understanding and Managing Leakage in Forest-Based Greenhouse Gas Mitigation Projects," Nature Conservancy, Washington, 2002.

16. K. M. Chomitz, "Baselines for Greenhouse Gas Reductions: Problems, Precedents, Solutions," World Bank, Washington, 1998.

17. Niesten and others, "Designing a Carbon Market."

18. D. Wheeler, "Beyond Pollution Havens," *Global Environmental Politics* 2 (2002): 1–10.

19. United Nations Framework Convention on Climate Change, Marrakech Accords and Marrakech Declaration, 2001.

20. Food and Agriculture Organization of the United Nations (FAO), "Global Forest Resources Assessment 2005: Progress towards Sustainable Forest Management," Forestry Paper 147 (Rome: FAO, 2006).

21. Greenpeace, "Sinks in the CDM: Do Not Change the 1990 Reference Year!" press release, COP 9, Milan, 2003 (www.greenpeace.org/international/press/reports/sinks-in-the-cdm-do-not-change).

22. Chomitz, "Baselines for Greenhouse Gas Reductions"; J. Ellis and M. Bosi, "Options for Project Emission Baselines," Organization for Economic Cooperation and Development, International Energy Agency, Paris, 1999. A well-known example is the setting of Kyoto emission targets for Russia relative to 1990-level emissions in order to induce the country's participation in the protocol, even though it was obvious that Russia's economic collapse had since led to a drastic drop in CO_2 emissions without any climate mitigation

efforts. S. V. Paltsev, "The Kyoto Protocol: 'Hot Air' for Russia?" Center for Economic Analysis, University of Colorado, 2000.

23. Rudel and others, "Forest Transitions."

24. J. Kill, "Sinks in the Kyoto Protocol: A Dirty Deal for Forests, Forest Peoples and the Climate," Forests and the European Union Resource Network (FERN), Brussels, 2001.

25. See S. Dessai and others, "Challenges and Outcomes at the Ninth Session of the Conference of the Parties to the United Nations Framework Convention on Climate Change," *International Environmental Agreements* 5 (2005): 105–24; P. M. Fearnside, "Saving Tropical Forests as a Global Warming Countermeasure: An Issue That Divides the Environmental Movement," *Ecological Economics* 39 (2001): 167–84.

26. J. McCarthy and S. Prudham, "Neoliberal Nature and the Nature of Neoliberalism," *Geoforum* 35 (2004): 275–83; P. Bond and R. Dada, eds., *Trouble in the Air: Global Warming and the Privatisation of the Atmosphere* (Johannesburg: Center for Civil Society, 2005).

27. S. Scherr, A. White, and A. Khare, "For Services Rendered: The Current Status and Future Potential of Markets for the Ecosystem Services Provided by Tropical Forests," ITTO Technical Series (Yokohama: International Tropical Timber Organization, 2004).

28. On the other hand, such activities can make it easier to demonstrate the additionality of a project and to gain host country approval.

29. E. Boyd and others, "The Politics of Afforestation and Reforestation Activities at COP-9," Briefing Note 12 (Norwich, U.K.: Tyndall Centre for Climate Change Research, 2004).

30. K. Brown and others, "How Do CDM Projects Contribute to Sustainable Development?" Technical Report 16 (Norwich, U.K.: Tyndall Centre for Climate Change Research, 2004); N. Landell-Mills and I. T. Porras, "Silver Bullet or Fool's Gold? A Global Review of Markets for Forest Environmental Services and Their Impact on the Poor," International Institute for Environment and Development, London, 2002.

31. R. A. Houghton, "Tropical Deforestation as a Source of Greenhouse Gas Emissions," in *Tropical Deforestation and Climate Change,* edited by P. Moutinho and S. Schwartzman (Belém and Washington: Instituto de Pesquisa Ambiental da Amazônia and Environmental Defense, 2005), pp. 13–22.

5

Forest Carbon and Other Ecosystem Services: Synergies between the Rio Conventions

JAN FEHSE

Sequestering carbon in forests by planting trees and reducing deforestation and forest degradation is an internationally accepted measure used to mitigate climate change. Projects that implement such measures may be awarded tradable "carbon credits." These carbon sequestration services, however, are not the only services forests can provide. They can simultaneously be harbors for biodiversity, protect watersheds from degradation, and provide natural beauty to local people and visitors.[1] Depending on their design, most carbon forestry projects provide one or more additional services that benefit the local, regional, and in some cases global community. In doing so they can also contribute to the objectives of the United Nations Millennium Declaration and of two United Nations Rio Conventions, the Convention to Combat Desertification (CCD) and the Convention on Biodiversity (CBD).[2] In the context of the Millennium goals, carbon forestry projects also have the potential to contribute to the reduction of poverty through income generation and to the reduction of opportunity costs by providing, for example, adequate supplies of bioenergy and water.

It is possible to attach economic values to the majority of these services.[3] The economic benefits may be either direct or indirect, through the avoidance of costs that would be incurred if the service ceased to be provided. For example, a deforested watershed can cause severe and expensive problems downstream, such as flooding, sedimentation, and reduced water quality. However, apart from their climate change mitigation benefits, forestry projects are in practice not valued and rewarded for their additional services. Experience also shows that in the current carbon policy and carbon market environment, revenues from the sale of carbon credits do not make a great contribution toward a project's implementation costs. Prices paid for forestry carbon credits are low relative to credits from other emission reduction sectors, and overall revenues on average raise the internal rate of return for commercial plantation projects by only a few percentage points, let alone cover the costs of conservation reforestation projects.[4]

As a result, carbon trading has so far not been a great incentive to implementing projects on a large scale. This may improve as markets become better established and the carbon policy context becomes more conducive to forestry in general and to avoided deforestation in particular. Nevertheless, considering that the global community has collectively agreed on the importance of sustainable development and has defined goals and objectives toward this end in the Rio Conventions, it would make sense to recognize, value, and reward not only the forest services that relate to climate change mitigation but also the other services forests provide. Doing so would create a more synergic approach to the implementation of the Rio Conventions and contribute to reducing the financing gaps that forestry projects currently face.

In what follows I discuss current and potential regulatory environments that could be created to facilitate the trading of noncarbon ecosystem services. I then offer an overview of carbon forestry project experiences that feature synergies with the objectives of the CBD and CCD.

Markets for Noncarbon Ecosystem Services: The Regulatory Context

The Kyoto Protocol, adopted under the auspices of the UN Framework Convention on Climate Change (UNFCCC), includes three carbon trading mechanisms that allow for the offsetting of unavoidable emissions by purchasing emission reductions or removals from elsewhere. These are the project-based Joint Implementation and Clean Development Mechanism and the International Emission Trading mechanism. They provide market incentives to generate lower-cost emission reductions, thereby converting emission reductions into business opportunities instead of only costs. Could similar incentives be used toward the objectives of the CBD and CCD? Could a scenario exist in which a contribution to desertification or loss of biodiversity would have to be offset, and consequently a contribution to the fight against desertification and loss of biodiversity be rewarded? Could there even be international targets to this effect? Such policies were not defined to this extent by the CBD and CCD, which indicates that doing so might not be as simple as for carbon trading.

There are several reasons for this. First, desertification and biodiversity lack an easily quantifiable unit of measure, such as the metric tons of CO_2 equivalent. This makes it impossible to design a straightforward offsetting system in which the offset equals the offense. What people mean by stating that ongoing desertification or loss of biodiversity is taking place is not exactly clear. Desertification generally involves the decline of vegetation, soil, and hydrological resources and thus includes the loss of a whole range of ecosystem services that would all have to be quantified differently, if they could be quantified at all. Even when focusing on, for example, hydrological services such as streamflow and streamflow distribution, one finds that they are dependent on many parameters, such as vegetation, soil, geology, and climate. Protection measures upstream might therefore not necessarily be detectable downstream, and conversely, a detected change in water flow might not be directly attributable to a specific protection measure. When researchers speak about loss of biodiversity, do they mean loss at the level of the gene, the organism, or the habitat? Is all biodiversity equal? For example, are rare species more valuable than common ones? How much biodiversity does the world lose if a certain area is destroyed, and how much does it gain if another area is restored? Although theoretically scientists should be able to answer these last questions, they simply lack the data with which to do so. The problems of precisely defining and quantifying the equivalence between damage and offset seem too challenging to resolve in the near future, but they might be resolved by using practical approaches such as proxy indicators (area of a habitat is often used) and multiplication factors—for example, the loss of one hectare of habitat might be offset by the restoration of three hectares, to guarantee that the principle of "no net loss" is followed.

Second, the variability in technical approaches to quantifying noncarbon ecosystem services offsets would require a lot of consensus-making between stakeholders before an internationally regulated offsetting system could be designed. Third, emission trading is based on the premise that the atmosphere is equally affected by the emission of one tonne of carbon dioxide in Africa and the emission of one tonne of carbon dioxide in Europe or North America. Noncarbon ecosystem services, with the exception perhaps of biodiversity, do not always (directly) provide global benefits but rather provide more local or regional benefits. Thus it would make sense to compensate those who suffer from the loss of ecosystem services by locating the offset nearby rather than on the other side of the planet.[5] Technical offsetting solutions may therefore need to be defined on a regional or even local level to ensure that offsetting remains meaningful and adequate, and this reduces the need for policies at the international level.

Nevertheless, increased interest seems to exist in setting up international payment schemes for ecosystem services, including biodiversity offsetting under the CBD. At a workshop in Geneva in September 2006, organized by the United Nations Environmental Program (UNEP), the World Conservation Union

(IUCN), and the CBD Secretariat, this topic was intensively discussed.[6] Various regulatory options were mentioned, varying from voluntary (that is, no regulation, only guidance) to a Kyoto-like international cap-and-trade system for biodiversity offsets, but no statement was made regarding the extent of any involvement by the CBD. A consensus does seem to exist, however, that a strictly voluntary approach, in which actions are driven mainly by public relations goals or altruism, would have a much smaller effect than a nationally or internationally regulated system. Similarly, the results of a stakeholder survey for biodiversity offsets showed that most business representatives believed regulation would provide more effective action than a strictly voluntary approach.[7] In this context it should be considered that the primary roles of regulation are to create the demand for ecosystem services either through offsetting or through command-and-control systems and to ensure the quality of desired activities through rule-making and the setting of criteria.

From the preceding discussion it seems that the best level for rule-making and criteria-setting is the (sub)national level at which the offset occurs (the "host country"), although for international offsets the CBD might provide general guidance. To ensure demand, national- and international-level commitments would be best. Indeed, many countries, including developing countries, already mandate offset measures when they oblige project developers to implement compensation measures for the negative effects that infrastructure, mining, and construction projects have on the environment. But although the general principle may already be anchored in environmental laws, problems usually exist, particularly in developing countries, related to the setting of equivalence criteria as well as to the enforcement and quality control of the compensation measures. As a result, those compensations that do take place are often in practice inadequate. Yet these existing obligations could form a good basis for the design of proper (sub)national biodiversity offsetting systems. One example in which this is being attempted is the San Nicolás project in Colombia, to which I return later.

It is unclear to what extent developing countries are prepared to burden their industries with offsetting requirements. Most countries also lack the capacity and the financial resources to design and enforce such a system. This strengthens the call for international offsets, which could be regulated by developed countries' governments, by the CBD, or by both. One lesson learned from the Kyoto Protocol is that it has massively accelerated the mainstreaming of climate change action in government policies, as well as the development of projects and of new technologies, and one could envisage a similar role for the CBD.

For combating desertification there are currently no targeted regulations toward offsetting or payments for ecosystem services. However, the National Action Plans drafted in each country as part of the CCD process define priority areas for action, and these may be used as criteria in the design of carbon or biodiversity offsetting systems.

Many cases exist in which a user of an ecosystem service would like to give the steward of the land that provides the service an incentive to adopt more sustainable management practices, but an offsetting system does not provide the correct policy instrument to do so. A good example of this situation is users of watershed-related ecosystem services. As its name describes, all services in this category are related to water flowing downstream, with waters from a larger catchment area concentrating eventually in one stream. Such services include erosion control, water quality, and peak and base flow distribution. The users of the upstream services invariably are located downstream, and it is their consumption of water in some form that defines the value they put on the service. Main groups of users include hydroelectric power generators, municipal water supply systems, irrigation systems, industrial users, and populations in flood-prone areas.[8] An additional group is populations in landslide-prone areas, who have been particularly affected by recent hurricanes in Central America and the Caribbean. If such users acknowledge the services provided to them, which often happens when costs are incurred in relation to their loss—for example, dredging costs resulting from sedimentation of reservoirs—they may be willing to make direct payments to upstream landowners to continue providing the services or to restore them.[9] However, in many cases users may become aware of the need for a payment scheme only when it is too late, or they may be too fragmented to take concerted action. In these cases local, regional, or national regulation is desirable. Such regulation could range from enabling legislation to obligatory direct payments or targeted and earmarked taxation.

Where then, on the regulatory level, do the synergies between forest carbon trading and other payment schemes for forest ecosystem services lie? First, it should be considered that one reforestation, forest restoration, or forest conservation activity may provide multiple ecosystem services. It therefore seems logical that the project design framework that has been established under carbon trading schemes, being in most aspects the most advanced, should also be used for the trading of other services. This framework includes aspects such as definition of the project boundary and project activities, methodologies for quantifying the project's benefits, and approval and quality control criteria and processes, among others. The income streams from these multiple service sales would complement each other to improve a project's capacity to attract capital for its implementation.

One project design criterion that is particularly important in this context is that of additionality. The additionality criterion is one of the cornerstones of the Kyoto Protocol's project-based mechanisms; it requires proof that a project activity takes place only because it is being enabled by the incentive of carbon trading—that is, it is "additional" to the business-as-usual scenario. It is difficult to see how any market-based environmental compensation scheme, using either offsetting or direct payment schemes, could function without using the additionality criterion. After all, an offset can be regarded as such only if a negative effect is compensated by a positive one, not by a neutral one (that is, business as usual). And a beneficiary of an ecosystem service would be prepared to pay for this service only if there were a threat of losing it or if it already had been lost and needed restoring. This means that there must be an activity that mitigates this threat or enables the restoration—in other words, an additional activity.

Why is the additionality criterion important for the design of a "synergic" project? The reason is that once the investment decision for a project activity has been taken, thus enabling its implementation, it cannot then be claimed that an additional payment scheme is required to provide the service, unless this scheme goes beyond what the current project is designed to achieve in terms of scope of activities, areas, and time frame. Therefore, all payments for a project's ecosystem services should be defined and agreed upon before the start of the activity. This is not easy to do, because it requires specific capacity and is likely to increase transaction costs and lengthen the project design period. But it is essential for trying to make income streams from multiple benefits work for projects.

Second, the provision and commercialization of multiple ecosystem services may not only be beneficial to the project itself but also serve larger political and regulatory objectives. In the context of current ideas about national baselines for avoided deforestation, in which carbon credits would be traded at the national level rather than the project level and credits would be issued only after an observed reduction in deforestation, facilitating payments for noncarbon ecosystem services could be a tool for a national government to provide extra incentives for concrete actions that reduce deforestation. Further interlinkages can be found with objectives or, in the future, perhaps even targets related to biodiversity conservation and combating desertification. Under the CBD and CCD, countries agree to take action, and they could use carbon forestry projects to leverage this by regulating or providing incentives to optimize other ecosystem services in the project design.

Finally, noncarbon ecosystem services offer strong benefits toward the other overarching objective of the climate change convention: adaptation. Nowhere are the links between the three Rio Conventions demonstrated so clearly. Biodiversity provides ecosystems with adaptive elasticity and resilience against climate change, thus preventing ecosystem collapse and desertification.[10] Deforestation and consequent desertification result in a region's increased vulnerability to climate change. As the potential double function of forestry projects for both climate change mitigation and adaptation becomes more valued, the role that ecosystem services other than carbon play in a project's design may also become better recognized within the climate change convention. This could perhaps lead to incentives such as exemption from the adaptation fund levy or even "adaptation payments" in one form or another.

Project Experiences with Synergies

In 2002, Landell-Mills and Porras identified twenty-eight projects with "bundled" approaches—that is, projects that either combined multiple ecosystem services in one product or marketed different services separately.[11] Many other projects have been developed since then. Rather than provide an overview of these projects, I give examples of projects that attempt to combine access to carbon markets with the marketing of other ecosystem services or other forms of pursuing the objectives of the Rio Conventions. My aim is to show that project designers have a wide range of possibilities for optimizing these synergies. Most of the examples are taken from the eleven years of experience working with carbon forestry projects worldwide gained by EcoSecurities, a carbon trading company that also provides carbon market consulting services to project developers and other clients.

Restoring Ecosystem Services

A project under the Clean Development Mechanism (CDM), one of the project-based carbon trading mechanisms under the Kyoto Protocol, may implement only afforestation and reforestation activities that enhance or restore an area's forest carbon pools. Therefore the main focus for the provision of other ecosystem services in a CDM project will also be on the restoration of these services. This offers great scope for activities that contribute to combating desertification through restoration of the vegetation cover on degraded and degrading sites that have lost their water and soil retention capacity. Indeed, the first projects to go through the approval and project registration process for CDM afforestation-reforestation (AR) methodologies have been implemented on degraded and further degrading soils. These projects were the first through the process because the simplicity of their baseline scenarios made their eligibility under the CDM relatively straightforward. This criterion proved in practice more relevant than their desertification-combating benefits, but they nevertheless made a clear contribution to the objectives of the CCD.[12] Unfortunately, none of these first CDM projects tried to secure an additional reward for this contribution. This will be reversed, it is hoped, by the Global Mechanism–EcoSecurities Partnership.[13] The Global Mechanism is a body subsidiary to the CCD that has the mandate to facilitate financing for activities that advance the convention's objectives. The Global Mechanism adopts an approach that integrates, with assistance from EcoSecurities, the climate change and desertification conventions by developing CDM AR and other relevant CDM projects (for example, projects that address the nonrenewable use of woody bioenergy resources) specifically within the context of combating desertification.

One example is the Julcuy project in the province of Manabí in coastal Ecuador, which is being developed in close cooperation with the Manabí provincial council.[14] This very dry region has suffered for decades from loss of the native dry forest and further degradation from overgrazing by goats. This has led to significant soil erosion and water scarcity, the latter accentuated by the fact that the region is fed by no external watercourses. The project seeks to restore around 10,000 hectares of the original forest vegetation, which includes trees that can provide high-quality fodder for the goats. In the longer term, the planted trees will provide local communities with a sustainable source of timber and fuel wood. The project area is important for the hydrological supplies of a number of urban centers, which have seen steep population growth in the last three decades. The project therefore seeks to market its hydrological benefits to the municipal water companies of these cities. Furthermore, the project area serves as a corridor between two important coastal nature reserves that form part of the larger Chocó-Manabí Conservation Corridor. In partnership with Conservation International (CI), the project's biodiversity benefits will be quantified and marketed. Similar projects are under development or being planned in Nicaragua, Morocco, and other countries in northern Africa and Central Asia.

In the same Chocó-Manabí Conservation Corridor in Ecuador, CI and the Fundación Maquipucuna are implementing a 500-hectare conservation reforestation project. The developers of this project submitted a specific baseline methodology that was approved by the CDM's Executive Board.[15] The project's focus is completely in line with CI's objective of conserving, restoring, and connecting ecological corridors that foster biodiversity through increased habitat size and access and improved genetic exchange. Although no direct efforts are being made to market the biodiversity benefits of this project specifically, CI's participation and financing efforts are indirectly a payment for biodiversity services, because its general marketing to donors and business partners is based on its biodiversity conservation strategy. The successful approach to using CDM AR for leveraging finance for its core biodiversity objectives is enticing CI to look for replication opportunities in both Ecuador and the rest of the world.

Further examples of CDM AR projects that fight land degradation are those that help rehabilitate degraded mine sites. Although in most countries mining companies have an obligation to rehabilitate decommissioned sites, in developing countries they rarely do in reality. These sites often continue to degrade, causing erosion and related problems of sedimentation or aerial dust. Even if some efforts are made toward rehabilitation, they usually involve only the establishment of grass cover. For mining companies, rehabilitation obligations are a cost they would prefer to avoid, but not rehabilitating is also a liability if it damages their public image. Generally the costs weigh more heaviliy than the public image, and minimal action, if any, is taken. The design of a CDM AR project can pull a company across the line by turning an environmental cost into an asset. It thus becomes the enabling factor to achieve the goal of rehabilitation of degraded sites. Without this other goal the CDM project would probably not have taken place.

An example of a direct combined benefit from carbon sequestration and hydrological services is the AES-Tiete project in Brazil. The project proponent is a hydroelectricity producer that tries to use the CDM AR project to restore riparian forests in the catchment area of a hydroelectric dam, and the primary benefit is securing sufficient water supply for the hydroelectric project.[16]

The issue of whether or not planting trees for the restoration of hydrological services causes detrimental effects warrants discussion. Studies have been published that show a statistically significant reduction in streamflow after reforestation, with particular reference to reforestation in the context of carbon sequestration (although the objective of the reforestation should make no difference).[17] However, the variability in hydrological responses to both deforestation and reforestation and the great number of factors that influence hydrological dynamics make it difficult to draw conclusions that are not site specific. Even so, the most significant changes affecting hydrology seem to be those inflicted on soils as a result of unsustainable management practices after deforestation. When an area is deforested, streamflow actually increases, because of reduced evapotranspiration. But when soils are compacted and erode—for example, in a scenario of overgrazing—their water-retaining capacity is reduced, and eventually this reduces streamflow and increases storm flow peaks. In addition, increased sedimentation reduces water quality. Therefore, when degraded sites are reforested, the longer-term benefits of soil recovery should be highlighted rather than the immediate hydrological effects. Moreover, when soil is not yet completely degraded, reforestation should also be seen in the light of conserving what is left, which in the long term may provide a greater hydrological benefit than the immediate loss of streamflow as a result of increased tree evapotranspiration.

Conservation of Ecosystem Services

Forest conservation is a holistic approach from the point of view of ecosystem services. Preventing deforestation of an area conserves the whole set of ecosystem services the forest offers. Although avoided deforestation as a climate change mitigation activity under the CDM fell out of grace with the Marrakech Accords in 2001, a great number of projects were initiated before that date.[18] Only one of these early projects, however, tried to market any of its other ecosystem services. This might change if an avoided deforestation carbon trading scheme emerges after 2012. The one exception was the national payment scheme for ecosystem services that Costa Rica set up in 1996. This first-of-its-kind scheme commissioned the national fund FONAFIFO to use funds from an earmarked fuel tax to make yearly per-hectare payments to landowners for conserving their forests.[19]

The idea was that in the long run the fuel tax funds would be replaced by payments for ecosystem services from third-party buyers. Predictably, the first service to be commercialized was carbon. The avoided deforestation benefits were certified by an independent auditor,[20] and in 1997 and 1998 Costa Rica delivered U.S.$2 million worth of Certified Tradable Offsets to a Norwegian consortium of buyers. Although it made no effort to quantify other ecosystem services, nor did it apply qualitative indicators or even differentiate between forest types, the system was clearly set up so that these components could be added at a later date. To date FONAFIFO has not managed to pass the stage of using the hectarage of forest as the single proxy indicator for the entire set of ecosystem services it provides. Yet it has managed to attract the investment of private companies. Those that stand out in particular are companies that sponsor the conservation of forest areas in watersheds important for their activities, notably a hydroelectric company, sugar manufacturers, and a cement company.

The San Nicolás project in Colombia (see case study, chapter 8) includes, alongside a CDM AR component, a non-CDM-eligible forest restoration and conservation component. This project has a clear mandate from its donors to develop innovative financing approaches toward sustainable forest management and conservation. Its goals include, next to carbon credits, the identification, quantification, valuation, and commercialization of noncarbon ecosystem services. Preliminary results confirm that for the identified target services—biodiversity, watershed management, and scenic beauty—quantification is indeed problematic. The most practical way of dealing with this is to use the area of forest conservation or restoration activity as a proxy, which can be refined by ranking the biodiversity or watershed management benefits of specific activities or of forest types. In the design of a payment scheme component of the project, a promising option is to set up a regional compensation system for the offsetting of damaging activities that require an "environmental license," such as mining. Applications for such a license are made to CORNARE, the public organization that manages the natural resources in the region.[21] CORNARE is also the project proponent of the San Nicolás project, which could provide the required offsets.

A final consideration should be made for CDM AR projects that seek to balance their project design by voluntarily adding a forest conservation or restoration component to a core of commercial reforestation. No commercialization of ecosystem services, including carbon, is foreseen from these activities, which should be seen as a form of "internal offsetting" of the apparent paucity in environmental benefits and, in some cases, the environmental risks of the core activities. Although this form of compensation is in no way quantified, neither is the environmental "offense," which in many cases is based mainly on perception. The project design of monoculture plantations (a valid climate change mitigation option, provided that host-country criteria for sustainable development are met) can be improved by incorporating conservation or restoration activities. A concrete example of such a project is the NFC project in Uganda.[22] A step further in this line of thinking is the certification scheme designed by the Climate, Community, and Biodiversity Alliance, which allows carbon forestry projects to demonstrate their community and biodiversity benefits in a ranking system according to a predetermined set of criteria.[23]

Final Remarks

Although many carbon forestry projects have the potential to leverage synergic benefits to all three Rio Conventions, the possibilities of receiving rewards for this in the form of marketing noncarbon ecosystem services are still limited. Much more work needs to be done to develop such markets further, and both national and international regulation should play important roles. If successful, this will help provide a more complete package of environmental finance incentives to those who make land management decisions. Sustainable land use works not only on the convention's drawing board; it must also be made to work in practice, and proper incentives are essential for this to happen.

Notes

1. De Groot provides a useful overview of the full range of hitherto recognized "ecosystem" or "environmental" services. R. de Groot, "Functions and Values of Protected Areas: A Comprehensive Framework for Assessing the Benefits of Protected Areas to Human Society," in *Protected Area Economics and Policy*, edited by M. Munasinghe and J. McNeely (Washington: World Bank, 1994).

2. United Nations Millennium Declaration, A/RES/55/2, September 18, 2000; United Nations Convention to Combat Desertification in Those Countries Experiencing Serious Drought and/or Desertification, Particularly in Africa, 1994, 33 ILM 1332, entered into force December 26, 1996 (hereinafter CCD); United Nations Convention on Biological Diversity, 1992, 31 ILM 818, entered into force December 29, 1993 (hereinafter CBD).

3. R. Costanza and others, "The Value of the World's Ecosystem Services and Natural Capital," *Nature* 387 (1997): 253–60.

4. T. Neeff and others, "Compensating Tropical Forestry for Environmental Services: Protecting Biodiversity, Mitigating Climate Change and Combating Desertification," *Quarterly Journal of Forestry* 101, no. 2 (2007): 135–44.

5. A comprehensive discussion of the current state of thinking regarding biodiversity offsets, including the previously mentioned technical issues, is provided by K. ten Kate, J. Bishop, and R. Bayon, "Biodiversity Offsets: Views, Experience, and the Business Case," IUCN and Insight Investment, Cambridge and London, 2004.

6. UNEP and IUCN, "Developing International Payments for Ecosystem Services: A Technical Discussion," summary report, Geneva, September 2006.

7. Ten Kate, Bishop, and Bayon, "Biodiversity Offsets."

8. S. Pagiola, N. Landell-Mills, and J. Bishop, "Making Market-Based Mechanisms Work for Forests and People," in *Selling Forest Environmental Services: Market-Based Mechanisms for Conservation and Development,* edited by S. Pagiola, J. Bishop, and N. Landell-Mills (London: Earthscan, 2002), pp. 37–62.

9. Landell-Mills and Porras provide an interesting overview and analysis of such direct-payment schemes that had been set up or were being designed at the time of their study. N. Landell-Mills and I. Porras, "Silver Bullet or Fool's Gold? A Global Review of Markets for Forest Environmental Services and Their Impact on the Poor," International Institute for Environment and Development, London, 2002.

10. Convention on Biological Diversity, "Biological Diversity and Climate Change, including Cooperation with the United Nations Framework Convention on Climate Change," document UNEP/CBD/SBSTTA/6/11, 2000.

11. Landell-Mills and Porras, "Silver Bullet or Fool's Gold?"

12. Ebeling and colleagues provide a comprehensive discussion of the potential of carbon forestry to contribute to ecological restoration. J. Ebeling, M. Virah-Sawmy, and P. Moura-Costa, "Using International Carbon Markets for Forest Restoration while Mitigating Climate Change," in *Ecological Restoration: A Global Challenge,* edited by F. Comín (Cambridge University Press, in press).

13. Global Mechanism, "Description of the GM's Strategic Programme for the Compensation for Ecosystem Services," 2007 (www.global-mechanism.org/about-us/strategic programmes/ces).

14. See www.global-mechanism.org/products-services/cesecuador.

15. The approved methodology with the code AR-AM0007 and the draft Project Design Document can be viewed at http://cdm.unfccc.int/methodologies/ARmethodologies/approved_ar.html.

16. The approved methodology with the code AR-AM0010 and the draft Project Design Document can be viewed at http://cdm.unfccc.int/methodologies/ARmethodologies/approved_ar.html.

17. R. B. Jackson and others, "Trading Water for Carbon with Biological Carbon Sequestration," *Science* 310 (2005): 1944–47. For overviews of the literature and evidence regarding the effects of deforestation and reforestation on the hydrology of a catchment, see M. Bonell and L. A. Bruijnzeel, eds., *Forests, Water and People in the Humid Tropics: Past, Present and Future Hydrological Research for Integrated Land and Water Management* (Cambridge University Press, 2005); and L. A. Bruijnzeel, "Hydrological Functions of Tropical Forests: Not Seeing the Soil for the Trees?" *Agriculture, Ecosystems and Environment* 104, no. 1 (2004): 185–228.

18. P. Moura-Costa and M. Stuart, "Forestry-Based Greenhouse Gas Mitigation: A Short Story of Market Evolution," *Commonwealth Forestry Review* 77 (1998): 191–202.

19. See www.fonafifo.com/paginas_english/invest_forest/i_ib_que_es_csa.htm.

20. This was in fact the first time a carbon offset project was verified by an independent auditor. It was carried out by Société Générale de Surveillance (SGS), and the verification system was designed by EcoSecurities.

21. See www.cornare.gov.co.

22. Neeff and others, "Compensating Tropical Forestry for Environmental Services."

23. Climate Community and Biodiversity Alliance, "Climate, Community and Biodiversity Project Design Standards (First Edition)," Washington, 2005 (www.climate-standards.org).

6

Forestry Projects under the Clean Development Mechanism and Joint Implementation: Rules and Regulations

SEBASTIAN M. SCHOLZ AND MARTINA JUNG

Under the Kyoto Protocol, nearly all industrialized countries and economies in transition ("Annex I countries") agreed to reduce their greenhouse gas (GHG) emissions to at least 5 percent below the level of 1990. Emissions from a variety of sectors including energy, industry, agriculture, and waste management are taken into account for compliance with Kyoto emission targets. Activities from land use, land-use change, and forestry (LULUCF) make up another category of GHG emissions and removals that also have to be accounted for, at least to a limited extent. Parties agreed on the emission reduction targets of the Kyoto Protocol before they decided whether and how LULUCF could be used to fulfill those targets. At the time of the negotiations of the Kyoto Protocol, scientific knowledge about the role of LULUCF was limited and negotiators were poorly informed about the estimation of emissions and removals in the land-use sector. It was clear, however, that the inclusion or exclusion of "sinks"—a shorthand term referring to the way terrestrial vegetation and soils take up carbon—would significantly affect the emission budgets of the parties, which made these negotiations difficult.

Some parties and NGOs saw the option to account for sinks as an unfortunate loophole that could be used to water down the actual emission reduction efforts of the protocol. For supporters, the inclusion of sinks was an economic necessity for achieving Kyoto targets, and some pointed to the massive quantities of carbon exchanged between the atmosphere, vegetation, and soils every year, an amount thirty times greater than the emissions from fossil fuels.[1] They were concerned that by omitting sinks, a major exchange of carbon, which could swamp any gains made through activities under the Kyoto Protocol, would be ignored. An eventful and at times controversial negotiation process followed that ultimately led to a complicated accounting system for LULUCF.[2]

It took almost six years of assessment and negotiation to agree on specific LULUCF rules for the project-based mechanisms. Negotiators finally reached a consensus at the ninth session of the Conference of the Parties (COP 9) in December 2003. At this conference the international community finalized an agreement on the definitions and modalities for the use of afforestation and reforestation (AR) projects in the Clean Development Mechanism (CDM). Before COP 9, in the so-called Marrakech Accords, parties had already limited the use of LULUCF in the CDM to AR projects only. In addition, the use of forestry projects in developing countries was restricted to meet no more than 1 percent of Annex I Parties' Kyoto obligations.[3] At present, LULUCF projects are in principle allowed under the two flexible mechanisms, Joint Implementation (JI) and the CDM. In the following sections, we first survey the general differences between LULUCF projects under JI and CDM and then discuss the rules and regulations in greater detail.

LULUCF under JI and CDM: The Most Important Rules and Modalities

Industrialized parties to the Kyoto Protocol measure progress toward meeting their emission reduction targets by preparing national inventories, which are required to account for LULUCF-related activities within the respective country. In general, parties account for emissions and removals from land use, land-use change, and forestry on the basis of different *activities,* not specific land areas. Articles 3.3 and 3.4 of the Kyoto Protocol define those activities. According to article 3.3, parties are obliged to account for afforestation, reforestation, and deforestation that started on or after January 1, 1990. Article 3.4 introduces additional LULUCF-related activities that parties can account for on a voluntary basis in the first Kyoto commitment period, which runs from 2008 to 2012. These activities are forest management, revegetation, cropland management, and grazing land management started on or after January 1, 1990.[4] Parties have to decide on the inclusion of activities under article 3.4 before the start of the commitment period.

Carbon dioxide removals accounted for in one commitment period can potentially be reversed at a later time. Therefore, the national inventory must reflect these changes and all activities a country elects to account for under article 3.4, not only in the first commitment period but also in future commitment periods. Furthermore, accounting of GHG removals from forest management activities under article 3.4 and credits acquired from LULUCF projects under JI in other countries are subject to country-specific limits.[5] A particular category of certificates, called "Removal Units" (RMUs), was created for removals by sinks in Annex I countries. RMUs are issued for net removals attributable to activities accounted for under article 3.3 (afforestation, reforestation, and deforestation) and article 3.4 (forest management, revegetation, cropland management, and grazing land management).

Joint Implementation (article 6, Kyoto Protocol) allows Annex I Parties to fulfill their commitments by participating in the development or financing of projects that reduce emissions or enhance sinks in other Annex I countries.[6] All LULUCF activities enhancing anthropogenic removals by sinks, which are set out in articles 3.3 and 3.4 of the Kyoto Protocol, are eligible under JI.[7] Credits generated by JI projects, so-called Emission Reduction Units (ERUs), can be used by Annex I countries for compliance with their Kyoto targets. JI projects create ERUs, which are converted either from Assigned Amount Units (AAUs) or RMUs held by the host country.[8] The number of ERUs equivalent to the emission reduction or removal attributable to the JI project is then transferred to the investor country (or entity).

Because JI is limited to Annex I countries, which are subject to emission reduction targets, and because JI ERUs are issued from the host country's assigned amount, the host party has no incentive to exaggerate the emission reduction attributable to the LULUCF project. Therefore, an external control and verification mechanism is unnecessary as long as both countries involved in a LULUCF-JI project fulfill their inventory and reporting obligations. These obligations require that the party (1) has ratified the Kyoto Protocol; (2) has calculated its assigned amounts; (3) has in place a national system for estimating all anthropogenic sources of GHG emissions and removals; (4) has in place a national registry; (5) submits its annual emission inventory; and (6) submits all supplemental information on its assigned amounts.

Parties fulfilling all these requirements are eligible for the so-called Track 1 under JI. This means they can issue, transfer, and acquire ERUs without having to undergo external certification. If a party fulfills only criteria 1, 2, and 4, it can still participate in Track 2, which involves a project cycle including third-party verification administered by an international body, the JI Supervisory Committee (JISC).[9] The JISC has the overall authority over the administration of the Joint Implementation mechanism and can accredit "Independent Entities," which conduct the verification and determination of the number of ERUs to be issued and transferred by the host country under Track 2. If a JI project releases previously stored carbon back into the atmosphere because of some kind of biotic or abiotic disturbance, the host country's cap still applies, and it is held liable for the loss as it appears in the country's accounting system. The ERUs issued for the project are not cancelled.

Contrary to JI, the CDM allows an Annex I country (or entity) and a non–Annex I country (or entity) to jointly realize a project leading to emission reductions or enhanced removals in the non–Annex I country (the so-called host country). CDM projects generate Certified Emission Reductions (CERs). These credits are transferred to the Annex I country (entity) and are valid for compliance with the party's Kyoto targets. Because CDM host countries are not subject to quantified emission targets under the Kyoto Protocol, the question arises as to who assumes liability if the removal of carbon in CDM projects is inadequately reported. Therefore, CERs generated under the CDM have to pass a stringent certification procedure.

Unlike in the CDM, the procedures for project registration and issuance of credits under JI are still in their infancy. We therefore focus in the following sections on the CDM while mentioning the most important differences between the two flexible mechanisms. Particularly, concerns about the "permanence" of credits from CDM forestry projects have led to a number of tailor-made rules for this project type.

Eligibility of Land for CDM Projects and the Definition of Forest

The eligibility of land for LULUCF-related CDM projects and the way the host country defines "forest" are closely intertwined. The host country communicates its particular forest definition to the Secretariat of the United Nations Framework Convention on Climate Change (UNFCCC) within a CDM-specific framework in which a forest is defined as a minimum area of land between 0.05 and 1.0 hectare with a tree crown cover of more than 10 to 30 percent. In addition, a forest consists of trees with the potential to reach a minimum height of 2 to 5 meters at maturity.[10] For each of the criteria, the host country has to report its lower threshold value to the Secretariat in order to determine which areas are considered "forest." Because LULUCF-related CDM projects deal solely with afforestation and reforestation, only areas that do not fall under the forest definition are eligible for such activities.

In addition, certain retrospective, time-related thresholds apply. For reforestation projects, the host country has to demonstrate that the area to be reforested was not a forest on December 31, 1989. For afforestation projects, it has to prove that the vegetation on the land to be afforested has been below the thresholds of the country-specific forest definition for at least fifty years before the project's start date. The bottom line of the threshold date for reforestation projects is that land deforested after December 31, 1989, for any reason, can currently earn no CDM credits for planned or ongoing reforestation. Project proponents shall apply the "Procedures to Demonstrate the Eligibility of Lands for Afforestation and Reforestation CDM Project Activities" in order to prove the land's eligibility.[11] These procedures formalize the land selection process according to the spatial and temporal criteria already outlined and oblige project proponents to verify the eligibility through remote imagery or ground-based surveys.

(Non)Permanence and Accounting

Projects based on activities in the LULUCF sector differ in one crucial aspect from projects based on activities in other sectors, such as energy projects. In the latter, a tonne of emission reductions, once achieved, remains a benefit to the atmosphere,[12] whereas in the former, a tonne of sequestered carbon is of benefit to the atmosphere only for as long as it remains in fact sequestered. Whether forests act as sinks, or reservoirs, for carbon from the atmosphere or eventually as sources of GHGs depends on whether or not disturbances such as insect invasions, forest fires, and man-made deforestation take place. This "nonpermanence" risk of sinks has been the basis of concern about the feasibility of LULUCF projects in general, both within and between Annex I Parties (articles 3.3, 3.4, and 6 for JI), and particularly in the CDM (article 12).

As outlined earlier, the potential loss of carbon in a JI project is covered by assigning liability to the host country. To counteract the problem of nonpermanence in the CDM, the negotiators agreed on two forms of temporary credits for LULUCF CDM projects: temporary CERs (tCERs) and long-term CERs (lCERs).[13] The reason for creating two types of credits was mainly political. Whereas tCERs expire at the end of the commitment period following the one in which they were issued, lCERs expire at the end of the project's "crediting period." The project developer chooses the length of the crediting period, the time over which the LULUCF project can generate new lCERs or tCERs. In general the crediting period is either thirty years nonrenewable or twenty years twice renewable, for a maximum of sixty years.[14]

The first verification for both forms of CERs takes place at a time chosen by the project proponent. Any subsequent verification follows at five-year intervals. For lCERs such verification is mandatory; for tCERs it is not. If the project owner decides not to have the corresponding tCERs issued again, then no subsequent verification is mandated. In the case of lCERs, the verifier confirms that the project retains at least the amount of carbon to cover the credits issued and verifies new GHG removals if they have occurred between two verifications. If there has been a net loss of sequestered carbon between two five-year verifications, then lCERs have to be replaced within a month by AAUs, CERs, ERUs, RMUs, or other lCERs from the same project activity.[15] For tCERs the verifier certifies a number of tCERs equal to the amount of carbon stored since the project's start date. Expiring tCERs have to be replaced only at the end of the commitment period following the period in which the credits were issued and used.[16]

Thus for tCERs the expiry, verification, and issuance of new tCERs corrects for any losses of carbon from a project, whereas with lCERs, if net carbon stocks decrease between two verification reports, the difference in lCERs needs to be replaced. In addition, COP 9 introduced a rule stating that at the end of the agreed-upon crediting period all tCERs and lCERs have to be replaced by credits from other projects. This created a liability to replace both forms of temporary credits at some future time, depending on the length of the crediting period. Whereas some parties feared that such a liability would deter the use of LULUCF in the CDM, others stated that liabilities at least twenty and as many as sixty years in the future would have little practical effect. For both forms of temporary credits, the liability to replace the credits after potential reversal rests with the Annex I Party in whose registries the unit is kept. Buyers of tCERs have either to replace them after five years or to acquire new tCERs if they wish to do so. Buyers of lCERs have to buy only once. From the seller's perspective, the regulation means that tCERs can be sold every five years, which is an economically interesting feature of tCERs. However, tCERs may incur greater transaction costs than lCERs because new certificates have to be issued every five years. Even though this is only a simple electronic transaction, it is expected that each transaction will carry an administrative fee and possibly additional fees contributing to special funds to assist developing nations.

The choice between tCERs and lCERs has to be made up front by the project proponent and cannot be changed retroactively.[17] Because tCERs and lCERs are temporary, their prices are lower than the prices of (permanent) CERs from other CDM project categories. In addition, Annex I countries can use tCERs and lCERs only in the commitment period in which they are certified and cannot carry them over to a subsequent commitment period (that is, no "banking" is allowed).[18] Removal Units (RMUs) are not bankable either.[19] However, it remains to be discussed to what extent LULUCF JI projects that generate ERUs from RMUs convert RMUs into bankable assets.

The Concept of Additionality and the Baseline

Additionality is a central concept with respect to the Kyoto Protocol's project-based mechanisms (CDM and JI). In order to be environmentally sound, it is essential that emission reductions and removals be "additional" to what would occur in the absence of the project. The basic idea of additionality is that the carbon sequestration or emission reduction of a CDM project would not have occurred without the incentives provided by the flexible mechanisms. Additionality is particularly important in excluding business-as-usual scenarios from the CDM. Given that the CDM is designed as a carbon-neutral process, it would be counterproductive if the additionality of a project could not be guaranteed.[20] The requirements for additionality, as COP 9 adopted them for LULUCF projects, state that an AR project is additional "if the actual net greenhouse gas removals by sinks are increased above the sum of the changes in carbon stocks in the carbon pools within the project boundary that would have occurred in the absence of the . . . project activity."[21] The CDM Executive Board, at its twenty-first meeting (EB 21), provided the first version of a stepwise tool for demonstrating the additionality of prospective LULUCF projects in the CDM. At EB 35 it further elaborated on the initial tool and issued a second version.[22] Developers of AR projects are strongly encouraged to make use of the tool, which screens a potential project for additionality in a stepwise approach.

The second element of additionality (to generate tCERs or lCERs) is to demonstrate that the GHG removals claimed by the project are more than would occur anyway, if the project did not exist. This "without-project scenario" is called the baseline scenario. Baselines for AR projects have to reflect the analysis of past trends and current situations for prevailing land-use patterns. Changes in land use are seldom random but are predictable on the basis of observable factors such as historically prevailing patterns of land use, presence of transportation networks, and access to markets. Sometimes land-use plans even exist.[23] The official text mentions three baseline approaches from among which project proponents must choose the most adequate. They may determine the baseline according to (1) existing or historical, as applicable, changes in carbon stocks in the carbon pools within the project boundary; (2) changes in carbon stocks in the carbon pools within the project boundary from a land use that represents an economically attractive course of action, taking into account barriers to investment; or (3) changes in carbon stocks in the pools within the project boundary from the most likely land use at the time the project starts.[24]

Option 1 establishes a business-as-usual baseline approach for AR projects. It describes a continuation of current land use and should therefore be chosen if the "without-project" scenario is most likely a continuation of current land-use practices. Option 2 describes a change in land-use patterns due to economic factors. The difference between options 2 and 3 is not self-evident at first view. Under economic considerations, "the most likely land use at the time the project starts" (option 3) is actually the one "that represents an economically attractive course of action" (option 2). By presenting these two alternatives, the negotiators distinguished between a more economically motivated land-use baseline scenario (2) and a more mandatory change in prevailing land use due to regulations and laws (3). Option 3 also captures any other nonfinancially motivated land-use options.

In general, the baseline scenario can either be established up front and remain fixed for the first phase of the crediting period or else be monitored on a representative basis (that is, through the monitoring of control plots) during the AR project implementation. In both cases an up-front estimation of baseline emissions and removals is necessary, but in the latter case the actual results from the control plots are used to calculate concrete emission reductions. The fixed baseline approach guarantees greater confidence about actual emission reductions generated by the project, whereas the second option may guarantee greater accuracy from a scientific point of view.

The determination of baselines for JI projects, especially for Track 2, is similar to that for CDM projects. To allow swift implementation of the JI, the JISC decided that appropriate existing CDM baseline methodologies could be used for JI projects as well (Track 2).[25] Under Track 1, the involved countries can decide on an appropriate baseline and determine the number of credits transferred to the investor country. Contrary to the CDM, JI projects can start to generate credits only after the beginning of 2008, and the crediting period has to be determined by the project developer. If it extends beyond 2012, the project has to be approved by the host party and is conditional on a second commitment period.[26]

Leakage

The issue of leakage plays an important role in assessing the number of credits that can be claimed for any CDM project. Leakage refers to the partial or even total offsetting of a project's climate benefits by activities or behaviors outside the project area that reduce sequestration or increase emissions.[27] For example, access restrictions imposed by plantations created for sequestration could displace rural people who then deforest areas elsewhere. In the case of avoided deforestation, usage restrictions might force people to clear forests elsewhere, which would counter the carbon reservoir protection.

From early in the negotiations, leakage was a major concern surrounding the issue of sinks, and so far approaches to assessing leakage lack the right mixture of scientific rigor and practicability. Seen from a purely economic perspective, leakage is a biologically inspired name for general equilibrium effects. Such effects are neither intrinsically condemnable nor avoidable. Indeed, it might be costly for a project to try to eliminate potential leakage entirely. This implies that project developers have to thoroughly address leakage in the project design and account for it by subtracting potential unintended GHG emissions due to the project activity (that is, leakage) from the project's performance. According to the definition of leakage for AR in the CDM, leakage has to be accounted for if it "is measurable and attributable" to the AR project activity.[28] Also, only leakage that *increases* emissions is to be taken into account, thereby precluding any claims for "positive leakage" or "spillover effects," as when a project leads to additional forest planting or growth outside the project boundaries.

Social and Environmental Safeguards

The CDM pursues a twin objective of climate protection and sustainable development. It is the host party's prerogative to confirm whether a CDM project activity contributes to achieving sustainable development.[29] It is expected that host governments will deny approval of projects that do not further their country's sustainable development goals, the establishment of which lies within the sovereign power of each country. Furthermore, CDM projects have to take potential environmental effects into account, including effects on biodiversity and natural ecosystems, as well as socioeconomic effects of the project activity.[30] Project participants have to submit documentation on the analysis of socioeconomic and environmental effects to the designated operational entity (DOE) as a prerequisite for project validation. This documentation has to include an assessment of "impacts on biodiversity and natural ecosystems, and impacts outside the project boundary" (that is, leakage) of the proposed AR project.[31] If mandated by law, or if the project participants or the host country government consider potential effects to be "significant," then an impact assessment (environmental, socioeconomic, or both) must be carried out in accordance with the host country's procedures. Parties introduced these provisions in order to respond to the concerns of certain parties, as well as some NGOs, who expected large-scale plantations with negative side effects to become eligible under the CDM. Clearly, ignoring these concerns would have eroded public support and undermined the general CDM framework. However, the failure to establish an internationally agreed-upon minimum standard by which to judge whether certain project effects were acceptable or not has been heavily criticized by environmental NGOs.

Furthermore, parties made the use of genetically modified trees and potentially invasive species in AR CDM projects subject to agreement between the host country and the investor. Finally, individuals, groups, and communities affected by the proposed CDM project activity have to be consulted. The time frame for the reception of stakeholder comments after validation is forty-five days for AR projects.[32] According to the Marrakech Accords, "'stakeholders' means the public, including individuals, groups or communities affected, or likely to be affected, by the proposed clean development mechanism project activity."[33]

Small-Scale Afforestation and Reforestation Projects in the CDM

The tenth Conference of the Parties, which met in Buenos Aires in 2004, decided on simplified modalities for small-scale AR CDM projects. Initially, projects that resulted in the removal of fewer than 8,000 tonnes of carbon dioxide equivalent (CO_2e) per year were considered to be small scale.[34] However, the sequestration rate limit constituted an additional barrier discouraging project proponents from proposing small-scale projects. Research on the topic revealed that even under optimistic assumptions about carbon price and transaction costs, small-scale AR projects would be infeasible under this size limit.[35] COP 13, meeting in Bali in 2007, raised the threshold to 16 kilotonnes of CO_2e per year.[36]

Because of the limit on the carbon sequestration rate of small-scale AR projects, the concrete project size depends on the actual project type, the species planted, and the natural conditions of the area concerned. For fast-growing species, project size will be small. For species or conditions with lower carbon uptake rates, projects will be relatively larger. Depending on these parameters, the maximum project area will vary between approximately 500 and 4,000 hectares, although a project may be made up of separated land units and thus affect a much larger land area.[37] However, small-scale AR CDM projects cannot be the result of a "de-bundled" operation of larger scale. If a project has the same project participants, has been registered within the previous two years, and has boundaries within 1 kilometer, then it is considered to be a de-bundled larger-scale operation.[38]

Small-scale AR projects are subject to simplified methodologies for the establishment of a baseline and for project monitoring.[39] To date, one approved baseline and monitoring methodology for small-scale AR CDM project activities exists for the conversion of grassland or cropland to forested land.[40] Another simplified methodology for small-scale AR CDM projects implemented on wetlands has been proposed by the AR Working Group.[41] The stepwise additionality tool mentioned earlier does not apply to small-scale AR projects. Additionality is established according to a simplified methodology.[42] Furthermore, small-scale projects must generally target low-income communities or individuals as defined by the host country. The latter aspect links those projects directly to an intended improvement of rural livelihoods. It remains to be seen whether the increase of the sequestration limit can promote the development of more small-scale AR projects in the future.

Conclusion and Outlook

In the early days of the carbon market, long before the international community adopted rules and regulations for forestry projects under CDM and JI, the majority of all carbon transactions were forest-related transfers. Particularly in developing countries in which agriculture and forestry accounted for a good portion of GDP, the number of carbon contracts outside of formalized international agreements was significant.[43] One reason was the anticipated positive influence of carbon-finance-related land-use systems on rural development and biodiversity conservation. Mainly NGOs administered these early projects and provided carbon-monitoring services for voluntary buyers from the private sector. In the following period, interest in forestry projects and forest-related credits declined, for a number of reasons. These included the controversial negotiation history of land-use-related project categories under the UNFCCC, the late agreement on and complexity of the rules and modalities for forestry projects under the CDM, and the relatively high (perceived) risk of this project category. Thus LULUCF assets, as a share of the volume contracted on the international carbon market, have dropped steadily in the last few years.[44]

Recent developments indicate a change of attitudes. Today most governments and the private sector take a more pragmatic position toward forestry projects. After COP 13 in Bali it is certain that forests will play a more prominent role in the post-2012 climate regime. The private sector is very supportive of LULUCF projects. Some companies are still worried about what at first glance appear to be cumbersome rules and regulations for the implementation of such projects under the UNFCCC framework. But an increasing number of them are prepared to buy carbon credits from forestry projects even beyond the Kyoto scheme on the voluntary market.[45] The major stumbling block discouraging private sector demand is clearly that forestry credits cannot be used for compliance under the European Union Emissions Trading Scheme (EU ETS).[46] A recent survey showed that 40 percent of the participating private sector entities would purchase forestry credits if they were recognized under the EU ETS.[47] Although there may have been good reasons for excluding LULUCF credits at the time the trading scheme was developed (for example, uncertainties about the accounting of emission removals by carbon sinks), the international framework of rules and regulations has clearly evolved since then. The COP and the CDM Executive Board adopted regulations for monitoring and accounting for forestry projects, and the first ten large-scale baseline and monitoring methodologies for AR projects have been approved.

In addition, it is worth noting that no other sector of the carbon market sees as many *voluntary* transactions that create credits outside the UNFCCC-regulated context as the LULUCF sector. And the trade of those voluntary carbon offsets is growing.[48] Transactions in this market sector are characterized by individual agreements between buyers and sellers that define the respective carbon rights and control measurement, verification, and transfer of certificates. Examples are companies that invest in socially responsible environmental projects to "green" their corporate identity and others actively offering "carbon neutral" products.

Overall, the regulatory framework existing today offers an encouraging starting point for LULUCF project development. The established tCERs and lCERs create a broad scope for contractual solutions between developing countries and Annex I Parties. Legal security for participants registering projects today is provided even if decisions concerning forestry projects under the CDM are subject to revision in further commitment periods. For all projects that are registered before the end of the first commitment period, crediting is guaranteed even beyond that period, depending on the crediting scheme chosen by project participants.[49] Thus regulations existing today pave the way for a project category that shows substantial developmental benefits for rural communities, on the one hand, and provides a flexible and supplementary mitigation solution for Annex I Parties, on the other.

The central issue of anthropogenic deforestation and its significant climate effects, however, remains to be comprehensively addressed.[50] The rules and regulations acknowledge that planting forests—for example, through afforestation and reforestation under the CDM—provides an opportunity to sequester carbon in vegetation and soils. But the text of neither the UNFCCC nor the Kyoto Protocol includes a satisfying mechanism by which to reduce the substantial emissions from deforestation, which are responsible for about a quarter of global GHG emissions. The serious problem is that it takes decades to restore carbon stocks that have been lost as a result of land-use change. Therefore reducing the rate of deforestation is the only effective way to decrease carbon losses from forest ecosystems.

With negotiations on a post-Kyoto agreement already begun, it is obvious that a more complete and target-oriented post-Kyoto regime will not only have to expand existing regulations by allowing additional land-use-related activities under the CDM but aiso create a framework to encompass all land-use and forest-related changes in *carbon stocks*. Developing countries today administer the majority of the world's environmental resources, such as tropical forests, and in doing so provide a vital global public good. They will have to be integrated into a more comprehensive incentive framework that also rewards forestry conservation and sustainable forest management besides afforestation and reforestation. The challenge now is to create such an incentive system for the protection of forests and to include it in the structural design of a future climate regime.

Notes

1. R. T. Watson, ed., *IPCC: Climate Change 2001: Synthesis Report* (Cambridge University Press, 2001); R. T. Watson and I. R. Noble, *The Global Imperative and Policy for Carbon Sequestration: The Carbon Balance of Forest Biomes* (London: Garland Science/BIOS Scientific Publishers, 2004).

2. E. D. Schulze, R. Valentini, and M. J. Sanz. "The Long Way from Kyoto to Marrakesh: Implications of the Kyoto Protocol Negotiations for Global Ecology," *Global Change Biology* 8, no. 6 (2002): 505–18.

3. For the most important definitions and rules on LULUCF, see 16/CMP.1 (decision 11/CP.7 before adoption by COP/MOP1).

4. Whereas a gross-net approach is applied for afforestation, reforestation, deforestation, and forest management (emissions are not included in the base year), the activities under article 3.4 are accounted for on a net-net basis.

5. See Appendix of Decision 16/CMP.1, FCCC/KP/CMP/2005/8/Add.3.

6. In reality, not only countries but also private entities participate in JI and CDM projects in a variety of setups. For simplification, we often refer only to countries, because they are the ones having to comply with Kyoto targets. Relevant COP/MOP decisions regarding article 6 are 9/CMP.1 and 10/CMP.1.

7. However, only activities chosen by the host country to be accounted for in its national inventory under article 3.4 can be registered as JI projects in that country.

8. AAUs define the base unit for the accounting of GHG emissions under the Kyoto Protocol. Experts and parties are currently discussing whether AAUs and RMUs can be converted to ERUs. See B. Schlamadinger, C. Streck, and R. O'Sullivan, "Will Joint Implementation LULUCF Projects Be Impossible in Practice?" (www.climatefocus.com/newspubs/downloads/JI_LULUCF_in_practice.pdf), and the webcast of the respective Subsidiary Body for Scientific and Technological Advice (SBSTA) side event from May 8, 2007 (http://unfccc.meta-fusion.com/kongresse/SB26/templ/ply_sideevent.php?id_kongresssession=531&player_mode=isdn_real).

9. See Decision 9/CMP.1, Annex, D, Participation requirements.

10. See Decision 16/CMP.1, Annex A. 1(a).

11. See EB 35 Report, Annex 18.

12. Unless it leads to emissions elsewhere or in subsequent years, that is, leakage.

13. See decision 5/CMP.1, Annex, Section A, Definitions, paragraph 1(g), "'Temporary CER' or 'tCER' is a CER issued for an afforestation or reforestation project activity under the CDM which . . . expires at the end of the commitment period following the one during which it was issued," and paragraph 1(h), "'Long-term CER' or 'lCER' is a CER issued for an afforestation or reforestation project activity under the CDM which . . . expires at the end of the crediting period of the afforestation or reforestation project activity under the CDM for which it was issued," FCCC/KP/CMP/2005/8/Add.1.

14. Whereas CDM projects were eligible for early crediting during the period from 2000 to 2008, JI projects can generate credits only after 2008.

15. The replacement obligation lies with the acquiring entity. Decision 5/CMP.1, Section K, paragraphs 49 and 50, FCCC/KP/CMP/2005/8/Add.1.

16. Ibid., paragraph 49.

17. Ibid., paragraphs 38 and 39.

18. Ibid., paragraph 41.

19. Annex to decision 13/CMP.1, section F, FCCC/KP/CMP/2005/8/Add.2. Carry-over of ERUs (not converted from RMUs and not used for compliance), as well as carry-over of CERs, is limited to 2.5 percent of the party's Assigned Amount Units.

20. Because CDM host countries do not have targets, the result of the CDM transaction is an unchanged atmospheric concentration of GHGs. The CERs created in the project are transferred to the Annex I country, which is allowed to emit an equivalent amount of carbon.

21. Decision 5/CMP.1, Section G, paragraphs 12(d) and 18, FCCC/KP/CMP/2005/8/Add.1.

22. CDM Executive Board, EB 35 Report, Annex 17, "Tool for the Demonstration and Assessment of Additionality in A/R CDM Project Activities" (http://cdm.unfccc.int/EB/035/eb35_repan17.pdf).

23. D. Kaimowitz and A. Angelsen, "Economic Models of Tropical Deforestation: A Review," Center for International Forestry Research, Bogor, Indonesia, 1998; S. Brown and others, "Changes in the Use and Management of Forests for Abating Carbon Emissions:

Issues and Challenges under the Kyoto Protocol," in *Carbon, Biodiversity, Conservation and Income: An Analysis of a Free-Market Approach to Land-Use Change and Forestry in Developing and Developed Countries,* edited by I. R. Swingland and others (London: Royal Society, 2002), pp. 42–55.

24. Decision 5/CMP.1, Section G, paragraph 22, FCCC/KP/CMP/2005/8/Add.1.

25. See Appendix B of Decision 10/CMP.1, Implementation of Article 6 of the Kyoto Protocol, FCCC/KP/CMP/2005/8/Add.2.

26. See "Guidelines for the Users of the Joint Implementation Land Use, Land-Use Change and Forestry Project Design Document Form," version 1 (http://ji.unfccc.int/Sup_Committee/Meetings/004/Reports/JISC04report_Annex_15.pdf).

27. Decision 5/CMP.1, Annex, Section A, Definitions, paragraph 1(e), "'Leakage' is the increase in greenhouse gas emissions by sources which occurs outside the boundary of an afforestation or reforestation project activity under the CDM which is measurable and attributable to the afforestation or reforestation project activity," FCCC/KP/CMP/2005/8/Add.1.

28. Decision 5/CMP.1, Section G, paragraph 12(d), FCCC/KP/CMP/2005/8/Add.1.

29. Ibid., paragraph 12(c).

30. Decision 5/CMP.1, "Modalities and procedures for afforestation and reforestation project activities under the clean development mechanism in the first commitment period of the Kyoto Protocol, Appendix B: Project design document for afforestation and reforestation project activities under the clean development mechanism," FCCC/KP/CMP/2005/8/Add.1.

31. Decision 5/CMP.1, Section G, paragraph 15(c).

32. Ibid.

33. Decision 3/CMP.1, "Modalities and procedures for a clean development mechanism as defined in article 12 of the Kyoto Protocol," Annex, Section A, paragraph 1(e), FCCC/KP/CMP/2005/8/Add.1.

34. Decision 5/CMP.1 Annex A. (i).

35. B. Locatelli and L. Pedroni, "Will Simplified Modalities and Procedures Make More Small-Scale Forestry Projects Viable under the Clean Development Mechanism?" *Mitigation and Adaptation Strategies for Global Change* 11, no. 3 (2006): 621–43.

36. FCCC/SBSTA/2007/L.18/Add.1.

37. See Decision 5/CMP.1, Annex, Section A, Definitions, paragraph 1(b), "The 'project boundary' geographically delineates the afforestation or reforestation project activity under the control of the project participants. The project activity may contain more than one discrete area of land."

38. Decision 6/CMP.1, Appendix C.

39. FCCC/KP/CMP/2005/4/Add.1, Annex II, "Simplified baseline and monitoring methodologies for selected small-scale afforestation and reforestation project activities under the clean development mechanism."

40. EB 33, Sectoral scope: 14, AR-AMS0001/Version 04.1.

41. EB 35, Sectoral scope: 14, AR-AMS000X/Version 01.

42. CDM AR Working Group, Sixth Meeting Report, Annex 2, Attachment B, "Assessment of additionality."

43. See, for example, K. C. Nelson and B. H. J. de Jong, "Making Global Initiatives Local Realities: Carbon Mitigation Projects in Chiapas, Mexico," *Global Environmental Change* 13, no. 1 (2003): 19–30.

44. K. Capoor and P. Ambrosi, "State and Trends of the Carbon Market 2007" (Washington: World Bank, 2007).

45. C. Dannecker, "Evaluation of a Survey about the Market for Certified Emission Removals from Forestry Projects: A Report by EcoSecurities, Ltd," Oxford, 2006; K. Hamilton and others, "State of the Voluntary Carbon Market 2007: Picking Up Steam," New Carbon Finance and Ecosystem Marketplace, London and Washington; Voluntary Carbon Standard, "Guidance for Agriculture, Forestry and Other Land Use Projects," 2007 (www.v-c-s.org/docs/AFOLU%20Guidance%20Document.pdf).

46. "Directive 2003/87/EC of the European Parliament and of the Council (13 Oct. 2003), establishing a scheme for GHG emission allowance trading within the Community and amending Council Directive 96/61/EC," *Official Journal of the European Union,* October 25, 2003: L 275, 25/10/2003 P. 0032–0046. On July 23, 2003, the Commission of the European Communities released the first draft of a "Proposal for a directive of the European Parliament and of the Council, amending the directive establishing a scheme for greenhouse gas emission allowance trading within the Community, in respect of the Kyoto Protocol's project mechanisms." COM, 2003, 403 final.

47. In addition, a workshop held in March 2006 in Brussels, organized by the World Bank, showed an increasing number of countries supporting the opening of the EU Emissions Trading Scheme for forestry credits. See www.carbonfinance.org.

48. Although some providers on the voluntary market specialize in energy-based projects, the majority of retailers are focusing on forestry projects. It is argued that trees are easier to sell to the general public because they are a more tangible and understandable countermeasure to global warming. See N. Taiyab, "Exploring the Market for Voluntary Carbon Offsets," International Institute for Environment and Development, London, 2006.

49. Decision 5/CMP.1, paragraph 3, FCCC/KP/CMP/2005/8/Add.1.

50. C. Streck and S. M. Scholz, "The Role of Forests in Global Climate Change: Whence We Come and Where We Go," *International Affairs* 82, no. 5 (2006): 861–79.

CASE STUDY

The Humbo Community-Based Natural Regeneration Project, Ethiopia

PAUL DETTMANN, TONY RINAUDO, AND ASSEFA TOFU

The Humbo Community-Based Natural Regeneration Project offers an opportunity to combine natural resource management, carbon sequestration, biodiversity, and poverty alleviation. The project is seeking registration as a CDM project activity. In 2005 the World Bank's Biocarbon Fund indicated interest in purchasing the carbon offsets generated by the project, which was developed and is being implemented by World Vision Ethiopia and Australia (WV), a humanitarian development organization, with the support of the government of Ethiopia. Following is a summary of the project and the key lessons learned to date.

Project activities include the restoration of indigenous, biodiverse forest species to a mountainous region of southwestern Ethiopia. These activities complement the natural resource management goals of the Ethiopian Ministry of Agricultural and Rural Development. The project offers significant social development outcomes in line with the goals of the government of Ethiopia and project proponent World Vision. It is expected that the project will sequester approximately 165,000 metric tons of atmospheric carbon dioxide by 2017 through the reforestation of 2,728 hectares of steep mountainside terrain that contribute to spring water recharge. The area has remained in its cleared state for more than two decades, and as a result, mudslides, floods, and erosion are common occurrences.

Core project goals include carbon sequestration, biodiversity protection and enrichment, income generation, protection and increased recharge of drinking water supplies, and improved livelihoods for current land users. Seven cooperatives have been established as nonprofit entities and have received land-use rights. All proceeds from carbon sales will be directed to projects benefiting all community members, such as projects in education, health, and agriculture, in accordance with group by-laws and constitutions.

Lessons learned from developing this CDM carbon sequestration project include the following. First, investment in advocacy and capacity-building is required. When this project concept was first proposed, the government of Ethiopia had not ratified the Kyoto Protocol and there were no CDM projects in Ethiopia. Significant time was required to sensitize potential stakeholders to the issues and opportunities associated with carbon projects. This CDM capacity-building for key stakeholders and government agencies will benefit other carbon projects currently being prepared and implemented in Ethiopia.

Second, small projects are costly to develop, implement, and manage. This project would be more cost-effective if it were larger; 20,000 hectares would be ideal. Under the current CDM framework, project development costs are substantial. As a result, small and medium-size projects (fewer than 5,000 hectares) without additional revenue streams (for example, from agroforestry, nontimber products, or tree crops) find it difficult to generate workable economies of scale. A conservative but simplified CDM methodology and monitoring process, if possible using remote sensing, would facilitate the replication of afforestation-reforestation (AR) projects. Implementation of many smaller projects with greater community ownership, by applying a more programmatic approach, might then also become feasible. In turn, the nonpermanence risk associated with AR projects would be spread over a larger number of communities and a greater geographical range.

Third, an existing strong working relationship between the implementing agency and the community facilitates project implementation. WV has more than 1,200 staff members working in some sixty-four area development programs in Ethiopia and has worked in the Humbo region for eighteen years. Experience in providing development assistance in health, agriculture, emergency relief, education, and community empowerment has given WV considerable understanding of local needs and earned it respect as a development partner from both the community and the local government. Difficulties experienced during implementation could have derailed the project had it not been for the strong relationships in place. Land-use and ownership issues are volatile topics in Ethiopia, and there is suspicion of anyone who proposes changes to the status quo. The overriding response to surveys on the potential role of WV in developing this project was one of strong endorsement based on the trust relationship that had been built over time. Additionally, the establishment of community use and ownership rights through these types of projects can be a substantial contribution in itself.

Fourth, afforestation projects can provide multiple benefits beyond carbon sequestration, and these should be recognized and rewarded. AR projects offer many services in addition to carbon sequestration, including the protection and management of biodiversity, maintenance and improvement of water quality, soil protection, and poverty alleviation. AR projects would greatly benefit from the quantification of at least some of these benefits in a standardized way (such as by applying the Climate, Community, and Biodiversity Standards), whereby projects may attract buyers seeking the provision of these services. The project team believes AR projects are unique among greenhouse gas removal projects and need to be recognized and rewarded for the additional services they provide. This additional service provision should offset the discount applied for their perceived lack of permanence. The Humbo project, for example, will provide additional benefits in the form of poverty reduction through managed harvesting of forest products, including firewood, fodder, edible fruits and leaves, medicines, and dyes; protection and enhanced recharge of drinking water supplies; protection of land systems from environmental problems such as erosion and flooding; protection of Lake Abaya from siltation and turbidity; protection and management of biodiversity; and ecotourism development.

Fifth, poor communities cannot engage with the current CDM process without significant technical assistance. The process the Humbo project had to undertake to secure CDM carbon credits was highly complex and likely would have been unsuccessful without the considerable resources of an organization such as World Vision. Some of the many challenges that confronted the project included the following:

—Legal challenges: understanding Ethiopian laws and regulations on property ownership and land-use rights; organizing seven local communities into incorporated cooperatives; negotiating with the World Bank to reach agreement on key issues such as appropriate forestry management practices; and identifying staffing needs and allocating responsibilities. In dealing with these, WV had the generous pro-bono support of Clayton Utz, a major Australian law firm.

—Financial challenges: development of a sophisticated financial forecast that built elements of commodities futures trading into the project model. The development of such a model requires advanced skills in financial analysis and modeling.

—Commercial challenges: understanding the carbon market in various jurisdictions and the difference in price between agricultural carbon sequestration and other forms of carbon emission reductions. It was challenging to reconcile the commercial and development aspects of the project and the way resources should be applied to the project.

—Project management challenges: analyzing project implementation and management issues in order to develop a robust forestry management plan.

—CDM challenges: preparation of a Project Design Document, appropriate application of a CDM methodology and monitoring plan, and preparation for an independent validation.

7

How Renewable Is Bioenergy?

BERNHARD SCHLAMADINGER, SANDRA GREINER, SCOTT SETTELMYER, AND DAVID NEIL BIRD

At forty-six exajoules per year, bioenergy constitutes more than 10 percent of global primary energy. Traditional solid biomass fuels (fuelwood, charcoal, dung, and straw) constitute about 80 percent of global biomass use. Lately interest in biofuels and bioenergy has increased significantly as a result of higher fossil fuel prices, concerns about energy security, and the need to reduce greenhouse gas (GHG) emissions to mitigate climate change. The drive to increase bioenergy consumption is raising concerns among nongovernmental organizations (NGOs) and in both the mainstream and the alternative press.[1] These concerns are not only that increased use of bioenergy will produce no net benefit for climate mitigation but also that it will negatively affect biodiversity, water, and food security.[2]

In this chapter we address the question, Under what conditions can biomass energy be seen as CO_2 neutral or renewable? We pay special attention to trade-offs and synergies with carbon sequestration on agricultural and forestry land. We discuss the relevance of emissions associated with the production and use of biomass, indirect effects through agricultural markets, and bioenergy's role in the Kyoto Protocol and especially in the Clean Development Mechanism (CDM).

The Greenhouse Gas Benefits of Bioenergy

The main interest in bioenergy as a climate change mitigation strategy is that it is a CO_2 neutral energy source, provided that the bioenergy is sustainably produced—

Figure 7-1. *Closed Carbon Cycle Associated with Biomass Use for Energy*

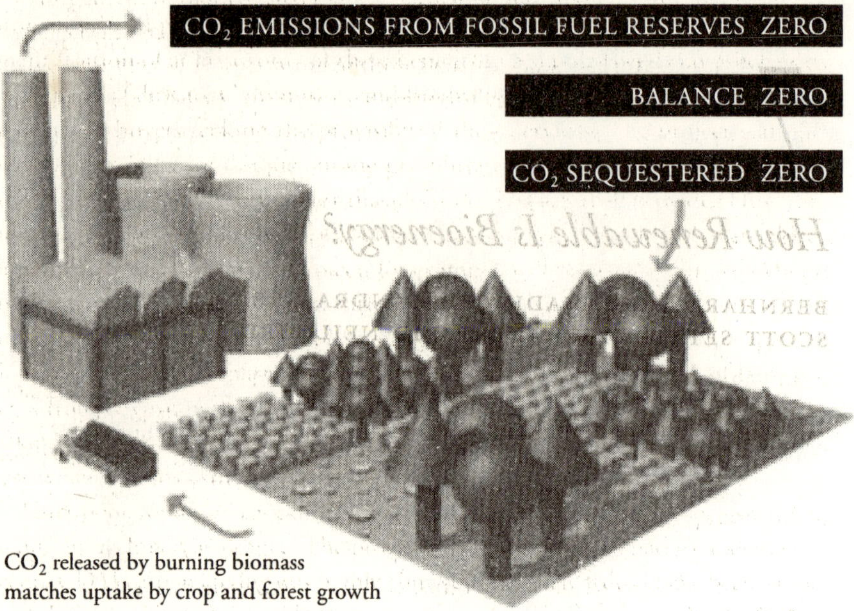

Source: IEA Bioenergy Task 38, Greenhouse Gas Balances of Biomass and Bioenergy Systems, "Answers to ten frequently asked questions about bioenergy, carbon sinks and their role in global climate change" (www.ieabioenergy-task38.org/publications/faq/). Figure compiled by R. Matthews.

that is, it comes from a renewable biomass source. This view is based on the premise that the same amount of CO_2 is released at the point of use as is removed from the atmosphere via photosynthesis (figure 7-1).

In countries with national GHG emission targets under the Kyoto Protocol (Annex I countries), it is assumed that sustainable agricultural and forestry practices are in effect and that if they are not, then any loss of biomass, particularly through deforestation, is captured in the land-use part of the national GHG inventory.

For countries without national GHG emission targets (non–Annex I countries), no assumption of sustainable agricultural and forestry practice is made, and the renewability of the biomass for energy use must be demonstrated. This has been a significant stumbling block for acceptance of methodologies for biofuel projects under the CDM. Net CO_2 emissions from the use of biomass can be ignored only if biomass is renewable.[3] Woody and nonwoody biomass is defined as renewable under the CDM if (1) there is no change in land use or the land-use change is from cropland or grassland to forest (that is, afforestation or reforestation), (2) sustainable management practices are in effect to ensure that there is no systematic decrease in carbon stocks on the land, and (3) the practice is in compliance with all national and regional forestry, agriculture, and environmental regulations.[4] Residual biomass from agricultural or forestry operations is also considered renewable if use of the residues does not cause a decrease in carbon stocks on the land where the biomass originated.

In practice, biomass energy is not always from renewable sources. Land management associated with production of biomass may result in decreases of carbon stocks in the five relevant carbon pools: above-ground biomass, below-ground biomass, deadwood, litter, and soil. For example, the production of biofuels from palm oil plantations is not renewable if the land was deforested to establish the palm oil plantation. Similarly, retrieving biomass may result in the decrease of deadwood, litter, or soil carbon stocks. A project that increases the collection of deadwood in an existing forest, for example, would not be considered renewable if this practice depletes the pool of deadwood in the forest. Another example is the planting of an annually tilled bioenergy crop such as rapeseed on grassland. The tillage of the soil could cause a systematic decrease in soil carbon stocks, so the practice would not be considered renewable. If the bioenergy system is not renewable, then one has to include the CO_2 emissions from direct land use in calculating net GHG emission benefits.

On the other hand, bioenergy production schemes exist that may increase terrestrial carbon stocks. For example, reforestation with energy crops such as oilseed-bearing trees may enhance carbon storage in plants and soils. Nevertheless, the use of land for producing biomass fuels may reduce the land's ability to store more carbon, as we discuss in the following section.

Bioenergy Production versus Carbon Sequestration

The authors of a recent article argued that forestation of land would sequester two to nine times more carbon over a thirty-year period than the carbon emissions avoided by replacing fossil fuels with biofuel grown on the same land.[5] They went on to suggest that only the use of woody biomass for energy is compatible with the retention of forest carbon stocks and that emissions avoided by substituting fossil fuels might be comparable to the carbon sequestered in an alternative, forest-restoration scenario that does not produce biomass for energy.

Figure 7-2 compares the merits of using land for bioenergy production with the merits of using it for carbon sequestration. The results are dependent on the following:

—The efficiency with which biomass energy replaces fossil fuel energy. This efficiency is high if biomass is produced and converted efficiently, if the replaced fossil fuel was used with low efficiency, and if a carbon-intensive fossil fuel is replaced.

—The time period under consideration. The longer the time frame of the analysis, the more attractive biomass energy is in comparison with carbon sequestration,

Figure 7-2. *Relative Benefits of Using Land for Substitution Management versus Sequestration Management*[a]

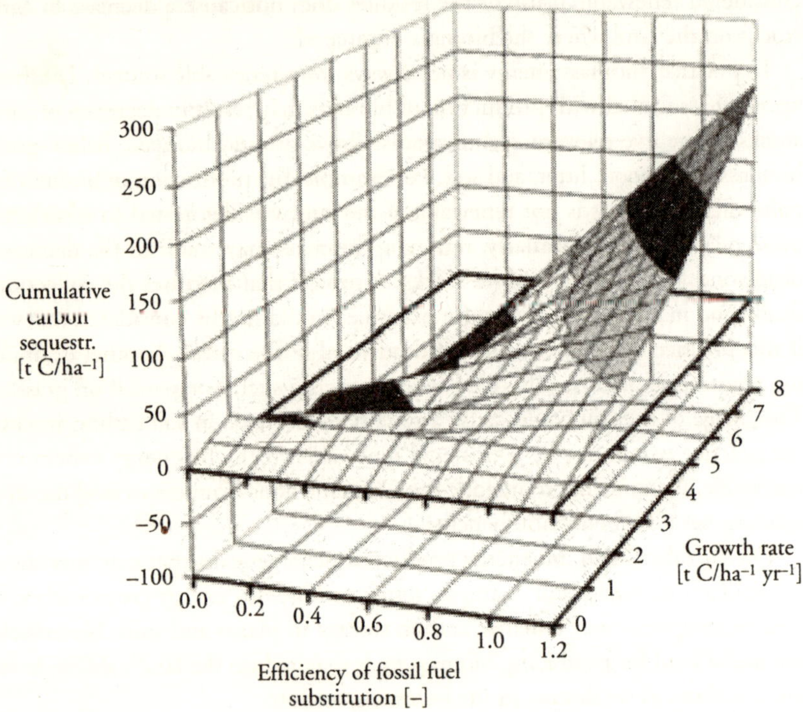

Source: B. Schlamadinger and G. Marland, "Forests for Carbon Sequestration or Fossil Fuel Substitution? A Sensitivity Analysis," paper presented at the eleventh World Forestry Congress, Antalya, Turkey, 1997 (www.fao.org/forestry/docrep/wfcxi/PUBLI/V1/T4E/3-3.HTM#TOP).

a. Difference after forty years between a scenario in which land is reforested with fast-growing species to produce biomass energy and a scenario in which land is reforested with the main purpose of storing carbon. The surface (vertical axis) depicts the cumulative carbon benefits of substitution over sequestration as a function of the efficiency of bioenergy use and the growth rate of the plantation. Positive values indicate that management for biomass energy is the better choice.

because the latter is constrained by saturation (only a limited amount of carbon can be stored on a hectare of land), whereas bioenergy can be produced repeatedly from harvest cycle to harvest cycle.

—The growth rate of the forest on the site. The higher the growth rate, the sooner the saturation constraints of carbon sequestration are reached.

The figure shows that a combination of high-yielding species and efficient use of the biomass to replace fossil fuel makes substitution management the preferable option over sequestration management. In the back right corner of the diagram the benefits of substitution management exceed those of sequestration management by almost 250 tonnes of carbon per hectare after forty years. On the other hand, low-efficiency biomass use, independent of growth rate, means that the land is better used for carbon sequestration. Where biomass is used efficiently but growth rates are low, the relative merits of substitution management are limited.

Harvesting versus Conservation

So far we have looked at the best use of nonforest land, comparing the carbon benefits of sequestration to the carbon benefits of biomass energy that can be derived from planting new forests. In other cases the question is whether existing forests should be used for bioenergy and timber production or protected for maximum carbon storage.[6] The modeling of results, an example of which is shown in figure 7-3, demonstrates that substitution management yields greater benefits if initial carbon stocks are low, growth rates are high, biomass is used efficiently, and a long-term view is taken. Figure 7-3 shows that the conversion of densely stocked forests into biomass-producing plantations—a scenario depicted in the front part of the illustration—can take a very long time to yield positive greenhouse gas benefits. For example, a forest stocked with 300 tonnes of carbon per hectare (admittedly an extreme case) can require a payback period of up to 100 years. With 150 tonnes per hectare the payback period is still around 30 years.

Emissions from Biomass Production

Direct land-use emissions may result from the cultivation of biomass for bioenergy. These emissions can be caused by the clearing of vegetation, including forests, to establish the bioenergy crop, the application of synthetic and natural fertilizers, and the use of fossil fuels during the cultivation and harvesting of the bioenergy crop. Emissions from the clearing of vegetation are predominantly CO_2 from the loss of biomass but may include methane (CH_4) and nitrous oxide (N_2O) emissions if the vegetation is burned during clearing. The use of fertilizers produces N_2O emissions.

Indirect emissions, generally termed *leakage,* are those that occur outside the project boundary as a result of biomass production for bioenergy. These emissions come from three main sources: the consumption of fossil fuels outside the project boundary during the establishment and management of the bioenergy system, the production of fossil fuels, fertilizers, or other soil additives used during cultivation, and the displacement of land-use activities. The first two tend to be small components of total project emissions. The last is more significant and is the primary concern of NGOs that oppose the rewarding of carbon credits for biological sequestration under the climate regime.

To examine the significance of the displacement of activities, we consider a typical jatropha plantation that produces one tonne of oil per hectare per year and

Figure 7-3. *Carbon Balance of a Forest Harvested at Time Zero and Regrown*[a]

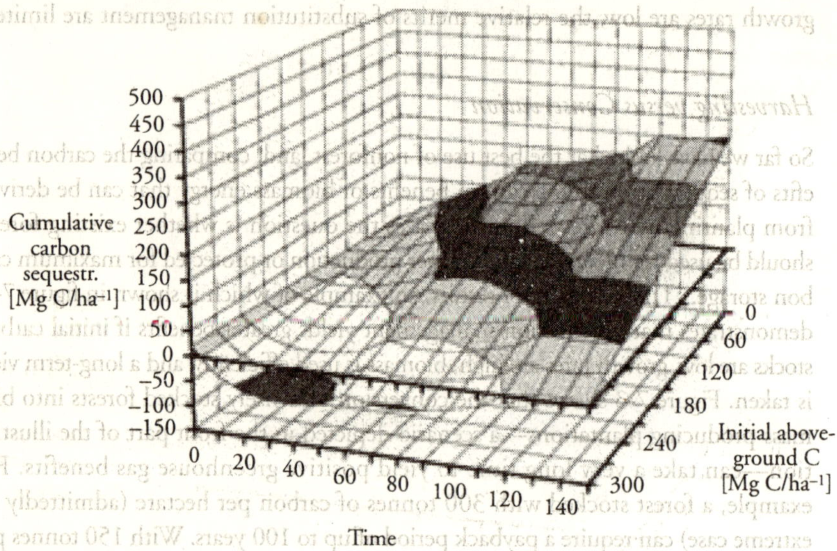

a. A forest with an initial above-ground carbon stock between 0 and 300 tonnes per hectare (axis on right), is harvested for timber and biomass fuel and then replanted. The surface (vertical axis) depicts the cumulative carbon benefit of substitution over forest protection ("no harvest") as a function of time and initial carbon stock. It includes on-site carbon storage as well as storage in wood products and credits for fossil fuel displacement. In this case it is assumed that biomass is used at high but not extraordinary efficiency (efficiency factor equals 1, meaning that one tonne of carbon in harvested biomass replaced one tonne of carbon in fossil fuel).

results in the reduction of emissions from diesel of approximately three tonnes of CO_2 equivalent per hectare (t CO_2e/ha) per year. If this plantation causes a shift in land-use practices that deforests an equal area of land, then approximately 300 t CO_2e/ha may be emitted as a result, and there is no net GHG mitigation benefit for 100 years.

Deforestation is an extreme example, but emissions will result from nearly any displacement of agricultural activities. For example, if the plantation causes an area of land that was grassland to be cultivated or grazed, then loss of soil carbon and emissions could be as high as 45 t CO_2e/ha.[7] Consequently, there will be no net GHG emission benefit for 15 years.

Liquid Biofuels and Agricultural Markets

The two principal liquid biofuels today are ethanol and biodiesel. Ethanol is produced from sugars and starches contained in a variety of crops, primarily sugar cane (in Brazil) and corn (in the United States). Biodiesel is primarily produced from vegetable oil, mainly from rapeseed (Germany), soybeans (United States and Brazil), and oil palms (Indonesia and Malaysia).

In recent years, investments in the production of biodiesel and ethanol have reached an unprecedented high. Between 2000 and 2005, global ethanol production more than doubled, and the production of biodiesel increased almost fourfold.[8] Factors spurring the development of the biofuels industry are rising oil prices and policies such as tax exemptions and fuel-blending targets that governments around the world are putting in place. In 2003 the European Union adopted a biofuels directive with a target of increasing the share of biofuel to 5.75 percent by 2010.[9] The EU is currently discussing the introduction of a 10 percent binding target by 2020. In the United States the 2005 Energy Policy Act contained incentive programs for biofuel investments, including a renewable fuel standard, tax incentives, and research and development support.[10] The advantages of biofuels most often cited include independence from insecure energy supplies, promotion of rural development, reduced local air pollution, and, last but not least, development of a new climate-friendly source of energy.

Despite significant investment and government support for biofuels, ethanol, the most commonly used biofuel, still represented a mere 1.2 percent of global gasoline supply in 2005.[11] And although biofuels have only a marginal effect on the gasoline market, they have a significant effect on the prices of agricultural crops. For example, in 2006 ethanol represented about 3.5 percent of gasoline supplies in the United States, but about 20 percent of the corn from the 2006 crop year went to ethanol production.[12] The U.S. Department of Agriculture estimates that by 2010 more than 30 percent of the U.S. corn crop will be used to produce ethanol.[13] The growing significance of biofuels to the agricultural markets has a number of important implications for agricultural prices and land use.

First, the prices of many agricultural crops are increasingly determined by the prices of their energy equivalents. The prices of corn and crude oil, the main feedstocks for ethanol and gasoline, respectively, have now converged when measured in their energy equivalent, after taking into account subsidies for ethanol production.[14] Since mid-2004 both crude oil prices and the prices of many biofuel crops have risen sharply. With increasing demand for crude oil, led by demand in developing countries such as China and India, many market observers believe that the price of crude oil and thus the prices of biofuel crops are likely to continue to rise in the coming years.

Second, as biofuel crop prices rise, the prices of other agricultural commodities will follow. Agricultural commodity prices are highly correlated with biofuel crop prices because cropland can be used to produce different commodities. Many commodities are substitutes in consumption. Agricultural commodities are internationally traded and have a single price after allowing for transportation and quality differences.[15] For example, higher corn prices encourage farmers to increase corn acreage. Because cropland used for soybeans can also be used for corn, soybean acreage shifts to corn production. Reduced production of soybeans, along with increased demand for soybean oil for biodiesel, leads to higher soybean prices. The prices of other vegetable oils that can be substituted for soybean oil also increase, and the prices of other grains, such as wheat, that are used as feedstock replacements for corn also rise. Meat and dairy prices are affected because higher feedstock costs must be recovered.

Third, rising agricultural prices driven by biofuel production provide incentives for farmers to convert more land to agricultural production. In some instances, additional cropland comes at the expense of carbon-dense forests, resulting in significant emissions from the conversion of land to be used for biofuel crops. Expansion of soybean production in Brazil and palm oil production in Malaysia and Indonesia are cases in point. In other instances, additional cropland comes from less carbon-dense land, but the shifting of preexisting activities from that land to forested land can lead to significant emissions. For example, the expansion of sugar cane production in Brazil has come largely at the expense of pasture, leading to worries that the grazing of cattle, with beef being another booming export product, could be shifted to the Amazon and result in greater deforestation.[16]

For policymakers, the linkage between land use and biofuels has important implications. First, biofuels, even if produced domestically on existing cropland, can result in land-use emissions elsewhere, because of price effects on other agricultural commodities caused by substitution and international markets. Second, the climate change benefits of biofuels, as measured on a life-cycle basis, must be evaluated against land-use emissions resulting directly or indirectly from bringing new land into agricultural production. And third, efforts to curb deforestation must recognize that the opportunity cost of forest land is influenced by the price of oil and government subsidies for biofuel as translated into agricultural commodity prices.

Albedo Effects

Finally, another potential climate change effect related to changes in land use is the reduction of albedo, or the reflectivity of the earth's surface. By inducing a shift from light land cover to darker land cover, a project may increase the absorption of sunlight and thus cause local warming. Some authors have suggested that reduction of albedo in regions that receive winter snow may contribute more to global warming than the cooling that results from the decrease in CO_2 concentration due to carbon sequestration.[17] Our own recent modeling (unpublished) indicates that this effect is not as strong as suggested, because the afforestation or reforestation is at high latitudes that in winter do not receive intense sunlight.

Moreover, the winters are often cloudy, and the clouds reflect a large proportion of the sunlight.

This may not be the case for bioenergy plantations, particularly those growing jatropha, that are planted on marginal land in sunny climates in middle latitudes. The marginal land will be light colored while the bioenergy plantation is darker, and the change in surface albedo will not be muted by the presence of clouds.

Another important consideration is the flux of latent heat due to increased evapotranspiration, which has a local cooling effect. Modeling of albedo and latent heat fluxes is still in its infancy and will need to be improved before policymakers are able to take these factors into account in formulating climate change mitigation policies.

Bioenergy, Forestry, and Agriculture in the Kyoto Protocol

In the context of the Kyoto Protocol, which focuses on short-term emissions and removals in the first commitment period, 2008 to 2012, the short-term optimization of carbon stocks in existing forests unduly favors sequestration over bioenergy. This is because the merits of bioenergy relative to those of carbon sequestration increase with longer periods of analysis—that is, beyond the commitment period.

On the other hand, the afforestation, reforestation, and revegetation of nonforest land not only sequester carbon but also contribute to a future biomass resource that will be needed if more emission reductions and greater shares of renewable energy are desired. Bioenergy helps overcome the saturation constraints of such reforestation activities and helps address the issue of nonpermanence. Incentives for carbon-stock-enhancing activities are needed to build the resources for modern biomass energy and to reduce pressure on existing forests. In the long term, bioenergy will be CO_2 neutral only if policies exist to enhance photosynthesis by the same amount that the use of biomass is increased. Simply increasing the use of biomass may lead to net depletion of carbon stocks and even to deforestation, degradation, and devegetation.

Overall, an integrated view of land use, land-use change, and forestry (LULUCF) and bioenergy is needed, both for LULUCF activities in which bioenergy is a coproduct and for bioenergy activities that require new and additional land resources or otherwise have a depleting effect on carbon stocks. So far the rulebook under the Kyoto Protocol has focused on LULUCF with a rather skeptical view ("cheap way out of having to reduce emissions") and failed to address bioenergy projects in developing countries, where the bulk of global biomass use takes place. Examples of such projects include improving the efficiency of cooking stoves and replacing nonrenewable fuelwood or charcoal with other, renewable energy sources.[18]

Biomass and Liquid Biofuels in the Clean Development Mechanism

The massive growth in biofuel investment in recent years has so far had no correspondence under the CDM. Not a single biofuel project has yet been successfully registered, and the situation is identical under the Joint Implementation (JI) mechanism. This is astonishing for a category that, by public and experts' accounts alike, is one of the key technologies for reducing emissions from transportation.[19] It is apparently not the lack of additionality that is hindering biofuels' participation under the CDM. With the notable exception of Brazil, the use of biofuel is far from a common practice and still faces a fair number of technological, logistical, and market barriers. The market share of biofuels in global fuel consumption is still only about 1 percent for ethanol and much less for biodiesel.

The obvious barrier with regard to the CDM is the lack of approved methodologies. Two issues in particular are holding up the development of methodologies for this asset class: the treatment of upstream emissions and concerns about double counting. These issues are by no means unique to biofuel projects, but they are particularly relevant in a biofuel context.

Status of CDM Biofuel Methodologies

As of November 2007 the CDM Executive Board (EB) had approved only one methodology for biofuel projects and had rejected all other submissions (table 7-1). The approved methodology, AM0047 version 2, is limited to projects using waste cooking oil or waste fat from biogenic origin as feedstock and supplying biodiesel directly to the end user. By restricting the scope of application, the methodology avoids the concerns of upstream emissions and double counting, but at the price of ruling out most biofuel projects. Recent methodology approvals by the board's Methodology Panel (MP), however, may open the door for biofuel projects based on cultivated biomass, with the important restriction that only biomass cultivated on severely degraded or underused agricultural areas is eligible.

Upstream Emissions under the CDM. Upstream emissions are emissions that arise in the production of an energy resource or commodity. They comprise, among others, the emissions from energy used during production, transport, and construction of the facility. Although upstream emissions are involved in every CDM project activity, they are generally not considered to have large effects and have little affected the approval of methodologies.[20] The situation is different for bioenergy projects. Emissions associated with the planting of an energy crop are considered substantial, particularly if the cultivated land had previously been forested. Bioenergy projects that have been registered under the CDM (table 7-2) rely almost exclusively on waste biomass from agricultural and industrial activities in which upstream emissions are simply transport-related emissions.

Table 7-1. *Biofuel Methodology Submissions to the CDM Executive Board (EB) and Methodology Panel (MP) as of November 21, 2007*

Submission number	*Short title*	*Biomass input*	*Status of approval*[a]
NM0228	Agrenco Biodiesel	Locally grown sunflower, soy, castor beans	Approved by MP 30; rejected by EB 36
NM0224	Biodiesel from palm oil and jatropha	Imported palm oil and locally grown jatropha oil	Rejected by EB 35
NM0223	Western Cape Biodiesel	Soybean imports and locally produced edible oils	Approved by MP 30; to be reviewed by EB 36
NM0180/ AM00047	Biolux Biodiesel Beijing	Waste oil	Approved by EB 29
NM0142-rev	Biodiesel Thailand	Palm oil	Rejected by EB 31
NM0129-rev	Biodiesel Thailand	Sunflower seeds	Rejected by EB 30
NM0108-rev	Biodiesel Andhra Pradesh	Jatropha, pongamia	Rejected by EB 30
NM0082-rev	Khon Kaen fuel ethanol	Sugar cane molasses	Rejected by EB 24

a. Numbers refer to the meeting of the board or panel, such as "thirtieth meeting of the Methodology Panel."

Upstream emissions from biomass production, on the other hand, are difficult to assess. It is not easy to quantify direct emissions arising from clearance of land, fertilizer use, and fossil fuel consumption, because it requires rigorous monitoring and measurements of the agricultural activities.

Quantification of indirect emissions is even more complex, particularly as it relates to the loss of sequestered carbon outside the project boundary. The issue is that cultivation of biomass on a piece of land may displace other, preexisting activities such as cattle grazing, which in turn causes degradation of grassland or forest elsewhere.[21]

Estimating the effects that a CDM project will have on deforestation can principally be done in two ways. One can take a project-specific approach, estimating the net decrease in above- and below-ground carbon from cultivating renewable biomass on a dedicated land area. Or one can take a regional approach, developing regional default factors to capture deforestation trends in different regions.

The difficulty with using a project-specific approach is that deforestation effects are not easily attributable to a specific project. Even if feedstock is supplied from an existing plantation, increase in demand may be met through the establishment of a new plantation elsewhere. Regional approaches, on the other hand, may be better suited to capturing macroeconomic trends, but they do not allow one to distinguish individual land management practices. One category in which a project-specific approach seems appropriate is cultivations that are established on degraded or waste land where no competing use is crowded out. Biofuel plantations established on

Table 7-2. *Biomass Energy Projects Submitted for Registration under the CDM as of June 29, 2007*

Project subtype	*Being validated*	*Requesting registration*	*Registered*	*Total*
Bagasse power	75	19	51	145
Palm oil solid waste	18	3	9	30
Rice husk	46	5	28	79
Agricultural residues: other kinds	58	4	44	106
Forest residues	12	1	13	26
Gasification of biomass	8	0	0	8
Biofuels	1	0	0	1

Source: United Nations Environment Program (UNEP) Risø Centre, CDM Project Pipeline, projects listed on June 29, 2007.

wasteland, such as jatropha cultivations, may well be the first category to succeed under the CDM. Recent approvals by the Methodology Panel of methodologies for severely degraded or underused agricultural land point in this direction.

The discussion of upstream emissions from bioenergy projects also reveals a severe bias. Whereas upstream emissions from the CDM project have to be factored in (in the case of fertilizers, direct emissions from the application of fertilizer as well as indirect emissions from the production of fertilizer should be considered), the upstream emissions avoided by the project are not. The substantial emissions associated with the extraction, refining, and transportation of fossil fuels are disregarded in the calculation. Furthermore, projects are generally not awarded credits for the generation of by-products such as glycerine and organic fertilizer that result in avoided emissions outside the project boundary. The asymmetric treatment of upstream emissions leads to a distortion of the emission balance of biofuel projects under the CDM and makes them appear less beneficial.

Early guidance on the treatment of emissions from biomass cultivation shows a rigorous approach. According to the Methodology Panel's draft "Tool to estimate the emissions associated with the cultivation of lands to produce biomass," every imaginable source of emissions has to be evaluated and accounted for by the project if deemed significant.[22] However, in an unconventional decision, the CDM EB recently rejected a methodology for biodiesel based on cultivated biomass (NM00228), which the MP had previously approved, noting "the excessive treatment of the estimation of project emissions from the cultivation of raw material used in production of biofuel compared to the calculation of the baseline emissions from the production of petrodiesel."[23]

Double Counting Concerns. Double counting occurs if a biofuel production plant constitutes a CDM project while the fleet of vehicles that uses the biofuels also becomes a CDM project. This issue has been discussed controversially by the CDM EB and its Methodology Panel without a final resolution. Whereas the EB gives priority ownership of Certified Emission Reductions (CERs) to the consumers of biofuel, the MP and public commentators favor the producers.[24] They point out that the bottleneck lies in the production of biofuel and that in the case of biofuel blended with normal fuel, consumers may not even be aware of what they are consuming. Furthermore, only the producers can monitor upstream emissions.

It is striking that the discussion of double counting occurs vividly in relation to biofuel while being a non-issue in other, similar contexts. For instance, in the case of electricity production from renewable energy it is common practice to allocate CER ownership to the producers.

Conclusion

Bioenergy is already a significant source of energy worldwide, particularly in the form of traditional fuelwood and charcoal. Although biomass is often considered to be CO_2 neutral, a refined way of assessing this has now been developed in the CDM, requiring that biomass be "renewable" in order for its CO_2 emissions to be counted as zero. However, problems in defining suitable CDM methodologies for biofuels also show that the CDM is ill equipped to resolve some of the larger questions at hand, particularly that of the effect of bioenergy on the availability of land. While the CDM focuses on the effects of individual projects, the land-use issues we have discussed can hardly be attributed to a single activity but tend to be the results of macroeconomic development. Land-use issues play an important role in bioenergy projects. The GHG balance is influenced not only by direct emissions from nonrenewable biomass use and displacement of land use resulting from the implementation of bioenergy production but also by indirect emissions and alternative land uses. And the issues extend far beyond climatic effects, to biodiversity, water, food security, land tenure, and social issues. It is these aspects of increased bioenergy use that are particularly in the public eye.

Notes

1. World Wildlife Fund for Nature (WWF), "WWF Contribution to the European Commission Energy and Transport Directorate," public consultation on the review of the EU Biofuels Directive, 2007 (http://assets.panda.org/downloads/wwf_response_to_the_ec_consultation_on_biofuel_certification.pdf); D. Adam, "World's Great Apes Face Disaster, Says Leakey," *Guardian*, May 31, 2007; P. Pontoniere, "Deforestation: The Dark Side of Europe's Thirst for Green Fuel," *New America Media,* February 28, 2006 (http://news.newamericamedia.org/news/view_article.html?article_id=37e104044bb11d19d566ac8f3621c63f).

2. Subsidiary Body on Scientific, Technical and Technological Advice (SBSTTA) to the Convention on Biological Diversity, "Recommendations Adopted by the Subsidiary Body on Scientific, Technical and Technological Advice at Its Twelfth Meeting," UNESCO, Paris, July 2–6, 2007 (www.cbd.int/doc/meetings/sbstta/sbstta-12/official/sbstta-12-xx-en.doc); J. Vidal, "Global rush to energy crops threatens to bring food shortages and increase poverty, says UN," *Guardian,* May 9, 2007.

3. Non-CO_2 emissions, such as methane (CH_4), and nitrous oxide (N_2O), produced during combustion, are still considered.

4. EB23, Annex 18, Definition of renewable biomass (http://cdm.unfccc.int/EB/023/eb23_repan18.pdf). There may be temporary decreases in carbon stocks due to harvesting.

5. R. Righelato and D. V. Spracklen, "Carbon Mitigation by Biofuels or by Saving and Restoring Forests?" *Science* 317 (2007): 902.

6. G. Marland and S. Marland, "Should We Store Carbon in Trees? *Water, Air and Soil Pollution* 64 (1992): 181–95; G. Marland and B. Schlamadinger, "Forests for Carbon Sequestration or Fossil Fuel Substitution? A Sensitivity Analysis," *Biomass and Bioenergy* 13 (1997) 389–97.

7. Calculation based on soil factors from the Intergovernmental Panel on Climate Change's "Good Practice Guidance for Land Use, Land-Use Change and Forestry" (tables 3.3.4 and 3.4.5) and an average soil carbon stock of 50 t C/ha before land-use change.

8. Worldwatch Institute, "Biofuels for Transportation: Global Potential and Implications for Energy and Agriculture," 2007, p. xiv.

9. "Directive 2003/30/EC of the European Parliament and of the Council of 8 May 2003 on the promotion of the use of biofuels and other renewable fuels for transport" (http://ec.europa.eu/energy/res/legislation/doc/biofuels/en_final.pdf).

10. See www.doi.gov/iepa/EnergyPolicyActof2005.pdf.

11. Worldwatch Institute, "Biofuels for Transportation."

12. National Corn Growers Association, "World of Corn: Consumption and Production" (www.ngca.com/WorldofCorn/main).

13. P. Wescott, "Ethanol Expansion in the United States: How Will the Agricultural Sector Adjust" (Department of Agriculture, Economic Research Service, FDS-07D-01, May 2007).

14. "Biofuelled," *Economist,* June 21, 2007.

15. M. Kojima, D. Mitchell, and W. Ward, "Considering Trade Policies for Liquid Biofuels," Energy Sector Management Assistance Program, June 2007, pp. 29–30 (http://siteresources.worldbank.org/INTOGMC/Resources/Consideringtrade_policies_for_liquid_biofuels.pdf).

16. L. Rohter, "With Big Boost from Sugar Cane, Brazil Is Satisfying Its Fuel Needs," *New York Times,* April 10, 2006.

17. S. Gibbard and others, "Climate Effects of Global Land Cover Change," *Geophysical Research Letters* 32 (2005): 237–40; G. Bala and others, "Combined Climate and Carbon-Cycle Effects of Large-Scale Deforestation," *Proceedings of the National Academy of Sciences* 104 (2007): 6551.

18. The reasons for this failure are that the emission reductions take place in the LULUCF sector, which is excluded from the CDM unless new forests are created (afforestation and reforestation).

19. For an overview of the GHG reduction potential of biofuels, see International Energy Agency and Organization for Economic Cooperation and Development, "Biofuels

for Transport: An International Perspective," Paris, 2004 (www.iea.org/textbase/nppdf/free/2004/biofuels2004.pdf).

20. The exception being methodologies for hydropower projects involving the construction of large reservoirs.

21. In CDM terminology this issue has been referred to as a "shift in pre-project activities."

22. The tool was proposed by the twenty-seventh meeting of the Methodology Panel (MP 27) but has never been approved by the CDM EB.

23. CDM EB 35th Meeting Report, October 19, 2007.

24. EB 26, Annex 12, "Guidance on double counting in project activities using blended biofuel for energy use."

PART THREE

Practical Experiences

8

Design Issues in Clean Development Mechanism Forestry Projects

BRUNO LOCATELLI, LUCIO PEDRONI, AND ZENIA SALINAS

The design of an afforestation and reforestation (AR) project under the Clean Development Mechanism (CDM) of the Kyoto Protocol is a two-stage process. The first stage includes the definition of a project idea, the evaluation of its eligibility under CDM rules, and preliminary estimations of carbon removals, among other things. The second stage involves the preparation or application of a baseline and monitoring methodology and the production of a Project Design Document (PDD). This second stage must be implemented in strict compliance with the modalities and procedures (M&Ps) of the CDM and any guidance provided by the CDM Executive Board (EB).

As a consequence of the decisions on M&Ps, the design of a CDM AR project deals with several technical issues more complex than those for CDM activities in other sectors.[1] In part this is the consequence of the complex political process that produced the M&Ps. Forestry projects were debated extensively because of concerns about their real contribution to climate change mitigation and their possible negative effects on host countries' sustainable development. It was feared that too lax regulations would jeopardize the environmental integrity of the Kyoto Protocol and create negative effects on local communities or biodiversity and that too stringent regulations would discourage project development.

In this chapter we review technical, legal, and market issues that proponents of CDM AR projects have to address when they design their projects. We analyze the tasks, types of information, and capacity required to address the project design issues, as well as the problems that project developers may face. We identify which design issues represent the most significant barriers for project proponents and propose recommendations to overcome these barriers. Our analysis is based on experience we gained during the first year of the FORMA project, which assisted in the preparation of ten CDM AR project initiatives in Latin America.[2]

In February 2006 FORMA launched a call for CDM AR project ideas and received forty-seven responses, which we refer to as the "large portfolio." These projects originated in fifteen Latin American countries. Some countries, such as Colombia and Bolivia with eight projects each, were well represented. After a first filter based on simple criteria, twenty-two projects were selected for further examination. Among them, seven were in early stages of development (substantiated by a general concept note or prefeasibility study), eleven had a Project Idea Note (PIN)—a short description that is not mandatory but has proved to be a useful marketing document—and four were already drafting PDDs. We refer to these twenty-two projects as the "small portfolio."[3] For this chapter, we performed a simple analysis on the large portfolio and a more detailed analysis on the small portfolio.

Technical Issues

During the development of a CDM project idea, project developers must first define project characteristics such as place, scale, activities, time frame, business plan, and partners. Many technical problems encountered during this step are present in any kind of forestry project, but other problems are specific to the CDM and are related to carbon estimations and the assessment of additionality—the reason the project would not happen without the CDM.

Carbon Estimations

Initial carbon estimations are intended to quantify the amount of carbon dioxide that will be removed from the atmosphere during the crediting period of the project. The estimations should define a simple baseline scenario and quantify the carbon stock changes that are likely to occur in both the project and the baseline scenario. Tools used during these initial estimations range from simple spreadsheets to comprehensive software programs such as CO_2Fix.[4] Input data can range from default values and generic equations to data specific to the site.

In FORMA's large portfolio, 46 percent of the projects were using no tool, 35 percent used a spreadsheet, and 20 percent used a specific tool. CO_2Fix was the most common tool, being used by 78 percent of the projects that used a specific tool. The use of tools became more widespread when the project design stage was more advanced: only 37 percent of projects in the early stage used a tool, but

Figure 8-1. *Use of Tools according to Project Design Stage in FORMA's Large Portfolio*

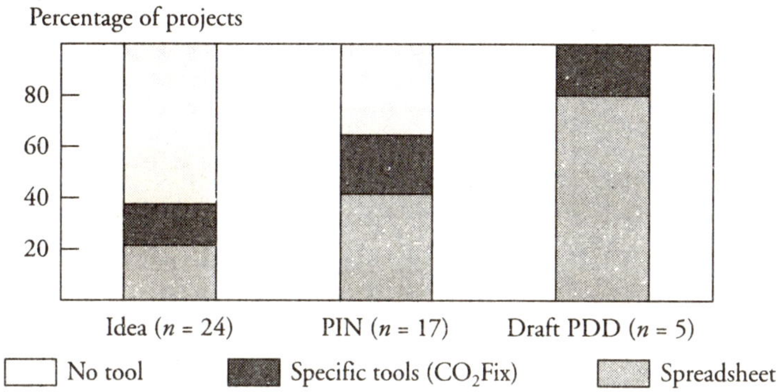

65 percent of those with PINs and 100 percent of those drafting PDDs used a tool (figure 8-1).

If significant leakage and project emissions are expected, they should be estimated broadly. In FORMA's small portfolio, 41 percent of the projects considered leakage (29 percent for projects in early stages, 36 percent for projects with PINs, and 75 percent for projects drafting PDDs).

A lack of good data and advice often hinder simplified but credible carbon estimations at early stages of project development. In many projects, baseline, leakage, and project emissions are underestimated or ignored, leading to overestimations of the expected carbon benefits. Even if there is no need to use an approved methodology for these first carbon estimations, the use of guidelines or specific tools and the application of the principles of transparency and conservatism are recommended. Of the forty-seven projects of the FORMA large portfolio, 71 percent of those in an early stage of design had a very basic carbon estimation or none at all. The quality of the estimation increased with advances in the project design. However, it is worrisome that 20 percent of the projects at the stage of drafting a PDD still relied on a basic estimation without referring to any methodology or using any tool (figure 8-2).[5]

Additionality

Initial additionality proofs are intended to screen out projects that may not comply with CDM additionality requirements. Even if the use of the additionality tool is not mandatory at this stage, it is highly recommended for guiding the analysis.[6] Even with the tool, some developers appear to face difficulties in understanding the notion of additionality and do not correctly follow the procedures proposed in the tool.

Figure 8-2. *Quality of Carbon Estimation according to Project Design Stage in FORMA's Large Portfolio*

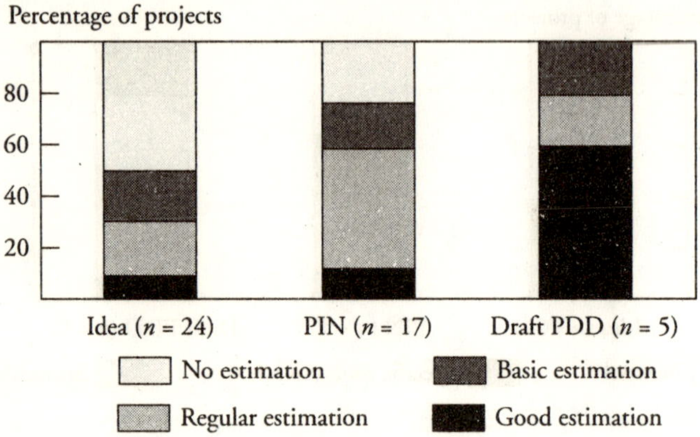

Many project developers have trouble following the logic of the additionality test. Normally, in order to find investors and partners, project developers try to demonstrate that their projects will be profitable and face no major barriers. In the case of the CDM, they must do the contrary: demonstrate low profitability or the existence of barriers to show the effect the CDM has on the project's viability. This may cause inconsistencies, as when an AR project without Certified Emission Reductions (CERs) is described as highly profitable in a document for investors and unprofitable in a CDM-related document. In FORMA's large portfolio, six out of forty-seven project developers, when asked about the barriers that might impede the implementation of their AR project, answered that there were no barriers. Four of these projects were at an early stage and two had PINs.

Demonstrating that the CDM will help to increase financial attractiveness so that the project is feasible is a necessary step to demonstrate additionality. Beyond the issue of additionality, this step is important for project viability. Project developers must evaluate whether accessing the CDM is worthwhile. If the revenue obtained from selling CERs is insufficient to offset CDM transaction costs, then the project would be better off without the CDM. Depending on projects and CDM costs and benefits, we have shown that on average the minimum project area required is 2,300 hectares if carbon trading is to be profitable for projects relying on temporary crediting (temporary CERs [tCERs] or long-term CERs [lCERs]).[7]

In FORMA's large portfolio, the area of the proposed projects ranged from 10 to 253,000 hectares, with a median of 4,625 hectares. In their project descriptions, five projects had fewer than 1,000 hectares (figure 8-3) and did not esti-

Figure 8-3. *Areas of Project Included in FORMA's Large Portfolio*

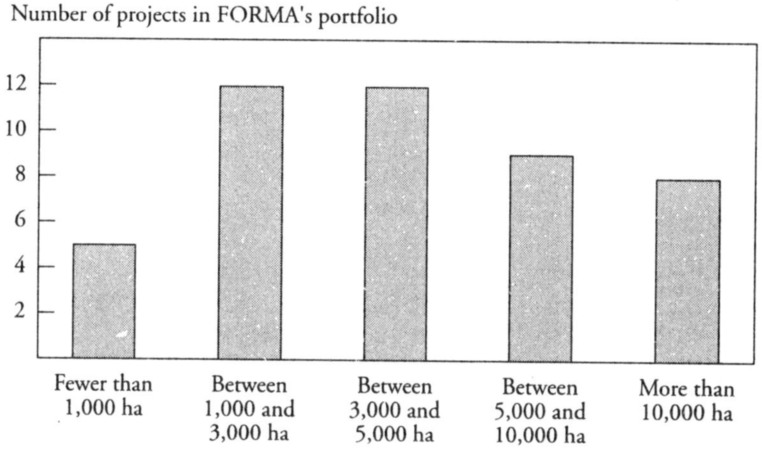

mate the transaction costs associated with the CDM. Three project ideas were for fewer than 30 hectares. Because of CDM transaction costs, projects of such small size will not benefit from carbon trading, even if the simplified modalities and procedures developed for small-scale projects are applied to them.

When evaluating whether the CDM is worthwhile, some project developers encounter difficulties in calculating the numbers of tCERs and lCERs expected, because such calculations are less straightforward than those for CERs of emission reduction projects. Moreover, almost no information exists about the prices of tCERs and lCERs and transaction costs for forestry projects. Most of the available information about costs and prices refers to industrial emission reduction projects, which are inapplicable to CDM AR projects.

Effects

Socioeconomic and environmental effects should be considered from the moment the project idea is developed. However, when these issues are addressed at early stages, they are usually addressed with very general considerations, such as unconvincing statements about the positive effects of forestry activities on livelihoods, biodiversity, or watersheds. In FORMA's large portfolio, many project developers proposed a list of general effects when they were asked to describe project effects. Ninety-six percent of projects in early stages, 76 percent of those with PINs, and 60 percent of those drafting PDDs gave a general and undocumented argument (figure 8-4). At early stages of project definition, using tools appears to be too cumbersome for most project developers. In FORMA's large portfolio, only one project description out of forty-seven mentioned the use of a standard

Figure 8-4. *Quality of Argument about Socioeconomic and Environmental Effects, by Project Design Stage in FORMA's Large Portfolio*

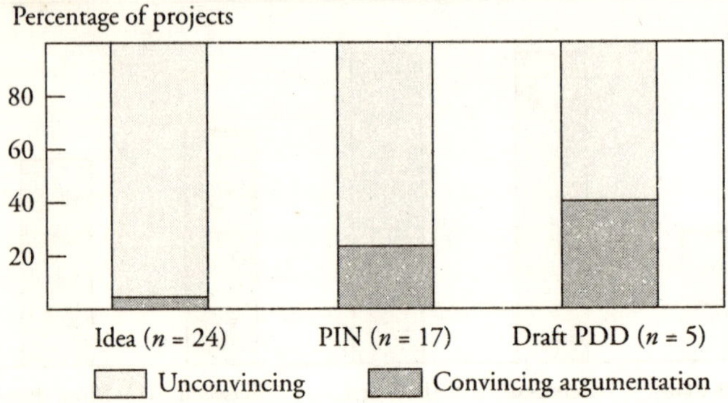

for evaluating the effects of the project on sustainable development. The project referred to the standards formulated by the Climate, Community, and Biodiversity Alliance (CCBA).[8]

Developing the PDD

During the development of the PDD, many project developers face serious difficulties related to the selection and use of baseline and monitoring methodologies. If none of the already approved methodologies is applicable to the project, developers must build their own. By August 2007, eight methodologies for AR projects had received final approval.[9]

Most approved methodologies were prepared by international consultants or with the support of specialists hired by carbon funds. Available tools and documents, such as the EB's guidelines for completing a new methodology for AR, will not help in developing a new methodology unless project developers possess the required analytical capacities, scientific knowledge, and drafting skills.[10]

In addition to problems related to applying a methodology, the development of the PDD raises technical issues. Project developers are required to delineate the project area geographically by identifying and mapping the limits of each piece of land that will actually be planted during the crediting period of the project. In FORMA's small portfolio, only 56 percent of the projects in an early stage or with a PIN had an idea of the project's boundaries, whereas all projects drafting PDDs had more precise definitions of their boundaries (even if in most cases the exact georeferencing still had to be done).

Identifying and mapping the limits of the project can be difficult if uncertainty exists about the list of land plots and participating farmers or if the eligibility of the land is still unclear, because the areas included in the project boundary must comply with the land eligibility criteria, which we discuss later. Technical problems may arise if the project encompasses several discrete areas, if few spatial data are available, or if the project has weak geographical information system (GIS) and remote sensing capacities. In FORMA's large portfolio, 35 percent of the projects mentioned GIS, satellite images, or aerial photographs (21 percent for early-stage projects, 47 percent for projects with PINs, and 60 percent for projects drafting PDDs). Only 17 percent mentioned these tools in connection with the issue of land eligibility (8 percent for early-stage projects, 24 percent for those with PINs, and 40 percent for those drafting PDDs).

In order to be eligible for the CDM, project developers must demonstrate that the land within the project boundary was (1) not a forest on December 31, 1989 (for reforestation), or during the past fifty years (for afforestation), (2) is not a forest at the moment the project starts, and (3) will not become a forest without the CDM. "Forest" should be defined according to the definition chosen for the CDM by the host country. According to the tool prepared by the EB (under review at the time of writing), projects can demonstrate eligibility by using remote sensing, aerial photographs, ground survey, or, in the absence of other data, testimony produced during a participatory rural appraisal.

Project developers must further select an approach for addressing nonpermanence—that is, selecting between temporary CERs (tCERs) and long-term CERs (lCERs). The selection depends on market factors (buyer preference, credit prices), project-specific financial needs, and project risk (high-risk projects should go for tCERs). The selection may be complicated if there is still no identified CER buyer or if information about market preference is lacking. Project developers must also define the project's duration and select a crediting period with a starting date. The crediting period can be renewable (twenty years each, renewable twice at most) or fixed (thirty years). The starting date for the crediting period must be the same as the project's starting date. The crediting period should be selected according to project duration, plantation turnover, and project risk.

As part of the project design, the project developer has to indicate the estimated net anthropogenic greenhouse gas (GHG) removals by sinks (project scenario minus baseline minus leakage) based on applying an approved methodology, followed step by step with adequate data, and applying principles of transparency and conservatism. But because of the large number of variables and equations used in any methodology, it seems impossible to present all steps in a transparent way; this would require that a spreadsheet be enclosed with the PDD, with a template adapted to each methodology.[11]

Finally, project developers must present in the PDD an analysis of possible effects on the environment (biodiversity and natural ecosystems, soils and hydrology, and perturbations such as fires, pests, and disease) and society (local communities, indigenous people, land tenure, local employment). If negative effects are considered to be significant, then project developers must present an environmental or socioeconomic impact assessment according to host country procedures and a plan to monitor and mitigate effects.

Carbon Market and Legal Issues

Many issues arising during project design are related to market and legal matters. A project can be accepted under the CDM only if the host country complies with the participation requirements listed in decision 3/CMP.1. The host country must have ratified the Kyoto Protocol and established a Designated National Authority (DNA) for the CDM. In addition to general CDM requirements, CDM AR is possible only if the DNA has chosen and reported to the EB a definition of *forest.*

Before proceeding to validation, projects must receive approval from the host country's DNA in the form of a Letter of Approval (LOA). Approval from the DNA must be obtained before presenting the PDD to the Executive Board. Obtaining the LOA is a legal requirement that may cause delay or refusal of approval of the project if the DNA has not been consulted at an early stage of the project design about the national procedures and criteria, if any, used in project evaluation. In FORMA's large portfolio, 22 percent of the projects mentioned contacts with the DNA (4 percent for projects in early stages, 35 percent for projects with PINs, and 60 percent for projects drafting PDDs).

Other major legal issues are not CDM requirements but must be resolved before the implementation of the project. Some are related to the contractual relationships between project participants, such as the project developer, the project manager, the project owner (if they differ), landowners, farmers, and contractors. Legal issues regarding the relationships between project participants are similar to those encountered in any forestry project, especially regarding land tenure. In FORMA's large portfolio, land tenure was clear in 70 percent of the cases, and no difference was observable among projects at different design stages. Landownership and tenure must be described in the PDD. Not clarifying these ownership issues may delay the registration of the project and induce conflicts and marketing problems.

In addition to the legal issues present in any forestry project, CDM project owners face the challenge of clarifying the legal ownership of CERs. It is possible that three different entities may own the CERs, the trees, and the land, respectively. Legal ownership of the CERs depends on the national legislation of the host country. In some cases governments may claim their rights to CERs on the grounds that an emission reduction or a carbon removal is a public natural asset or national property.[12] Many countries have no specific regulation dealing with the ownership of CERs. In these cases general legal principles regulating ownership in the respective jurisdiction apply. The LOA and a host country authorization to sell will help make the buyer comfortable.

Many uncertainties also persist with respect to domestic taxes on CERs. Depending on whether a CER is considered a good, a service, a financial instrument, or a business asset, it may be taxed differently. A levy on CERs may also be applied by host countries to finance their DNAs and is in addition to regular taxes: value added tax, consumer tax, sales tax, or income tax. For project developers, an unclear legal framework for CDM AR projects complicates the project design.[13] In FORMA's large portfolio, only 7 percent of project developers had an idea of national legislation about CDM projects and CER ownership.

Carbon contracts deal with the legal and financial issues common to any AR project as well as specific CDM risks and products (CERs). The allocation of risk among buyers, projects, and other entities involved (for example, banks), the definition of CER prices and volumes, and the responsibility for transaction costs are crucial items in negotiations leading to the sale of CERs.[14] In the forms received during FORMA's call for proposals, only 2 percent of the projects mentioned purchase agreements (none for early-stage projects and projects with PINs; 20 percent for projects drafting PDDs). This observation underscores the fact that very few projects consider legal arrangements.

As a consequence of failure to give the required importance to legal issues and of lacking juridical knowledge and information about carbon transactions, many project developers face problems when negotiating contracts. In many host countries, project developers do not have the financial capacity to hire good legal advisers to help negotiate and evaluate contracts internal to the project or with buyers and investors. Moreover, few lawyers have deep knowledge of the CDM and its modalities and procedures. Legal guidebooks specific to the CDM and model contracts are worthwhile for project developers, as are capacity-building training sessions.[15]

Design Issues: Barriers for Project Developers?

Project developers replying to FORMA's call for proposals were asked to identify the barriers they faced in the design of their CDM projects. A majority (74 percent) acknowledged the existence of barriers. The perception of barriers differed between design stages; for instance, all project developers drafting PDDs perceived at least a technical barrier, in comparison with only 33 percent of developers of projects in early stages. Conversely, perception of legal and market barriers increased between the stages of project idea (38 percent) and PIN (59 percent) but did not increase during advancement toward the PDD. This observation may be interpreted optimistically (it results from an alleviation of legal and market barriers) or pessimistically (the lack of awareness about legal and market issues is persistent).

Because few projects in the portfolio showed progress in market and legal aspects, the second interpretation appears to be more likely. Among technical issues, baseline methodology and estimation of emission removals were perceived as major barriers by 30 percent and 26 percent of developers, respectively. The next four perceived technical barriers, in descending order, were additionality (15 percent of developers perceived this), technical data (13 percent), monitoring methodology (13), and land eligibility (11 percent). This result shows that emphasis is placed on the estimation of additional CO_2 in a first step (baseline and estimation) and then on specific technical sections of the PDD. Socioeconomic and environmental effects were considered barriers by only 7 percent of the developers.

Among market and legal barriers, lack of funding for preparing the PDD and covering the costs of the project cycle was perceived as a barrier by 37 percent of the project developers, making this issue the most frequently mentioned barrier. This perception was more acute for projects drafting their PDDs (60 percent) than for projects at an early stage (29 percent). Next in this category, 9 percent of developers cited a shortage of information about the market and CER prices as a barrier. It is notable that developers' perceptions of market and legal barriers involved mainly short-term issues such as immediate funding for design. Few project developers mentioned barriers related to the search for CER buyers or investors (4 percent) or legal issues (4 percent), and none mentioned negotiations with buyers.

The foregoing result might suggest that a majority of projects had already overcome legal or market problems, but this was not the case. Regarding market issues, only 13 percent of projects had succeeded in guaranteeing total funding, 22 percent had partial funding, and 28 percent mentioned a possible CER buyer. Regarding legal issues, only 22 percent referred to contacts with DNAs, only 7 percent had clear information about national legislation on CERs, and 2 percent mentioned legal arrangements with market actors. Almost none of the small-scale projects in the portfolio mentioned small project size as a market issue, nor did they recognize that buyers might not be interested in small projects with high transaction costs.[16]

Particularities of CDM AR Projects

Relative to CDM activities in other sectors, CDM AR projects appear to face more barriers. First, because the CDM modalities and procedures were decided later for AR activities than for activities in other sectors, forestry projects are suffering a delay in development and implementation that is prejudicial in a market interested in buying CERs for the first commitment period of the Kyoto Protocol (2008–2012). This delay is an important issue because the sale of CERs from forestry projects depends on tree growth. In the case of energy projects, a technological change will produce immediate additional emission reductions, but forestry projects deliver most of their carbon removals when the trees reach their maximum growth, several years after the project starts. This specificity makes it impossible for AR projects to deliver a significant number of credits in the first commitment period.

In addition to time frame differences, forestry projects are disadvantaged relative to industrial emission reduction projects with respect to methodological issues. For instance, determining land eligibility, calculating preexisting carbon stocks, conducting GPS surveys, developing GIS databases, and performing other measurements, surveys, and evaluations required by the methodologies are time and resource consuming. The nonpermanence issue and the current rules that cause all temporary CERs to expire at the end of the crediting period also create a disadvantage relative to emission reduction projects. The problem comes not from the procedures for calculating tCERs or lCERs but from the market risk associated with expiring credits.

Regarding legal issues, forestry projects are disadvantaged by more complex regulations. For instance, the DNAs must choose a definition for *forest* and may define rules about exotic, transgenic, and invasive species. In the PDD, AR projects must include sections that are not required for other CDM projects, one about socioeconomic effects and another about ownership issues.

Non-CDM-Related Barriers

Difficulties encountered by project developers are not only related to the CDM. The success of a project depends on the developer's ability to overcome the traditional barriers of the forest sector. Financial barriers were faced by 72 percent of the projects in FORMA's large portfolio and included lack of funding for plantation activities or lack of interest by banks and international funding agencies, which can be explained by the high risks and long-term financial return associated with slow tree growth.

The second major barrier, faced by 52 percent of the developers in FORMA's portfolio, is a cultural one. In many places a lack of forestry culture impedes the implementation of plantation activities. This barrier is particularly important for projects working with farmers planting trees on their own land. The third type of barrier, faced by 46 percent of the projects, is institutional. It includes the lack of priority given to the forest sector by national policies and institutions or the lack of stability or transparency of organizations related to the forestry sector. Technological barriers were mentioned by 30 percent of the project developers and included deficiencies in information sharing and lack of experience with plantations of certain species, such as native species. Natural barriers (pests, soil degradation) and human resources (lack of capacity-building, resistance to change toward a modern vision of forestry) were also mentioned as barriers by some project developers.

Capacity as a Barrier

The quality of the CDM component of a project determines whether a proposed CDM AR project achieves validation. Analyzing the twenty-two projects in FORMA's small portfolio, we found that quality depended heavily on the capacity of the project developers. In our analysis, each project was valued by a pair of experts, and paired evaluations were confronted. The evaluation was organized according to the following three criteria:

—the technical capacity of the developer (indicators: experience in similar forestry activities measured in number of projects or in years, knowledge of technical forestry issues, use of relevant technical forestry information)

—the organizational capacity of the developer (indicators: capacity of entities involved in the project, clarity of roles in the implementation of the project, clarity of the project concept, clarity of the work plan and the chronogram)

—the technical quality of the CDM project (indicators: knowledge of CDM cycle and requirements, knowledge about needed and available data, perception of barriers, awareness of the need for a methodology, a clear baseline scenario, clear selection of carbon pools in the baseline, good estimation of carbon removals in the baseline, a clear project scenario, clear selection of carbon pools and sources in the project scenario, good estimation of emissions and carbon removals in the project scenario, and good assessment of leakage)

A regression analysis showed that the technical quality of the CDM project was positively correlated with the technical capacity of the developer ($r = .66$; $p = .001$) and with organizational capacity ($r = .45$; $p = .035$). This result shows that CDM design issues do not represent the same barriers for all projects. A team with greater organizational and technical capacity will remove these barriers more easily.

Conclusion

The success of a CDM project transcends technical design issues. With the accumulation of experience in the design of CDM AR projects, the production of guidebooks about technical issues, the capacity-building sessions organized in host countries, and the technical assistance provided by specific funds (for example, the BioCarbon Fund) and projects (for example, FORMA and ENCOFOR, or "Environment and community-based framework for designing afforestation, reforestation and revegetation projects in the CDM"), project developers should recognize that technical issues can be overcome.[17] Although some guidebooks provide useful information about legal issues, the majority of documentation and capacity-building efforts deal mainly with technical issues.

The viability of CDM projects will eventually depend not on technical issues but on market and legal issues such as contract negotiation and project attractiveness in the carbon market. Until now, few developers have taken these market and legal issues into serious consideration, as was shown in FORMA's portfolio. Because of the delay in the takeoff of CDM AR projects, experience regarding market and legal issues is lacking. Only a few carbon contracts have been signed with AR projects, and many of them are confidential. Information about the prices of tCERs and lCERs is inaccessible.

Project developers should strive to make their projects attractive in the carbon market. For that they must be aware of the importance of project scale and CER volume for CER buyers, because many buyers may seek to avoid the high management cost induced by a portfolio of small projects. At the same time, some buyers might be reluctant to buy from very large-scale industrial projects because of the negative image possibly associated with them. Public image is an important factor explaining project attractiveness, and project developers could improve this by working transparently and demonstrating the absence of negative effects. Some projects may choose to be certified—for instance, by applying the CCBA standards for demonstrating positive effects on community and biodiversity and facilitating project marketing—even if, for the moment, it is unclear whether certification will result in higher CER prices.

Other important factors influencing project attractiveness in carbon markets are low complexity, good governance, level of delivery risk, and time horizon (the sooner the project will issue CERs, the better for buyers). Another important marketing factor is the possibility for a forestry project to propose innovative solutions to the problem of nonpermanence. For instance, the project may be included in a portfolio of projects or facilitate the purchase of nonexpiring credits in replacement of its tCERs and lCERs at the moment they expire.

After the creation of the CDM in 1997, many foresters were enthusiastic about this new mechanism. They envisioned the possibility of selling large quantities of carbon credits at high prices—for instance, from forest conservation projects. Since 1997 the modalities and procedures of the CDM have been progressively defined, and project developers have begun to realize that the mechanism has become increasingly restricted by complicated rules and requirements, such as the rules about eligible activities, permanence, land eligibility, project cycle, and associated transaction costs. As the technical issues became increasingly cumbersome, a good portion of the initial enthusiasm was lost. At the same time, capacity-building efforts were conducted, which helped remove the barriers created by the technical issues. Project developers with sufficient technical capacity may now be able to tackle the technical issues and reach project validation. However, the viability of a

CDM AR project reaches beyond the validation step and depends largely on its attractiveness in the market and the legal arrangements in which it is involved. Market and legal issues define the next challenge for CDM AR project developers.

Notes

1. The decisions 3/CMP.1, 5/CMP.1, and 6/CMP.1, adopted during the first Meeting of the Parties to the Kyoto Protocol (Montreal, November 28–December 10, 2005), represent the legal basis for forestry projects under the CDM for the first commitment period (2008–2012). These decisions had previously been adopted by the Conference of the Parties under the following references: decision 17/CP.7 (Marrakech, FCCC/CP/2001/13/Add.2), 19/CP.9 (Milan, FCCC/CP/2003/6/Add.2), and 14/CP.10 (Buenos Aires, FCCC/CP/2004/10/Add.2).

2. FORMA is the acronym for "Fortalecimiento del MDL en los sectores forestal y bioenergía en Ibero América," which translates as "Strengthening the CDM in the forestry and bioenergy sectors of Iberoamerica." FORMA was a joint effort of INIA (Instituto Nacional de Investigación Agraria y Alimentaria, Spain), CATIE (Tropical Agricultural Research and Higher Education Center, Costa Rica), CIFOR (Centre for International Forestry Research, Indonesia), ECOSUR (Colegio de la Frontera Sur, Mexico), and MGAP (Ministerio de Ganaderia, Agricultura y Pesca, Uruguay) to assist CDM AR project initiatives in Latin America and produce guidelines and tools to facilitate the preparation of these types of projects.

3. A multicriteria analysis was applied to the small portfolio to select ten projects for financial and technical support by the FORMA project.

4. G. M. J. Mohren and others, "CO_2FIX for Windows: A Dynamic Model of the CO_2 Fixation in Forest Stands," Institute for Forestry and Nature Research, Instituto de Ecologia de la UNAM, Wageningen, Netherlands, 1999.

5. For the purpose of our analysis, we defined a "regular estimation" as a consistent but preliminary estimation and a "good estimation" as an in-depth estimation providing details about methodology and presenting consistent results for the whole project lifetime.

6. "Tool for the demonstration and assessment of additionality in AR CDM project activities," report of the twenty-first meeting of the CDM Executive Board, September 2005, Annex 16 (http://cdm.unfccc.in).

7. B. Locatelli and L. Pedroni, "Accounting Methods for Carbon Credits: Impacts on the Minimum Area of Forestry Projects under the Clean Development Mechanism," *Climate Policy* 4, no. 2 (2004): 193–204.

8. Climate, Community, and Biodiversity Alliance (CCBA), "Climate, Community and Biodiversity Project Design Standards," first edition (Washington, May 2005) (www.climate-standards.org).

9. See http://cdm.unfccc.int/methodologies/ARmethodologies/approved_ar.html.

10. "Revised Guidelines for Completing the Project Design Document for AR" and "The Proposed New Methodology for AR, Version 05," report of the twenty-sixth meeting of the CDM Executive Board, September 2006, Annex 21 (http://cdm.unfccc.int); T. Pearson, S. Walker, and S. Brown, "Sourcebook for Land Use, Land-Use Change and Forestry Projects," BioCarbon Fund, Winrock International, Washington, 2005; S. Kamel, "Clean Development Mechanism PDD Guidebook: Navigating the Pitfalls" (Roskilde, Denmark:

UNEP Risø Centre, 2005); M. K. Lee, ed., "Baseline Methodologies for Clean Development Mechanism Projects: A Guidebook," 2005 (www.cd4cdm.org).

11. See "Tool for Afforestation and Reforestation Approved Methodologies (TARAM)," developed by FORMA and BioCarbon Fund (www.proyectoforma.com; http://carbon finance.org/biocarbon).

12. M. Wilder, M. Willis, and J. Carmody, "Legal Issues Guidebook to the Clean Development Mechanism" (Roskilde, Denmark: UNEP Risø Centre, 2004), p. 65.

13. N. Chowdhury and V. Kumar, "Legal Implementation of the CDM in India: Challenges and Opportunities," *CDM Investment Newsletter* 1 (2005): 3–7 (BEA International and Climate Business Network); M. Socorro and others, "Legal Aspects in the Implementation of CDM Forestry Projects," World Conservation Union (IUCN) Environmental Policy and Law Paper 59 (Gland, Switzerland: IUCN, 2005).

14. CD4CDM, "Legal Issues Guidebook to the Clean Development Mechanism."

15. Ibid.; Socorro and others, "Legal Aspects in the Implementation of CDM Forestry Projects"; C. Streck and B. O'Sullivan, "Legal Tools for the Encofor Programme," 2007 (www.joanneum.at/encofor/tools/Tools.htm); CER Sales and Purchase Agreement (www.cerspa.org).

16. B. Locatelli and L. Pedroni, "Will Simplified Modalities and Procedures Make More Small-Scale Forestry Projects Viable under the Clean Development Mechanism?" *Mitigation and Adaptation Strategies for Global Change* 11, no. 3 (2006): 621–43.

17. W. Kägi and D. Schöne, "Forestry Projects under the CDM: Procedures, Experiences and Lessons Learned," Forests and Climate Change Working Paper 3 (Rome: Food and Agricultural Organization, 2005); I. K. Murthy and others, "Methodological Issues in Developing a Community Forestry Greenhouse Gas Emissions Mitigation Project in Mancherial," *Environmental Science and Policy* 9 (2006): 525–37; Pearson, Walker, and Brown, "Sourcebook for Land Use, Land-Use Change and Forestry Projects"; Kamel, "Clean Development Mechanism PDD Guidebook"; Lee, "Baseline Methodologies for Clean Development Mechanism Projects"; M. K. Lee, ed., "CDM Information and Guidebook," second edition (Roskilde, Denmark: UNEP Risø Centre, 2004); A. Olhoff and others, "CDM Sustainable Development Impacts," UNEP Risø Centre, 2004 (www.cd4cdm.org). On ENCOFOR, see www.joanneum.at/encofor/.

CASE STUDY

The San Nicolás Project in Colombia

CARMENZA ROBLEDO AND PATRICIA TOBÓN

The San Nicolás Project in Colombia was designed to address the following questions, among others: How can CDM projects improve the living conditions of the rural poor? Are there examples that demonstrate this potential? And what makes a real difference for the rural poor?

The San Nicolás valleys lie in the northwest of the Antioquia region in Colombia. The region belongs to the area of influence of the Corporación Autónoma Regional de los Rios Negro y Nare, or CORNARE. The region's strategically important watersheds and hydrological resources have been subject to the construction of two hydropower dams, which generate more that 30 percent of Colombia's total electricity. Although natural resources, especially forest and water related, are vital to the development of the region, they have been seriously degraded in recent decades.

During the 1990s CORNARE recognized the increasing need to ensure sustainable management of its natural resources, especially forests. As a consequence, in 1998 CORNARE agreed to a legal instrument (Agreement 016) with all the municipalities of the region, establishing 40,000 hectares for forest conservation and 32,000 hectares for multiple forest uses, including plantations.

As a means of enforcing this agreement and promoting the sustainable use of forest resources, CORNARE partnered with the Swiss Federal Institute for Material Testing and Research to develop the project "Alternative financing model for sustainable forest management in San Nicolás." In 1999 a project proposal was presented to the International Tropical Timber Organization, and Switzerland, Japan, and the United States agreed to cofinance it.

The first phase of the project lasted from 2001 to 2004. It was clear to the project partners from the beginning that in addition to technical matters, the project needed to address the socioeconomic situation of the communities. More than 12,000 families live in the nine municipalities covered by the San Nicolás Project's 72,000 hectares, at elevations ranging from 800 to 3,000 meters. A process called the Regional Forum was thus created to promote capacity-building, discussion, and decisionmaking for the project. As part of the Regional Forum, more than 170 meetings and workshops were organized with the farmers and with industry partners and municipalities.

The Regional Forum was designed around three main phases: information, participation, and decisionmaking. The first phase, information, aimed to familiarize all potential local actors with the project, including communities, industry, and policymakers at the regional and local levels. On the basis of comments received from these groups, specific needs and requirements were assessed and future project activities were adapted accordingly.

The second phase of the Regional Forum, participation, was aimed at exchanging knowledge and expectations about future land uses according to the institutional agreements (including Agreement 016) and the needs expressed by the community. During this phase all forestry activities were analyzed by project experts together with members of the community, using a social mapping methodology. A clear need for community capacity-building related to management practices for new forestry activities and to entrepreneurship was identified. The project responded by designing and implementing a capacity-building program, which has been supported by local research and education institutions.

All the information collected during these first two phases allowed the project team to elaborate a detailed forest management plan. The plan, together with the analysis of forest cover and trends, enabled the carbon potential of the project to be quantified.

The third phase of the Regional Forum, decisionmaking, was aimed at ensuring a participative decisionmaking process among small and medium-size farmers, NGOs, municipalities, regional institutions, and the private sector. The main issues to be agreed upon in this phase were the forest management plan and the creation of an institution to leverage the implementation of the plan and facilitate the trade of all products and services, including carbon sequestration.

In 2008 the project region has a master forest development plan designed with and adopted by the local farmers. It integrates a CDM reforestation component that includes the use of timber and nontimber products, as well as a conservation component. The CDM component includes agroforestry, silvopastoral, and small plantation activities. The project prepared a baseline and monitoring methodology that has been already approved by the Executive Board of the CDM (ARAM 009).

As a response to the second objective of the Regional Forum, a public, private, and civil-society partnership was created to manage the project over the coming decades, as well as to facilitate CER purchases from international business. The Corporation for the Sustainable Management of the Forests (MASBOSQUES), established in September 2003, includes all sectors of society and focuses on strengthening civil society and ensuring the participation of communities in planning and decisionmaking regarding sustainable management of the forest beyond the Regional Forum. MASBOSQUES will take responsibility for implementing the forest management plan as well as putting in place the different schemes for payment for environmental services, including carbon sequestration and emission reduction. MASBOSQUES has already signed a contract with the BioCarbon Fund of the World Bank for the CERs of the project, and negotiations for including it in a REDD scheme are under way.

By 2008 more than 2,500 hectares of new forests had been established, creating new income for the community. To support this, appropriate forest management activities were designed, and research on traditional uses of nontimber products was conducted. The master plan also includes a capacity-building and an entrepreneurship program covering new products and markets. Beneficiaries of these programs are now selling nontimber products such as shampoo, soap, and essential oils. Timber and nontimber forest products are to be commercialized by local farmers, who remain the owners of all these products. The social returns from the project can be summarized as empowerment of local communities, creation of public-private partnerships, improvement of local capacity, improvement in family income, and improvement in food security.

High levels of community participation and empowerment, supported by high-quality documentation, made the project a success in San Nicolás. The question remains whether other, less well-supported communities in developing countries can also use the CDM for promoting poverty reduction and sustainable development in rural areas.

9

The Permanence Challenge: An Economic Analysis of Temporary Credits

FRANCK LECOCQ AND STÉPHANE COUTURE

Carbon sequestered in biomass or soils may be released accidentally because of fire, windstorms, or other natural hazards or because of conversion of the land to agriculture or pasture. The so-called nonpermanence risk constitutes a fundamental difference between biological sequestration projects and projects that reduce the emission of carbon into the atmosphere. The risk of nonpermanence matters because sequestering carbon temporarily does not have the same effect on global warming as avoiding emissions permanently.

To deal with the risk of nonpermanence, carbon sequestration through land use, land-use change, and forestry (LULUCF) projects under the Clean Development Mechanism (CDM) generates credits with finite lifetimes. Temporary Certified Emission Reductions (tCERs) last for one commitment period, with a new tCER issued at the next period only if verification shows that the carbon stock is still sequestered. Long-term Certified Emission Reductions (lCERs) may last longer, but they will expire before their final due date if periodic verification shows that the carbon is no longer sequestered. In addition, under current rules tCERs and lCERs cannot be renewed beyond the project's last crediting period (a maximum of sixty years), regardless of whether the carbon is still sequestered or not.[1]

The current rules thus create a double liability for the carbon buyer. First, if the carbon stock underlying the credit is no longer sequestered or is not reverified, then the holder of the credit suffers a debit that must be compensated for by the acquisition of a permanent credit or of another temporary credit based on a different carbon stock.[2] The second liability arises at the end of the project's crediting period when a replacement has to be found whether or not the underlying carbon stock is being maintained. However, the fact that tCERs and lCERs have finite durations may also provide an opportunity for buyers seeking to gain time until carbon is less expensive—for example, because they anticipate that their emissions will fall in the future.

Our purpose in this chapter is to evaluate how much carbon buyers would be willing to pay for temporary credits (or streams thereof) relative to permanent credits and to discuss the implications for the attractiveness of LULUCF projects relative to other CDM projects. In the following section we show that the price of tCERs depends on the relative prices of carbon today and at the next commitment period. We then extend the model to lCERs (and to streams of tCERs) and show that their price depends on both the expected price path of permanent credits and the risk of nonpermanence.

Next we provide empirical evidence that the risk of nonpermanence, though difficult to evaluate, is likely to be significant. As a result, the risk premium attached to nonpermanence is likely to be large. Finally, we discuss the implications of the preceding findings for LULUCF project developers. We suggest that few buyers are likely to be interested in temporary credits per se and that most will in fact demand quasi-permanent credits even from LULUCF projects. Given the magnitude of the nonpermanence risk premium, we explore some strategies that LULUCF project developers might adopt to make LULUCF-derived credits more palatable to buyers.

The Price of tCERs

In this section we discuss the value of tCERs relative to the value of permanent credits. Because tCERs expire at the end of the commitment period following the one for which they were issued, purchasing a tCER is equivalent to purchasing the right to defer the purchase of a permanent credit by one commitment period (not two, because the tCER must be replaced before it expires).

At the present time, markets for carbon assets during the second commitment period, let alone subsequent ones, are virtually nonexistent.[3] Thus the price of tCERs depends on the difference between the observed price of permanent credits today and the expected price of permanent credits during a second period.

Precisely, let p_1 be the price of permanent credits for the current commitment period, let $E(p_2)$ be the expected price of permanent credits for the next commitment period, and let r be the discount rate. A risk-neutral entity with obligations under the Kyoto Protocol will purchase a tCER if and only if its price p_{tCER} is lower than the difference between the current and the discounted future price of carbon (equation 1).[4]

$$p_{tCER} \leq p_1 - \frac{E(p_2)}{(1 + r)^5} \tag{1}$$

If the price of carbon is expected to increase at a rate higher than the rate of discount between the first and second commitment periods, then tCERs are worth zero. In this case, purchasing a temporary credit would amount to purchasing the right to buy carbon at a higher price (even in discounted terms) in the future.

If, on the other hand, the price of carbon is expected to increase less rapidly than the discount rate, then temporary credits have a positive value. For example, if the price of carbon remains constant in real terms over the two periods, and if the discount rate is 6 percent, then buyers will buy tCERs at, at most, one quarter of the value of a permanent credit.

The Price of lCERs

The price of lCERs or of streams of tCERs is slightly more complex to estimate than the price of tCERs. This is because lCERs have lifetimes ranging from a few years to a maximum of sixty years, for they may be issued at any time during the project and will expire at the end of the project's last crediting period. They will also expire before their theoretical duration if the carbon pool underlying the lCER disappears or the periodic verification process stops.

To see how the nonpermanence risk matters, we consider an lCER issued during commitment period 1 and planned to end during commitment period T. Let α be the probability that verification does not take place or that the carbon underlying the lCER is found to be missing during each commitment period.[5] We assume α to be constant here, but other risk profiles might also be considered depending on the tree species and project characteristics. Time-varying risks would not fundamentally alter the following discussion.

By definition there is a risk α that the lCER expires at period 2 and thus that the buyer will be forced to purchase a replacement credit at price $E(p_2)$. And there is a probability $(1 - \alpha)$ that the lCER remains valid until period 3. At that point there is again a probability α that the credit expires—and thus that a replacement credit will be purchased at price $E(p_3)$, and a probability $(1 - \alpha)$ that the lCER remains, and so on until commitment period T, when, if it has not already expired accidentally, the lCER expires as planned.

Summing up, there is a probability α that the lCER will expire at period 2, a probability $\alpha(1 - \alpha)$ that the lCER will expire at period 3, . . . , a probability

$\alpha(1-\alpha)^{T-3}$ that the lCER will expire at period $T-1$, and a probability $(1-\alpha)^{T-2}$ that the lCER will expire as planned during period T. A risk-neutral entity would thus purchase this lCER instead of a permanent credit at period 1 if and only if the price of the lCER p_{lCER} is such that

$$P_{lCER} \leq p_1 - \alpha \frac{E(p_2)}{(1+r)^5} - \alpha(1-\alpha)\frac{E(p_3)}{(1+r)^{10}} - \ldots - \alpha(1-\alpha)^{T-2}\frac{E(p_T)}{(1+r)^{5(T-1)}}. \quad (2)$$

If there were no risk of nonpermanence ($\alpha = 0$), then the lCER would guarantee that the purchase of a permanent credit would be delayed until commitment period T. The price of the lCER would depend only on the expected price of carbon at period T and on the discount rate. And the logic would be the same as in the case of a stream of tCERs: if the price of carbon is expected to increase at the rate of discount or higher between commitment period 1 and commitment period T, then there is no point in purchasing such a "risk-free" lCER ($p_{lCER} = 0$).[6] Conversely, if the price of carbon is expected to increase less rapidly than the rate of discount, then the price of the lCER is positive.

For example, if the price of carbon is expected to remain constant from period 1 to period T, and if the discount rate is again 6 percent, then a "risk-free," twenty-year lCER ($T = 5$) will sell at more than two-thirds (69 percent) of a permanent credit. This value is very sensitive to expectations about the future price of carbon: it falls to 53 percent if the price of carbon is expected to rise by 2 percent per year instead of 0 percent. The longer the duration of the lCER, the closer one gets either to zero (if the price of carbon is expected to increase at a rate higher than the rate of discount until the lCER expires) or to 100 percent of the value of a permanent credit (if the price of carbon is expected to rise less rapidly than the discount rate until the lCER expires). For example, with a constant price of carbon, a thirty-year risk-free lCER ($T = 7$) would sell at nearly 83 percent of a permanent credit, and a sixty-year risk-free lCER ($T = 13$) would sell at 97 percent.

When the risk of nonpermanence is taken into account ($\alpha > 0$), the price of lCERs depends not only on the price of carbon at expiration date but also on the expected *path* of the price between commitment period 1 and commitment period T. Everything else being equal, the price of the lCER diminishes when $\alpha > 0$ relative to the case $\alpha = 0$, and the difference is the risk premium attached to nonpermanence. The magnitude of the premium depends on the magnitude of the risk and on the expected price path of carbon.

If the price of carbon is assumed to evolve smoothly—for example, at a constant rate—over time, then the risk premium remains relatively small. For example, if the price of carbon is assumed to remain constant in real terms from period 1 to

Implications of the Sixty-Year Maximum Duration for the Prices of lCERs and tCERs

The rule stating that lCERs and tCERs cannot be renewed beyond a project's last crediting period even if the carbon remains sequestered is unnecessary from an environmental point of view. Indeed, one could imagine lCERs or tCERs being renewable indefinitely as long as the verification process is maintained and the carbon remains sequestered. Could the seller make much more money by selling such "indefinite" lCERs?

On the one hand, the seller would get a better price because the buyer would not need to plan for the certain replacement of the credit sixty years hence. On the other hand, the seller would have to accommodate the risk of nonpermanence for each commitment period from year sixty onward. The magnitude of each effect depends on the risk of nonpermanence, on the expected price of carbon at year sixty, and on the expected price path of carbon beyond year sixty.

Equation 2 can easily be extended to indefinite credits by pushing T to infinity. On this basis, numerical tests suggest that the seller would probably earn a net gain if the price of carbon were not expected to peak after year sixty. In other words, "indefinite" lCERs would sell at higher prices than sixty-year lCERs. But the magnitude of this effect is likely to remain modest. For example, if the price of carbon is expected to increase at 3 percent per year indefinitely, and if the risk of nonpermanence is 5 percent per commitment period, then an indefinite lCER would sell at 75 percent of a permanent credit, whereas a sixty-year lCER would sell at 68 percent. To increase the difference between indefinite and sixty-year lCERs, one must assume that the price of carbon increases more rapidly, but in that case neither indefinite nor sixty-year lCERs have much interest relative to permanent credits anyway.

period 5, then the price of a twenty-year lCER relative to the price of a permanent credit drops from 69 percent without risk to 65 percent with a permanence risk of 5 percent per commitment period. The risk premium represents less than 4 percent of the value of a permanent credit. Under the same assumptions, the risk premium is still small even for thirty-year (6.6 percent) or sixty-year lCERs (11.3 percent).

The risk of nonpermanence will matter greatly, on the other hand, if the price of carbon is expected to go through one or several peaks between commitment periods 1 and T. In that case the costs associated with replacement of the temporary credit at peak price would be high, and the risk premium can become significant. Let us assume, for example, two paths for the price of carbon over the next thirty years. In path 1, the price of carbon increases at a constant rate of 2 percent per year. In path 2, the price of carbon increases at the rate of discount (6 percent) during the first twenty years and then diminishes linearly to reach the same level as in path 1. If the risk of nonpermanence is 5 percent, then the risk premium is worth 6.3 percent of the price of a permanent credit under path 1 and 13.8 percent under path 2—more than double.

Estimating the Risk of Nonpermanence

In most instances—and to our knowledge, in all LULUCF projects currently in the CDM pipeline—LULUCF project developers intend their projects to be permanent. Yet natural hazards or mismanagement may result in a partial or total accidental release of the sequestered carbon at any time. In addition, land use is not necessarily stable in the long run.

The risk of land-use change is probably the easiest to estimate. Migration flows from rural to urban areas are expected to continue in the future, and rural areas that are subject to agricultural extension are usually the richest, most productive areas. As a result, remote areas with lower agricultural productivity are less likely to be claimed for agriculture.[7]

The key issue is the risk of accidental carbon release. It has two components, one related to project risk and the other to natural hazards.

First, carbon could be released if the project is scaled down or abandoned before the temporary credit (or stream thereof) was set to expire. For example, an agroforestry project may be abandoned before its due term and the land converted back to pasture or agriculture.[8] It is obviously difficult to estimate project risk in LULUCF activities, but project failures tend to occur early in project operations.[9] If a forestry project has been operational for one or two decades, it is probably more likely to continue.

Second, carbon could be accidentally released following natural disturbances such as wildfires, windstorms, droughts, diseases, insect outbreaks, frosts, and heavy snows. In this case the risk is for the most part independent of the project sponsor's actions (although preventive and coverage measures can be taken to reduce both the incidence of natural hazards and their effects when they occur). Wildfires and windstorms, in particular, are significant ecological disturbances, and their size and frequency appear to be increasing.

Table 9-1 provides estimates of the annual risk of wildfires for several regions of the world, based on the share of forested land affected by wildfires between 1990 and 2000. These data are complemented with country-level surveys conducted by the UN Food and Agricultural Organization. The point to be drawn from the table, which is only an illustration, is that the risk of wildfire can be relatively large in some parts of the world. However, three caveats must be made regarding these data. First, the overall quality of the data is poor, especially for developing countries. Records, when available, are often limited, with no assessment of the quality of the statistics. Second, regional-level averages mask large differences across countries, as the minimum and maximum numbers suggest. The same applies within countries, where fire incidence can vary dramatically by region and biome. Third, FAO data do not provide information about the amount of biomass that is left standing and alive after a forest fire. Table 9-1 should thus be understood only as

Table 9-1. *Estimated Annual Risks of Wildfires for Seven World Regions*[a]

Region	*Number of observations*	*Average*	*Minimum*	*Maximum*
Boreal Europe	3	0.00	0.004	0.006
Boreal non-Europe	2	0.12	0.094	0.205
Temperate Europe	18	0.06	0.003	0.225
Mediterranean	17	0.23	0.003	1.249
Latin America and Caribbean	4	0.52	0.121	1.343
Asia	6	1.40	0.016	6.876
OECD Pacific	4	0.29	0.004	0.343

Sources: Food and Agriculture Organization, *Global Forest Fire Assessment 1990–2000* (Rome, 2001); Food and Agriculture Organization, *Global Forest Resources Assessment* (Rome, 2005).

a. Based on country-level areas affected by wildfires and supplemental country data. Average, minimum, and maximum figures are annual chance of wildfire occurrence as percentage.

providing a range of plausible figures; it cannot be used for project-specific risk assessment.

Similarly, wind can cause serious damage to forests. For example, storms were responsible for 53 percent of the total damage to forests by natural disturbances in Europe between 1950 and 2000.[10] However, the literature regarding the risk of damage due to wind is less developed than the literature relative to fires. A major difficulty is that there is a large difference between the risk of occurrence of a windstorm and the risk that the wind will cause lasting damage to the stand. The latter depends on the height and age of the trees, crown size, health of roots, soil moisture, fertilization, thinning, topography, and position in relation to wind direction. Although ample meteorological data exist on the occurrence of windstorms, data on damage to forests are limited.

Nevertheless, the data suggest that the risk of nonpermanence in a single forest stand can be relatively large and that the 5 percent per commitment period figure used earlier is by no means an overestimation. These data also suggest that good management of a particular forest, although necessary, is not a sufficient condition for guaranteeing quasi-permanence.

How to Make LULUCF Attractive to "Bullish" Carbon Buyers

The preceding discussion shows that buyers with bearish views about the carbon market will be attracted to tCERs and lCERs. Buyers with mild or bullish views about the carbon market are unlikely to be attracted to tCERs or lCERs at all; they would not be ready to purchase such credits regardless of how cheap they were. Chomitz and Lecocq have concluded that (1) the longer the horizon, the more likely it is that the value of streams of temporary credits will be either zero or the value of permanent credits, and not an intermediate value; and (2) it takes bearish views of the carbon market for the value of streams of temporary credits to be greater than zero.[11]

As a result, only two groups of carbon buyers seem likely to be interested in temporary carbon assets. The first is actors with bearish views of the market, either in the long run (for example, because they believe that climate change will prove less dangerous than expected or that technical change will provide rapid solutions) or in the short run (for example, because they anticipate some pause or delay in the buildup of international climate regulations). The second is actors with optimistic views of their *own* mitigation potential, regardless of the overall price of carbon. For example, firms may use temporary credits to synchronize carbon constraints with their investment cycle and avoid replacing long-lived capital (say, a boiler) prematurely. Entities that do not meet these criteria may still engage in LULUCF projects and thus still purchase temporary credits, but they will be interested only in quasi-permanent assets and thus are likely to demand strong guarantees about the permanence of the sequestered carbon.

We discuss five strategies to help guarantee that carbon credits will be quasi-permanent.

The first is insurance, although insurance for natural hazards is not readily available, even in many developed countries. For example, less than 5 percent of the French forest surface is insured, and prices are widely considered prohibitive. Insurance is likely to be unavailable for most CDM LULUCF projects, although the recent development of weather insurance mechanisms may modify this situation in the future.[12]

A second possibility for the project developer is to self-insure and self-protect. The developer can take several actions to limit either the risk of occurrence of a hazard (so-called self-protection actions such as brush clearing to limit fire risks) or the effect of a hazard should it occur (so-called self insurance actions such as limiting stand height to minimize wind-related risks). An important point here is that not only carbon revenues but also timber revenues depend on risk limitation. The forester's incentives are thus aligned with those of the carbon buyer even if carbon revenues form a limited share of total revenues from the stand. And hazard management techniques are part of the forester's tool kit.

Third, self-insurance may involve the creation of carbon reserves that would be tapped only if a hazard occurred in the main stand. Ideally these stands should be distant from the main forest to limit the risk that both will be exposed to the same hazard. Larger project developers may also pool carbon credits from a wide range of forests, thereby dividing risks.

Fourth, project developers may bundle streams of temporary credits from LULUCF projects with low-price future permanent credits or options for the purchase of future permanent credits. For example, efficient buildings may generate emission reductions in the long run with some certainty, but developers may still want to cash in rapidly on those credits, even at cheap prices, because they are selling the building.[13] Projects that combine LULUCF and emission reduction components, typically bioenergy projects, can also rely directly on self-generated permanent credits from the energy component to back up the sale of the temporary credits from sequestration. This suggests that large, multisector project developers or funds may be at an advantage over small, LULUCF-only project developers.

Finally, the risk of natural hazards is uncorrelated with other risks to which buyers are exposed, such as price risks in their main domain of activity and stock market risks. Thus for risk diversification reasons, buyers may be willing to trade some exposure to their traditional risks for exposure to nonpermanence risks—just as, for risk diversification reasons, large financial institutions have an interest in holding some forestry assets in their portfolios. For example, a financial company might be willing to purchase tCERs or lCERs instead of buying (more expensive) permanent credits and use the differential to reduce risk exposure in other markets. This introduces some risks on the carbon side, but the combination of risk reduction and risk diversification may justify it.

Notes

1. The rules for accounting for tCERs and lCERs are set out in FCCC/KP/CMP/2005/8/Add.1 as Decision 5/CMP.1, "Modalities and procedures for afforestation and reforestation project activities under the clean development mechanism in the first commitment period of the Kyoto Protocol." See Annex, section J, "Issuance of tCERs and lCERs," and section K, "Addressing non-permanence of afforestation and reforestation projects under the CDM."

2. Carbon extracted from forests but stored in artificial carbon pools such as furniture or construction is not accounted for under current CDM regulations.

3. K. Capoor and P. Ambrosi, "States and Trends of the Carbon Market 2007" (Washington: World Bank, 2007). The Kyoto Protocol sets out quantified emission reduction and limitation targets for industrialized countries for the years 2008–2012, the first commitment period. It does not set out similar targets for additional periods after 2012. A possible second commitment period is currently being discussed by the international community.

4. Two caveats must be made on Equation 1. First, the tCER might be purchased any time during the first commitment period, and the permanent credit aimed at replacing the tCER might be purchased any time during the second commitment period, so there might be more or less than five years between purchase of the tCER and purchase of the replacement credit. Second, a risk-adverse entity would add a further discount on temporary credits because of the risk that p_2 might be higher than expected. Similarly, a risk-prone entity would value temporary credits more than in (1) because it would put a premium on the potential for gains at the second period if p_2 turns out to be low.

5. Verification in LULUCF projects is supposed to occur at least once every five years.

6. A stream of tCERs refers to the issuance and periodic reissuance of tCERs from a CDM project.

7. World Bank, *World Development Report 2003: Sustainable Development in a Dynamic World. Transforming Institutions, Growth and Quality of Life* (Washington, 2003); K. Chomitz, "At Loggerheads? Agricultural Expansion, Poverty Reduction, and Environment in the Tropical Forests," World Bank, Washington, 2007.

8. Abandonment or conversion may also be forced, as in the case of civil war or government nationalization of the land.

9. Of course risks exist in any carbon project, but they matter for a much longer time in the LULUCF case. For example, mismanagement that occurs even ten years after the end of the delivery of temporary carbon credits will still affect the renewal of those credits, whereas if a power plant is improperly maintained and ceases production the day after a delivery of a carbon credit has been completed, the validity of the credit and of the underlying emission reductions that have been sold is unaffected.

10. M.-J. Schelhaas, G.-J. Nabuurs, and A. Schuck, "Natural Disturbances in the European Forests in the 19th and 20th Centuries," *Global Change Biology* 9 (2003): 1620–33.

11. K. Chomitz and F. Lecocq, "Temporary Sequestration Credits: An Instrument for Carbon Bears," *Climate Policy* 4, no. 1 (2004): 65–74.

12. M. Brunette and S. Couture, "The Impact of Public Intervention on Self-Insurance and Insurance Activities in Risky Forest Management," LEF Working Paper 2006-07 (Nancy, France: Laboratoire d'Economie Forestière, 2006). The difficulty of providing insurance for forestry projects is compounded by the fact that some of the events are low-probability and high-consequence events, which are inherently more difficult to insure.

13. Chomitz and Lecocq, "Temporary Sequestration Credits."

10

Project-Based Mechanisms: Methodological Approaches for Measuring and Monitoring Carbon Credits

TIMOTHY PEARSON, SARAH WALKER, AND SANDRA BROWN

Projects based on land use, land-use change, and forestry (LULUCF), such as afforestation and reforestation (AR) projects, have the ability to produce real, significant, positive carbon effects. However, without scientifically tested and rigorously applied methods for measuring these effects, the actual carbon dioxide emission reductions will not be quantifiable or creditable. Fortunately, widely recognized standards exist for the quantification of land-based carbon credits. The responsibility of a project developer is to understand how these methods can be used to design a measurement plan that maximizes the verifiable, conservatively estimated carbon emission credits while minimizing the resources required for project implementation.

Robust methodologies for measuring and monitoring carbon stocks and carbon stock changes have been developed over many years of forestry expertise. These methods form the basis of approved methodologies for measuring emission removals under the Kyoto Protocol's Clean Development Mechanism (CDM).

In this chapter we discuss where measurement and monitoring are required in carbon projects in the land-use sector. We also discuss what represents "good practice" in measurement and outline ways to create measuring and monitoring plans that will inspire confidence. Although other documents exist that detail exactly how to measure the carbon stocks of LULUCF projects,[1] we discuss all elements of project measurement and monitoring and how to maximize creditable net carbon benefits through good practice. Our aim is to provide project developers and practitioners with the tools to create highly creditable carbon project activities.

We discuss measurement and monitoring generally, but where we introduce a focus on a specific offset mechanism, the focus is on the Clean Development Mechanism of the Kyoto Protocol, with mechanisms such as Joint Implementation (JI) and the voluntary market receiving lesser mention. The reasons behind this focus are that the CDM will be the most prescribed, with the highest standards to achieve in measurement, and is the most advanced of the mechanisms in terms of definition of rules, requirements, and regulations.

Key Concepts

The key concepts under measurement and monitoring are conservatism, accuracy, and precision.

Conservatism is an approach to uncertainty. Where a range of possible values exists, a conservative approach to project development and reporting would be to adopt the value that gives the lowest (most conservative) net carbon effect. For example, when 95 percent confidence bounds exist around a value, it is conservative to take the higher bound in the baseline and the lower bound during the project.[2] In each case the net carbon benefit will be reduced and will be more conservative. Other conservative approaches are to exclude carbon pools that will increase or remain unchanged during the project and to exclude project emissions that will be higher in the baseline than in the project.

Sampling involves examining a subset to understand the whole. Sampling is the basis of all measurement for LULUCF carbon projects, because destructive measurement of all trees is an expensive approach to evaluating carbon stocks and is counterproductive with regard to greenhouse gases (GHGs). For carbon sampling, the two key concepts are accuracy and precision.

Accuracy represents how close a measurement is to the true value. When sampling, one cannot know the true mean, but if statistically sound, rigorously tested scientific methods are used and carefully applied, then one can feel confident that the true mean is well represented. Precision represents the range of values between which the true value may lie, or the repeatability of a measure. In other words, accuracy is how well our measurements represent reality, and precision represents how confident we are in the measurements we have taken.

To illustrate the concepts of accuracy and precision, we use a hypothetical forest that has a carbon stock of 120 tonnes of carbon per hectare (t C/ha) (table 10-1). In scenario A, the mean of the measurements is identical to that of the known stock, and the 95 percent confidence interval (a measure of the variability of the repeated measurements) is small, even for just five samples. Thus scenario A is accurate and precise. Scenario A should be the desired outcome of a measurement plan.

Table 10-1. *Illustration of the Concepts of Accuracy and Precision*[a]

Sample	*Scenario A: Accurate and precise*	*Scenario B: Accurate but imprecise*	*Scenario C: Inaccurate but precise*
1	118	145	180
2	123	95	183
3	121	170	177
4	118	110	178
5	120	80	182
Average	120	120	180
95% confidence interval	2.6	45.8	3.2

a. Numbers are hypothetical estimates of carbon stocks, in tonnes of carbon per hectare, derived from samples in a hypothetical forest with a "known" carbon stock of 120 t C/ha. In reality the true mean for a forest can be known only if all trees are harvested and weighed.

In contrast, in scenario B the mean is accurate but the range is very large (low precision), with the consequence that we have less confidence in the mean. A designated operating entity (DOE) (or independent accredited entity, IAE) may require a deduction because of this lack of confidence. In scenario C, the range is small (high precision) but the mean is inaccurate (significantly greater than the true forest mean). Scenario C is the worst of the three possibilities, because the high precision indicates confidence but the low accuracy illustrates misplaced confidence.

A lack of accuracy arises from poor measurement methods or poorly applied methods; a lack of precision arises from poorly planned or insufficient sampling. Measurement and monitoring should always be planned to be accurate and precise, and when there is any lack of confidence, the effect is minimized by being conservative.

The Elements of Measuring and Monitoring

All project-based mechanisms have several components that affect carbon emissions. Therefore proper measurement and monitoring include more than tracking carbon stocks in trees as they grow. To quantify the effect of a carbon project properly, all project aspects that alter net carbon emissions must be quantified using established, transparent methods. The four elements of measurement and monitoring are monitoring project carbon stocks, establishing the baseline, monitoring leakage, and estimating project emissions. After considering each of these elements we examine the errors that sometimes arise through poor development or improper implementation of measuring and monitoring plans and suggest ways in which these errors can be minimized or avoided entirely.

Monitoring Carbon Stocks

Carbon flows from LULUCF activities are usually assessed using the stock-change approach, which typically estimates net carbon flow on the basis of measured changes in an inventory of carbon stocks. For example, the change in the biomass carbon stock in a community of trees is a result of the difference between inflows of carbon to trees through photosynthesis and outflows through respiration and leaf fall. Because carbon stocks generally change relatively slowly, annual measurements or estimations of the inventory of carbon stock are unlikely to capture significant changes. It is therefore common to conduct successive inventories at intervals of several years, often five for vegetation and more for soil. Carbon flows to and from the atmosphere are then estimated as the differences in carbon stocks between successive measurements. If the total change in the carbon stock is positive, it means carbon is being sequestered from the atmosphere (the forest is a sink); if negative, it means carbon is being emitted to the atmosphere (the forest is a source). Methods for inventorying and monitoring carbon stocks in the key LULUCF pools are based on commonly accepted principles of forest inventory, soil sampling, and ecological surveys, are well established after many decades of use and refinement, and are thus the preferred methods for LULUCF projects.

In general, the steps for estimating changes in carbon stocks are to (1) divide the project area into strata (groups) on the basis of similarities in estimated carbon stocks; (2) measure the area of each stratum; (3) estimate the average carbon stock changes in each stratum; (4) multiply the area of the stratum by the carbon stock change estimate; and (5) sum the estimated carbon stock changes of each stratum. Therefore the monitoring of project-level carbon benefits has two major components: monitoring any changes in the project area and monitoring changes in each of the monitored carbon pools.

Established, field-tested methods exist for measuring carbon stocks in the field and summing the carbon stocks within a project. *Good Practice Guidance for Land Use, Land-Use Change and Forestry,* published by the Intergovernmental Panel on Climate Change (IPCC) in 2003, provides an overall guide to LULUCF carbon measurements. The World Bank's 2005 *Sourcebook for Land Use, Land-Use Change and Forestry Projects* presents more targeted, specific field methodologies. These two sources can be used to assist project developers in implementing carbon monitoring plans.

MONITORING PROJECT AND STRATA AREAS. The project boundary delineates the area in which occur all emission changes that are attributable to the project. The project boundary can include one contiguous area or several discrete areas of land. For afforestation-reforestation, the project boundary will most likely be the planted area of each stratum. At the beginning of and throughout the life of the project, this area must be delineated, ideally using a global positioning system

(GPS) in combination with a map or a georeferenced, remotely sensed image. The area of each stratum may vary during the project (because of changes in, for example, the planting regime or the growth and mortality of trees in certain areas), so these changes must also be determined. As for all components of monitoring, the area must be estimated conservatively.

CHOOSING CARBON POOLS. The carbon in a landscape can be grouped into six "carbon pools": above-ground tree biomass, above-ground nontree biomass, below-ground roots, forest floor litter, deadwood, and soil. Depending on the project activity and environment, the change in each carbon pool and the proportion of this pool in total carbon stock measurements will vary.

The pools to be monitored should depend on several factors, including expected rate of change of carbon stocks in the pool, magnitude of change, accuracy and accessibility of methods for quantifying this change, and costs associated with measuring the pool to different levels of precision.

For all land-based projects, tree biomass will constitute the majority of the carbon stocks found in the system. Nontree biomass (for example, shrubs and grasses) may be important during initial stages of projects or in savanna-type systems, but it often does not form a significant carbon pool in mature forests. Carbon in deadwood usually accumulates slowly, and the litter pool in most forest types is small. The soil carbon pool will also change slowly or not all and therefore will show significant changes only over long periods of time.

APPROACHES TO CARBON STOCK ESTIMATION. The estimated carbon stock change within each pool must be a conservative estimate. This ensures that any emission removals or reductions certified have actually occurred. For afforestation-reforestation projects, this means that the credited *estimated* increase in a carbon stock must be equal to or less than the *actual* carbon stock increase and never greater than the actual carbon stock increase.

Each method used to estimate carbon stocks comes with its own level of certainty. An estimate with a low level of certainty means that the true value of the stock could sit in a wide range. Where there is low certainty, a conservative approach is warranted. For example, the lower bounds of the confidence interval could be taken.

Additionally, there are two approaches to estimating the carbon stock of any pool: (1) direct field measurement and (2) the field measurement of a parameter for which a relationship between the carbon stock and the parameter can be created with certainty. As an important example of the second approach for land-use projects, the biomass in trees within a project area cannot be measured directly, because to do so would require that all the trees be cut down and weighed. Instead, relationships (in this case a regression equation) between an easily measurable variable or variables (usually diameter at breast height or diameter and height) and the biomass of trees can be established by cutting down and measuring and weighing a small number of trees. Relationships can be used to determine biomass from tree diameter, tree height, or tree volume and to determine root biomass from aboveground biomass. The relationships take the form of default factors or regression equations. Such relationships to be applied by the project exist on a scale of increasing level of certainty, from default estimates to ecosystem-specific published relationships, locally or regionally derived published relationships, and project-derived specific relationships.

If the estimates or relationships used are not derived by the project, then limited field-based measurements can be taken of most pools to verify that the estimates used are credible and conservative and do not include a systematic bias.

Where project-level field measurements are taken to estimate carbon stocks, these measurements will likewise have an associated level of accuracy and precision. Generally, the precision of estimates can be improved by increasing the number of field measurements taken. Systems that are more variable require a greater number of field measurements to reach a level of precision equal to that for less variable systems.

MAXIMIZING RETURN ON INVESTMENT IN MONITORING. Project developers must analyze which approach is appropriate for each pool in order to maximize creditable sequestration or emission reductions while minimizing costs. For any increase in the number of pools monitored and in the certainty of the estimate, additional costs will occur. When developing the project, developers would find it a useful exercise to perform a cost-benefit analysis to determine the level of certainty that will maximize the return on their investment. By doing this, the carbon credit return on every increase in certainty can be estimated.

This desired level of certainty may vary for each part of the project. For example, if one of the project strata is highly variable, then it may make sense to use the lower bounds of a highly uncertain estimate. Similarly, if a carbon pool is extremely variable, so that a large number of samples would be required to produce a precise estimate, then this must be weighed against the expected number of carbon credits to be received from the changes in carbon stocks of this pool.[3] This is especially important to consider with respect to the tree biomass pool, because it will be the dominant pool. An analysis could also be conducted to determine whether producing a project-specific equation would increase the number of carbon credits and consequently the income from credits received beyond the initial costs of developing the equation (a cost-benefit analysis). For some pools, such as the deadwood pool in afforestation-reforestation projects, monitoring may not be cost effective, especially because this pool may accumulate carbon slowly during the life of the project in managed forest stands.

The concept of maximizing investments also extends to monitoring the project area. For example, if the boundary of the area planted is highly contorted, it may be more cost effective to reduce the credited area to a boundary that is more easily measurable than to take detailed GPS measurements, each with an associated error level.

The cost-benefit equation is affected by project scale. For very large projects with a high expected total carbon yield, it is more cost effective to invest early on in project-specific defaults and equations that will maximize reported carbon benefits. For very small projects, the cost of these initial measurements would be insufficiently offset by the added return in carbon credits.

Establishing the Baseline

Proper accounting for baseline, or "without project," carbon stocks helps to ensure that only additional carbon stocks attributable to the project activity are credited. Any stocks determined to have changed within the baseline would not be included in any credited carbon stocks. Baselines under the Kyoto Protocol have several distinct elements, each of which requires an initial evaluation (and, in limited cases, monitoring throughout the project). These baseline elements are additionality, eligibility, baseline land cover, and baseline carbon stock.[4]

The concept of additionality under the flexible mechanisms of the Kyoto Protocol is meant to ensure that all credited carbon represents a real positive climate effect. Specifically, the additionality test is designed to prove that without finances derived from the trading of carbon credits, disadvantageous economics or other barriers would not have been overcome and the project would not have proceeded.

Additionality in the LULUCF sector is best assessed using the CDM Executive Board–approved AR additionality tool.[5] It proves additionality through one of two tests: a financial test and a barriers test.

The methodological requirement of proving additionality is a one-time assessment at the beginning of each project crediting period. For projects with no harvest and hence no additional project income (for example, conservation, reforestation with mixed native species), the financial test will be passed automatically. For other project types, other barriers may need to be examined, such as the availability of financing and the local existence of the necessary skills to undertake the activity.

The second baseline element is eligibility. Under the CDM, projects are eligible only if the project area meets the criteria of lacking forest (according to the host country's definition) on December 31, 1989, and of not having been reforested and deforested in the intervening time. The main purpose of the eligibility criteria is to ensure that carbon finance does not create an incentive for deforestation, so that forests are not cut down in order to provide a site for subsequent reforestation.

The methods for proving eligibility are governed by the mandatory eligibility tool developed by the afforestation-reforestation working group of the CDM for CDM projects.[6] The methods involve a one-time, pre-project analysis of land cover in 1989 and a more recent year. This may involve government maps and documents or, perhaps most often, an analysis of satellite images combined with on-the-ground information.

The chosen approved methodology guides the selection of baseline land use, the third element. For most projects this will be the land use that is prevalent immediately before the start of project activities. In the situation where multiple land uses exist in the baseline condition, these lands should be stratified in the project.

Finally, the total baseline carbon stock is equal to the project area multiplied by the carbon stock per unit area (usually per hectare). Within all baseline strata, the carbon stocks per unit area need to be assessed to a high level of precision (usually to within 10 percent of the mean with 95 percent confidence). Alternatively, a reduced level of sampling could be used with the conservative approach of applying a baseline stock equivalent to the mean plus the 95 percent confidence interval. This is conservative because using a high baseline value gives a lower project benefit that equals the project carbon stocks minus the baseline carbon stocks. Using the mean plus the confidence interval gives a high and conservative baseline.

For the evaluation of baseline carbon stocks, the same considerations apply as for methods for monitoring project carbon stocks, including use of standard methods for measurement and evaluation of required sampling intensity,[7] consideration of whether to use project-specific or region-specific, national, or international data and allometric equations, and balancing the financial cost of high precision against the cost of lost carbon from taking a conservative value.

In rare cases, such as where baseline carbon stocks are increasing over time at an unknown rate, it could be necessary to monitor the baseline carbon stocks.[8] This would be achieved through proxy plots in sites outside the project boundary but identical in all other ways to the project area. If possible this should be avoided, because of the doubled costs of monitoring both the project and the baseline simultaneously to similar levels of accuracy and precision and because of questions about the representativeness of the proxy sites. That is, if the sites are under project control, then are they really independent proxies? And if they are not under project control, then will they be lost at some time during the project's lifetime?

Monitoring Leakage

Leakage, whereby emissions are increased outside the project boundary as a direct result of project activities, is a primary consideration for all carbon projects. Consideration of leakage should be among the initial steps of any project design. The aim should be to avoid leakage rather than merely to track it. Leakage can be avoided if it is anticipated and if programs are put in place to provide, for example, alternative livelihoods for displaced peoples and fuelwood plantations to replace fuel taken from within project boundaries. Whenever possible, project developers should identify possible sources and potential magnitudes of leakage in the design phase and plan the project so that leakage can be minimized.[9]

The methods for leakage evaluation, based on project type, are given within the CDM AR approved methodologies. Here we provide details on the types of leakage, when each type should be considered, and considerations for tracking leakage. For afforestation and reforestation projects, the most important forms of leakage are considered to be vehicle use and activity shifting.

It is likely that all projects will have leakage through vehicle use outside the project boundaries. Even though this will often be a small emission, the requirement exists for tracking it under the CDM. Typically this will require tracking of kilometers traveled, fuel used for different classes of vehicles, or both. A system needs to be devised to facilitate tracking in a transparent, cost-effective, and verifiable manner.

If project land is being used in any way before the start of activities, then activity displacement may occur as an effect of project implementation. The activity displaced could include grazing, growing crops, collecting firewood, or employment to carry out these activities. The leakage component would be the potential for the displaced users to shift their activities elsewhere, thus leading to devegetation or deforestation and to associated carbon loss outside the project boundaries.

For tracking activity displacement, a sampling scheme is needed that selects the sample community members, tracks their activities and movements, and records and analyzes the results from interviews. In addition, carbon sampling (with considerations identical to those for monitoring project carbon stocks) is required to evaluate the carbon stocks of areas actually or potentially decreasing in stocks due to the activity shifting.

To reduce the threat of activity shifting in a project, the developer could make sure that "alternative livelihood" measures are initiated, so that any activities originally taking place in the project area are replaced by other livelihood activities that do not require additional forested land. Examples of such measures include improved farming techniques that increase crop yields and income on existing "outside the project boundary" land.

Requiring leakage to be monitored creates a vested interest in the project to ensure that alternative livelihood strategies are adopted and maintained. If strategies initially attempted are unsuccessful, then the project must work with the community to find alternative strategies, or it will risk reductions in carbon credits via leakage.

Beyond vehicle use and activity shifting, other forms of leakage could exist. One is increased use of wood posts for fencing. A second is product production within the project area that leads to GHG emissions outside the project area. An example is forage production within the project boundary that increases the number of livestock and hence livestock emissions outside the project boundary. A third form of leakage is due to market effects. It occurs, for example, where timber produced by a project leads to a distortion in the market or where a reduction in crop production leads to a price increase and causes clearing elsewhere to fill the perceived gap in supply. This could be an important form of leakage for very large forest management activities. No currently approved methodology includes this form of leakage, and estimation methods could involve complex economics.[10]

A final form of leakage takes place as a result of the "crowding in" or "crowding out" of investment impacting carbon stocks outside the boundaries of the CDM project area. No methodology exists to account for this type of leakage.

Estimating Project Emissions

The final methodological consideration for project planners is the need to account for emissions that occur during project implementation. These emissions will decrease the project's total carbon benefits. Such emissions are emissions from site preparation, emissions from fires, emissions due to fertilizer use, especially in the form of nitrous oxide, emissions arising from wetlands or irrigation, and emissions from project vehicle and machinery use within the project boundary.

The methods for assessing such emissions are given within the CDM-approved methodologies. If no wetlands or irrigation exists within the project area and if no fertilizing is planned, then these forms of emissions can be excluded at the outset of the project. Otherwise, accounting for all emissions must occur, even if each is likely to be small.

In the project implementation documentation, systems for verifiably tracking the use of fertilizer, the distances traveled by project vehicles, and the fuel purchased for project vehicles and machinery are needed. Emissions from burning fuel are complicated by the fact that emissions within the project boundaries are considered project emissions, but emissions outside the boundary are considered leakage. Therefore the system used to track these fossil fuel emissions needs to be designed so that it can partition fuel burned by vehicles and machinery inside and outside the project boundaries.

Minimizing Errors

Errors can be divided into two categories: those that reduce accuracy and those that reduce precision. A good carbon project will minimize both types of errors. Here we introduce four sources of error and discuss whether each will affect accuracy, precision, or both and what steps can be taken to minimize the error.

The most obvious error in implementing a carbon project is choosing poor methods to assess project carbon benefits. However, many of these methods are excluded through the validation process. For example, the application of approved methodologies excludes the use of poor or inappropriate growth models or default tables.

Beyond the initial validation level, closely following the approved methodology and the more detailed guidance in *Good Practice Guidance for Land Use, Land-Use Change and Forestry* and *Sourcebook for Land Use, Land-Use Change and Forestry Projects* will ensure that appropriately designed and tested statistical methods will be used that will give accurate results to known levels of precision.

The second source of error is poorly applied methodology. Even with excellent methods, inaccurate measurements will result if they are collected incorrectly or out of context. For example, if the field crew does not fully understand which trees to include and exclude from plots, then errors will multiply and lead to systematic inaccuracy. Equally, poor data entry into computer software programs used for calculations could lead either to systematic errors in the results and hence inaccuracy or to greater variation and hence lower precision.

Inaccuracy due to poor application of methods is avoided through the design and firm implementation of a quality assurance and quality control (QA-QC) plan. A QA-QC plan defines the way project personnel should be trained and the way checks should be conducted in the field, in the laboratory, and in data entry to detect and correct errors. At the most basic level, a key step is to ensure full training of all personnel using a set of standard operating procedures that will be followed at all times. Random and systematic checks of methods used, results collected, and data entered will ensure consistent high standards.

As mentioned earlier, the use of inappropriate regression equations—for example, for calculating biomass from tree diameter—can create significant systematic errors. For carbon pools and emission sources that represent a very small proportion of the project sequestration (perhaps a couple of percentage points), the use of unverified equations and defaults, recommended in the Tier 1 level of the IPCC guidance, and other national or international equations is appropriate. For more significant steps in the estimation procedure, such as the use of regression equations linking tree diameter with biomass or factors linking stem biomass with total tree biomass, project data (such as limited harvests) should be used to check for systematic biases.

Finally, the greatest cause of a lack of precision in a project's reported results is insufficient sampling. If the 95 percent confidence intervals around the estimated mean greatly exceed 10 percent of the mean, then the sampling plan was faulty. Either more plots were needed to capture the variation more completely or additional stratification should have been done to partition the variation further.

The collection of preliminary, pre-project data is critical. These data will give project developers an indication of the necessary stratification and the number of plots needed to achieve the desired precision. This stated, a precision within ±10 percent of the mean with 95 percent confidence is unnecessary for each carbon pool individually. High variation can be expected in deadwood stocks and understory vegetation, for example, but these pools and their variation will form a small component of total stocks. The important end result is for *total* variation to be low enough that the targeted precision is achieved.

Conclusion

Methodological approaches exist that can track the carbon effects of land-use activities with a high degree of precision in a manner that is real, transparent, and conservative. The methodological approaches are divided among four elements: monitoring project carbon stocks, establishing the baseline, monitoring leakage, and estimating project emissions. With appropriate planning and the following of published methods and methodological advice, each can be tracked with a high degree of accuracy and precision.

Notes

1. T. Pearson, S. Walker, and S. Brown, *Sourcebook for Land Use, Land-Use Change and Forestry Projects* (Washington: World Bank BioCarbon Fund and Winrock International, 2005); S. Brown and O. Masera, eds., "LULUCF Projects," in *Good Practice Guidance for Land Use, Land-Use Change and Forestry,* edited by J. Penman and others (Kanagawa, Japan: Institute for Global Environmental Strategies for Intergovernmental Panel on Climate Change, 2003), pp. 4.89–1.120.

2. The 95 percent confidence bounds are a calculation, based on standard statistical methods, of the range in which the true value is likely to lie.

3. Tools exist for assisting project developers in determining the number of samples required to receive a certain level of certainty and in cataloging the potential costs of field measurements. See Pearson, Walker, and Brown, *Sourcebook for Land Use.*

4. For discussion of and guidance on these elements, see T. Pearson, S. Walker, and S. Brown, "Afforestation and Reforestation under the Clean Development Mechanism: Project Formulation Manual" (Yokohama, Japan: International Tropical Timber Organization, 2006).

5. UNFCCC, "Tool for the demonstration and assessment of additionality for afforestation and reforestation CDM project activities," 2007 (http://cdm.unfccc.int/methodologies/ARmethodologies/AdditionalityTools/Additionality_tool.pdf).

6. UNFCCC, "Procedures to demonstrate the eligibility of lands for afforestation and reforestation project activities," 2007 (http://cdm.unfccc.int/EB/035/eb35_repan18.pdf).

7. Pearson, Walker, and Brown, *Sourcebook for Land Use*; Brown and Masera, "LULUCF Projects."

8. An example is where natural succession may be taking place in the baseline.

9. L. Aukland, P. Moura-Costa, and S. Brown, "A Conceptual Framework and Its Application for Addressing Leakage on Avoided Deforestation Projects," *Climate Policy* 3 (2003): 123–36.

10. B. Sohngen and S. Brown, "Measuring Leakage from Carbon Projects in Open Economies: A Stop Timber Harvesting Project in Bolivia as a Case Study," *Canadian Journal of Forest Research* 34 (2004): 829–39.

11

Characterizing Sequestration Rights Legally in Chile

DOMINIQUE HERVÉ AND EDMUNDO CLARO

Forestry is an important economic sector in Chile. Tree plantations currently make up 14 percent of forest cover. Although forests are harvested primarily for industrial purposes and firewood, they are also used for animal grazing and are protected for conservation reasons. Since the 1990s the forestry sector's contribution to GDP has been growing; in 2003 that contribution was 3.4 percent. In economic terms, only the mining and industrial sectors contributed more than the forestry sector.

In view of this forestry-oriented economy, we analyze in this chapter the legal implications of the development of afforestation and reforestation (AR) projects in Chile in the context of the Clean Development Mechanism (CDM) of the Kyoto Protocol. Specifically, our aim is to determine the legal and economic scenario for developers of and investors in these kinds of projects in Chile. First we explain the basic legal and institutional issues surrounding forestry, land, and the environment. Next we analyze the legal nature of Certified Emission Reductions (CERs) under Chilean legislation and the implementation of the CDM in the country. We then identify the legal requirements for CDM AR projects in Chile, describe existing and projected activities in this area, and draw some conclusions.

This chapter is based on previous work done for IUCN and founded by GTZ and UNEP.

Institutional and Regulatory Background in Chile

The main government institutions involved directly in the forestry sector in Chile are the National Forestry Corporation and the Forestry Institute. The National Forestry Corporation, or CONAF (Corporación Nacional Forestal), is a private legal entity subordinate to the Ministry of Agriculture (MINAGRI). Its mission is to contribute to the conservation, enhancement, management, and good use of the country's forestry resources and to collaborate with the administration in developing state-protected wildlife areas, which consist of national parks, forest reserves, and state-owned forests.

The Forestry Institute, or INFOR (Instituto Forestal), is a private corporation subordinate to the Ministry of Economics (MINECON). Its purpose is to serve as an entity with technical personnel specialized in forestry matters who can collaborate in developing the forestry sector in every aspect. INFOR's mission is to research and generate information about the sustainable use of forestry resources and ecosystems, their derived products and services, and the transfer of those products and services to public and private entities, contributing in this manner to the economic, social, and environmental development of the country.

The term *forest land* is not used explicitly in Chile. Instead, the closest expression corresponds to "land preferentially suitable for forestry." This is defined as "all the land that, due to climate and soil conditions, should not be plowed permanently, independent of whether it is covered by vegetation or not, excluding land that might be used for intensive agriculture, fruit growing, or cattle ranching without suffering degradation."[1] A special Decree Law passed in 1974 (DL 701, "Statute of Forestry Promotion"), promoting forestry activity in Chile, defines a forest as "a piece of land with vegetable formations in which trees predominate and that extend for at least 5,000 square meters, with a minimum width of 40 meters, with a tree top greater than 10 percent in arid and semiarid areas, or greater than 25 percent in more favorable conditions."

Several pieces of legislation regulate aspects of the forestry sector in Chile such as the conservation and preservation of forests, the promotion and development of forestry activities, and environmental requirements. Only two are of major practical significance: the Forest Law of 1931 (last modified by Law 18.959 in 1990), which established regulations for forest use and protection, and DL 701 (last modified by Law 19.561 in 1998), with its various regulatory bylaws, which establish the bases for large-scale private investment in exotic plantation forestry and regulate the management of native forests (technical bylaw 259, 1980). Because of pressure on native forests coming from plantations, substitution, and other activities, a Native Forest Law was presented for discussion in 1992. Since then, attempts have been made to pass a law whose aim would be to encourage the conservation and sustainable use of native forests. In August 2007 the Native

Forest Law was approved by the Chilean Senate, and by the beginning of 2008 it had met all requirements to be enacted. The absence of this law has been seen by many as one of Chile's major deficits in terms of environmental management.

The general rule in Chile is to allow the development of AR activities without requiring any permit or management plan. Nevertheless, in practice most AR projects go through the voluntary incentive program established by DL 701, which acts as a central permitting system for this activity. DL 701 regulates some of the most important forestry activities: management, cutting, exploitation, and AR. Specifically the statute was designed to regulate forest activities on "land preferentially suitable for forestry" and on "degraded land," promoting such activities especially on the part of small landowners.[2] It also attempts to protect, prevent the degradation of, and recover forest land within the national territory. AR activities executed in conformity with DL 701 receive important subsidies.

This incentive mechanism establishes a differentiated system of bonuses and tax exemptions for forestry activities developed on certain land. The bonuses consist of the payment of a percentage (75 to 90 percent) of the total costs of these activities by the government. The tax exemptions apply mainly to land tax (applicable to land with plantations benefiting from this subsidy) and income tax (applicable to income generated by the sale of forestry products).

In order for AR activities to be eligible under DL 701, a CONAF-approved management plan is required. The qualification of certain land as "preferentially suitable for forestry" is the basic condition required to benefit from the subsidy scheme defined by the law. The owner of the land must request this qualification from CONAF, which approves the submission in order for it to be considered under the Statute for Forestry Promotion. The land qualification from CONAF creates rights and duties for the owner. The rights include tax exemptions and the possibility of obtaining support for the development of AR activities. The support consists of the government's payment of a percentage of the total costs of the activities. The corresponding duties are to submit a management plan to CONAF and to comply with it. Noncompliance may give rise to sanctions.

Land

In Chile, the Ministry for Housing and Urban Development (MINVU), MINAGRI, the Ministry of National Assets, and the regional governments and municipalities deal with the use, administration, and management of land. MINVU, together with regional and local governments, is in charge of the planning and regulation of land use, mainly regarding land in urban zones. Regarding non-urban land, including agricultural land and land suitable for forestry activities, power is vested in MINAGRI, together with MINVU. The Ministry of National Assets administers all the public land of the country.

The legal basis for the system of landownership in Chile is established by article 19, number 24, of the Constitution, which includes "the right of property" as one of the fundamental rights of every person. Besides the Constitution, the Civil Code regulates property rights in their different variations. A property right is defined as a "real right" (as opposed to a "personal right") over a corporeal thing, which can be used and disposed of arbitrarily as long as it is not against the law or the right of any other person.[3] There are three inherent attributes of this right: to use, to benefit from, and to dispose of property.

The owner of land also has the right to own the fruits of the land through "accession." Under this right the owner of a piece of land is also the owner of the forests on the land. Forests belong to the category of "immovable goods," or real property, because they cannot be moved from one place to another. They can become "movable goods," however, when they are harvested. Accordingly, the owner of the land can separate it from its forest and sell the right to the forest independent of rights to the land.[4]

Considering the aforementioned regulations, the owner of the land could also be considered the owner, through accession, of CERs that might be issued as a consequence of emission removals and reductions produced on the land. They would be considered "civil fruits" as opposed to "natural fruits," which are given by nature (Civil Code, articles 643–48). If the landowner has transferred the usufruct right over something that might generate emission removals (as in the case of forests), then the CERs would be owned by the usufruct holder. Under these conditions, CDM AR investors would have to negotiate the allocation and assignment of CERs with forest owners, who might or might not be the landowners.

Chilean legislation establishes a system of preexisting rights over the land, represented mainly by the rights of indigenous people, long-time occupants, and users of mining property. Law 19.253 of 1993 regulates indigenous people's rights, including the legal status of indigenous land. This land includes both historically occupied land and land currently owned by indigenous people. Indigenous land is protected by legislation and cannot be sold, seized, mortgaged, or acquired by prescription. The only exception is when the transaction is between indigenous communities or between persons of the same ethnic group. Moreover, land owned by indigenous communities cannot be rented or given to third parties for their use or administration.

Chilean legislation also gives long-time occupants of small parcels of land the opportunity to acquire property rights to that land. Such people are beneficiaries of a special statute (DL 2695, 1979) that aims to regulate the possession of small parcels of land when there is no legal title by allowing acquisition of the property through "prescription."[5] The statute allows the responsible authority to register such land in favor of its irregular possessors, giving them the ability to acquire land through a short prescription of one year.

In Chile, mining activities are considered under a special law, the Mining Code, which is based on the principle of freedom and protection of such activities. The Constitution and the Mining Code establish that the state has absolute, exclusive, inalienable, and imprescriptible ownership of all mines, including those mining natural guano deposits, metal-bearing sands, salt deposits, coal and hydrocarbon deposits and fields, and other fossil substances except surface clays, regardless of the property rights of natural or legal individuals over land on which these resources may be found.[6] Consequently, CDM AR investors will have to consider the possibility that mining activities might take place on land where their AR projects are being developed.

Chilean legislation allows someone other than the owner of the land to undertake AR activities. It is important to distinguish between two cases, those of state-owned land and privately owned land.

Chilean legislation allows the use of public land by private persons. Decree Law 1939 establishes the applicable rules for the management of the state's property and considers, besides the transfer of the property, four legal concepts that can be relevant for AR projects: concessions, rents, destinations, and alienation. The Ministry of National Assets can grant concessions of public properties to Chilean legal persons for predetermined objectives. The concessions are allocated through a national or international, public or private bid or, in certain cases, directly.[7] Regarding rents, private persons can use and benefit from state property through lease agreements. The maximum term to lease state property is five years for urban and ten years for rural property, except in the case of educational institutions or other legal persons that assign the property to national or regional interest objectives, in which cases the term can be up to twenty years.

The concept of destination applies when a public institution requires the use of state property in order to execute its functions. The Ministry of National Assets assigns the property to an institution, which must use it for the specified purpose. In the case of AR projects this concept could be applicable if the project developer is a public institution and the development of the project is part of its function. Finally, state property that is not considered essential for the fulfillment of the state's objectives can be subject to alienation. Such property can only be sold, except when the transfer is in favor of municipalities (local towns), public institutions, or nonprofit private entities.

For privately owned land, Chilean civil law considers several legal rights, different from property rights, that can be applied in order to develop AR projects. These are usufruct (or "enjoyment") and leasing or renting.

Usufruct is a real right or entitlement to enjoy a thing with the obligation to conserve its original form and substance and return it to the owner at the end. This right enables one to enjoy natural and civil fruits of the thing. Its maximum term is until the death of the person who is entitled to enjoy the right or thirty years if the right holder is a legal person. Therefore, this right is not subject to succession. It can be created by contract or will. If it relies on real property (land), then it includes the right to enjoy forests, with the only obligation being to conserve their substance.[8]

A lease is a contract in which one person is obliged to grant the use of a thing and the other is obliged to pay a price for this use. Chilean legislation distinguishes between the renting of urban and rural land. For rural land there are special requirements such as the obligation to protect and conserve existing natural resources.[9]

We can conclude, then, that Chilean legislation regarding land property allows the implementation of AR projects by the owner of the land and by third parties through other rights over the land, different from ownership, on both private and public land.

The Environment

The environmental authority in Chile is CONAMA, created in 1994 by Law 19.300, or the Environmental Framework Law (EFL; Ley de Bases Generales del Medio Ambiente). CONAMA is a functionally decentralized public authority, originally accountable to the ministry of the Presidential General Secretariat. In March 2007 a legal reform established that the highest political authority of CONAMA is the minister of the environment, who is now the president of the Board of Ministers (the highest body within CONAMA) and is accountable directly to the president of the republic.[10]

Besides the minister of the environment, the administration of CONAMA lies with the executive director, who is also appointed by the president. CONAMA has a regional director in each region of the country who provides technical support for the corresponding regional political entities, the COREMAs.

The EFL includes several provisions that apply to the forestry sector. The most relevant for the purpose of this chapter are provisions that establish the obligation to undertake environmental impact assessments (EIAs) for certain projects that might cause "environmental impact"—the so-called Environmental Impact Assessment System (EIAS).[11]

The EIAS is the procedure used to determine whether the environmental effects of an activity or project are consistent with the applicable legislation. Projects belonging to categories listed in the EFL can be executed only if they have been previously environmentally assessed, by the submission of either an environmental impact study (EIS) or an environmental impact declaration (EID), and afterward approved.[12] Because an EIS is reserved for projects of greater magnitude or that have greater potential to adversely affect the environment or any of its components, the general rule is to submit an EID.[13]

CERs under Chilean Legislation

CERs are not defined under Chilean law. Nevertheless, important precedents exist or are under consideration in Chilean legislation regarding rights with similar characteristics. The first is Supreme Decree 4, 1992, from the Ministry of Health, which establishes air emission standards for particulate material and is applicable to stationary sources in the Metropolitan Region. This regulation created an offset program based on agreements between the sources in which one of them obliges itself to reduce its level of emissions by the same amount that the other increases it. This system is managed by the regional health authority, which is in charge of authorizing the agreements and keeping a registry of the sources and their emissions.

The second precedent, a system of tradable emission permits, is not yet in force, but a bill establishing such a system has been under discussion in the National Congress since 2003. The bill, called the Tradable Emission Permits Law (Ley de Bonos de Descontaminación), would create a legal framework for the development of a new market in which the reduction of emissions of certain pollutants could be traded. Under this mechanism the government would be able to allocate obligatory emission quotas to different sources of emissions in a determinate air (and eventually water) basin.[14] These sources would be able to generate emission credits if, at the end of the year, their emissions fell below their allocated quotas, and they would be able to sell credits to sources whose emissions have surpassed their quotas. The system would recognize the right to these reductions, although it would not determine the legal nature of that right.[15] These credits are similar to CERs, their major difference being the scale of application: whereas for CERs the basin corresponds to the atmosphere and it is possible to trade and use them internationally, for Chilean credits the basin is locally demarcated and trades and uses are local as well. Although no plans exist to define a special right to CERs through national legislation in Chile, we believe that legal definitions contained in this bill of law are antecedents that permit interpreting the legal status of CERs in Chile.

Therefore, the right to a CER under existing law arguably may be characterized as a private property right that is enforceable by its holder against all parties and is exercised over an intangible and movable good, namely, the credit for the emission reduction or carbon removal.[16] Therefore, the right to a CER could be a "real right" (it can be exercised *erga omnes*) over an incorporate and movable good (the credit) that gives its holder the right to use it, enjoy it (receive its benefits), and dispose of it (sell it). If this right were sufficiently and precisely defined by law, then contracts would have to respect the requirements established by the corresponding law. Under Chilean private law, parties can agree on concepts different from the existing legal categories of rights if and only if those concepts do not oppose the law, the public order, and good custom. Therefore it would be possible for the parties to revise or modify this right as long as they do not oppose these concepts.

Legal Requirements for CDM AR Projects in Chile

Developers of AR projects that involve the execution of activities in national parks, national reserves, natural monuments, wildlife reserves, nature sanctuaries, marine parks, marine reserves, or any other areas under official protection must submit their project proposals to the EIAS and wait for a positive Environmental Qualification Resolution before they can proceed.[17] If an AR project does not fall into any of the foregoing categories, it may be executed without further requirements. If the developers of an AR project wish to receive the benefits of DL 701, the project can be implemented only on land classified as "preferentially suitable for forestry" and "forestable," as defined earlier.

Also, because most AR projects benefit from subsidies established under DL 701, they usually have to comply with the law's requirements, which are directed toward recovering degraded land and helping small farmers. Specifically, in order for AR projects to qualify for DL 701 incentives, the land must meet one of the following conditions: it must be fragile, be in a process of desertification, be degraded with the potential of being eroded by wind, or be any kind of land "preferentially suitable for forestry" in the hands of small farmers.

All AR projects in Chile, as long as they submit to the system of DL 701, require the preparation of a "Forestry Management Plan" and its approval by CONAF. These plans are aimed at maximizing the benefits of using the natural resources provided by a piece of land, making sure that these resources and their ecosystems are protected. The plans basically consist of detailed information about the management of the AR project, along with a basic and limited description of how the environment will be protected. There are no guidelines for an environmental impact analysis, and there are no social elements in a Forestry Management Plan.

The Designated National Authority (DNA) for Chile—the entity that grants approval for AR projects under the Kyoto Protocol's CDM—is located in the Board of Ministers (the highest-ranking body) within CONAMA. The board created a steering committee specifically for this purpose, and it exercises all CDM functions. The steering committee includes representatives of the Ministry of Foreign Affairs, MINAGRI, the National Energy Commission, and the National Council for Clean Production and, if needed, a representative of a ministry with specific powers over a particular project. The DNA is concerned only with the sustainability and voluntary character of projects, limiting its review to aspects mandated by the Kyoto Protocol and decisions adopted under the protocol. It does not create additional political or legal institutions.

If a project has to go through the EIAS, its procedure includes public participation. There are two different kinds of requirements for public participation in the EIAS. Projects submitting an EIS invite comments from citizen organizations and individuals for a period of sixty days. For projects submitting an EID, the EFL requires the publication of a list of the projects or activities subject to EIDs in a periodical with regional or nationwide circulation and in the corresponding municipalities, to keep the public duly informed. In this case there is no opportunity for public comment.

If projects do not go through the EIAS, they need comply only with the sector-specific permits that are legally required. In these cases there is usually no public participation. If an AR project registers for DL 701 benefits, it needs the corresponding Forestry Management Plan, which is given by CONAF and whose procedure does not include public comment. This means that project developers who are interested in participating in the CDM and do not have to go through the EIAS have to comply with Kyoto Protocol public comment requirements through procedures that must to be determined by the project developer on a case-by-case basis.

Overview of CDM AR Projects in Chile

Before the Forest Law was passed in 1931, forest plantation rates were approximately 150 hectares per year in Chile. From 1931 to 1974, when DL 701 was introduced, the annual plantation rate was 16,000 hectares. As a result of the incentives provided by DL 701, annual plantation rates since 1974 have averaged 90,000 hectares.[18]

Between 1994 and 2004, approximately 560,000 hectares were afforested and 540,000 hectares reforested in Chile, giving an AR annual average of 100,000 hectares (figure 11-1). Forty-seven percent of the AR land was planted in radiata pine, and 40 percent in varieties of eucalyptus. It can be expected that in the coming years AR projects will result in the planting of approximately 100,000 hectares annually, mostly in radiata pine and eucalyptus.

On July 4, 2007, CONAMA issued the first letter of approval for an AR project under the CDM. The project was developed by the Canadian firm Mikro-Tek and aims to increase carbon sequestration in forestry projects through the inoculation of soils with naturally occurring fungi called mycorrhizae. A number of additional projects have been presented as suitable for obtaining CERs under the CDM scheme (table 11-1).

Despite the increasing interest in AR projects in Chile, there are two basic obstacles to their development. One is methodological uncertainty concerning plantation dynamics, including growth rates, environmental effects, and fire risks. The other is the issue of additionality. Because Chile has been afforesting and

Figure 11-1. *Afforestation and Reforestation in Chile, 1994–2004*

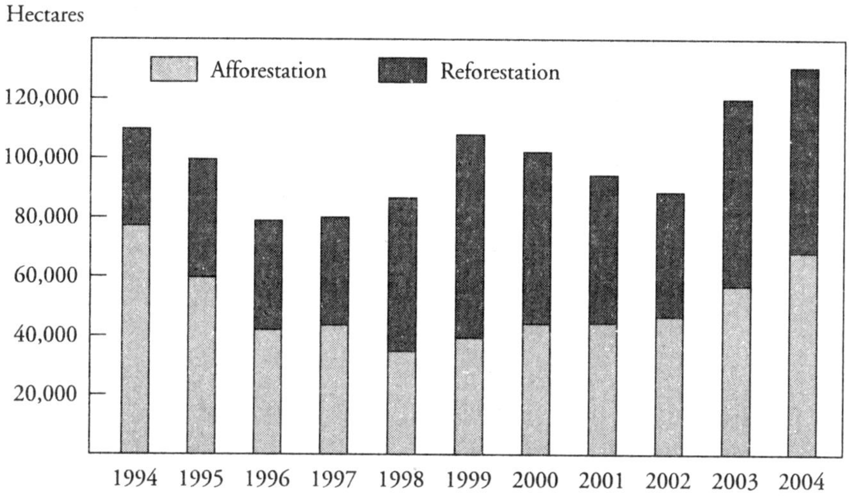

Source: www.infor.cl.

reforesting approximately 100,000 hectares annually for the last fifteen years, it may be difficult to prove that CDM AR projects would not have happened in the absence of the CDM. Considering these obstacles, and stressing that the incentive for forestation backed by the legal instrument DL 701 is strong, we think it probable that the effect of the carbon market on forestry rates in Chile will not be considerable and will not modify the structure of the activity in a significant way.

Conclusion

It is unlikely that the CDM mechanism will have a significant effect on AR practices in Chile. Since 1974 a forestation subsidy has existed that can cover up to 90 percent of forestation costs and eliminates territorial taxes. Nevertheless, in areas where the national incentive has been insufficient for forestry projects to develop, the CDM could have a positive effect. It could provide an incentive for the reforestation of vast expanses of land in Chilean Patagonia that were burned by colonizers during the early twentieth century and have not been reforested to date.[19] If guided well by Chilean forestry authorities, the CDM might also act as a pivot for the establishment of native forest plantations and plantations based on forestry systems other than monoculture. Thus, although we think the CDM will not have a significant effect in Chile concerning AR practices, we are of the opinion that the existing incentive system will not impede project participants from demonstrating additionality if the case is well articulated.

Table 11-1. *CDM AR Projects in Chile*

Project name	*Project type*	*Project status*	*Stakeholder*	*Area (ha)*	*Species*	*Expected t CO_2e*	*Source of information*
Sociedad Inversora Forestal	Afforestation	Planted between 2003 and 2005	Fundación Chile	3,063	Radiata pine, *Eucalyptus globulus*	1.1 million up to 2017	www.prochile.cl/servicios/medioambiente/total_pin_ chilenos_2006.pdf
Bundle of small-scale projects relying on mycorrhizae inoculation	Reforestation	Under implementation	Mikro-Tek	2,400	Radiata pine, eucalyptus, *Quillaja saponaria* (a native species)	1.2 million over 20 years	www.mikro-tek.com
Forestación Campesina	Afforestation	Started planting in 2007	INFOR	6,000	Ponderosa pine	7 million in 24 years	www.infor.cl/web_proyectos/web_seminario/documentos/servicios_ambientales/Paulo_Moreno.pdf
Afforestation of degraded land in Aysén, Chile	Afforestation	Expected to start planting in 2008	Forestal Mininco	10,000	Ponderosa pine	100,000 per year	www.prochile.cl/servicios/medioambiente/ideas_proyecto/forestation_01mininco.php

Sources: Sociedad Inversora Forestal: www.prochile.cl/servicios/medioambiente/total_pin_chilenos_2006.pdf. Small-scale projects: www.mikro-tek.com. Forestación Campesina: www.infor.cl/web_proyectos/web_seminario/documentos/servicios_ambientales/Paulo_Moreno.pdf. Afforestation of degraded land in Chile: www.prochile.cl/servicios/medioambiente/ideas_proyecto/forestation_01mininco.php.

Chilean contract law allows parties to determine freely the characteristics of the rights they transact. Because there is no legal definition of the right to a CER, there are no limits to the will of the parties except what the Civil Code calls "respect of the law, public order, and good customs."[20] This has permitted the development of CDM contracts in other areas, such as energy and waste. We think the absence of a legal definition in Chilean law poses no problem for the materialization of CDM contracts in the area of AR projects. In any case, the eventual contractual conflicts would be similar to conflicts over other types of contracts, without additional complications.

The owner of the land in Chile is the owner of its products, including forests, and may separate forests and their products and services from the land and sell them independently. Thus an AR project proponent in Chile who intends to produce CERs will have to own the project forest or receive the rights to the CERs from forest owners. In cases where land titles are not always regularized, especially in the cases of small and medium-size landowners, it seems appropriate to recommend that potential AR investors in Chile, as a matter of standard due diligence practice, look for land with its property documentation in order and updated. At the same time, it seems reasonable not to locate CDM AR projects on indigenous land, because it cannot be sold to persons of a different ethnic group or used or administered by third parties. It is also possible that CDM AR projects in Chile will have to resolve problems related to people's using or living on project land when those people are not the owners. These issues will have to be resolved case by case.

With the exception of AR projects involving the replacement of native forest or taking place in areas under official protection, which are rare and have to be submitted to the EIAS, AR practices face no environmental legal barriers or restrictions in Chile.

Chilean mining law allows the granting of mining concessions for exploration or development on any land regardless of property rights over surface tenements. Nevertheless, it limits the concessionaire's rights in order to prevent damage to the owner of the land. In the case of forests these requirements imply authorizations, reports, and licenses analogous to those required to develop mining activities in protected areas, such as national parks, national reserves, and national monuments. There is no way for a CDM AR project to avoid the possibility of a mining activity's being established on land destined for the AR project. Again, due diligence would aim to identify potential conflicts between mining interests and potential forestry land at an early stage.

Chilean law tacitly assumes that AR projects are good, and so AR regulation is framed basically as an incentive to promote the activity. Although in 1998 legislation was modified to focus on small and medium-size property owners, big projects can still benefit from incentives if they are developed on degraded land. If the developers of a CDM AR project want to register for DL 701 benefits, they must submit a management plan to CONAF and have it approved before developing the AR activities.

AR activity in Chile is dynamic and has a long history. This is evidenced by the straightforwardness of the enabling legislation and by a regulatory framework that subsidizes AR practices. Thus we think there is little space or opportunity for the implementation of new legal instruments. The only major new piece of forestry legislation is the recently approved Native Forest Law, which will address the management and conservation of native forests but will not interfere with AR activities.

Two issues need to be addressed regarding CDM processes and current AR procedures in Chile. The first has to do with the requirement for public comment under the Kyoto Protocol and its probable absence for most AR projects. This might be resolved by the project developer's adopting certain requirements similar to those established for the EIAS. Similarly, the second issue is related to the requirement for socioeconomic impact assessments under the Kyoto Protocol and its potential nonexistence for most AR projects in Chile. Again this gap might be bridged by the project developer, who might, for example, follow the Chilean national certification scheme, CERTFOR, in its principles for dealing with social issues.

Notes

1. The legal act that provides this definition is the Forest Law of 1931 (Supreme Decree 4363).

2. Article 2 of DL 701 defines degraded land as "those dry lands, and those lands classified as Irrigation Class IV by the Tax Authority [Servicio de Impuestos Internos] when determining their agricultural value, ranging from moderate to severe erosion and that are susceptible of being repaired through soil conservation practices or activities."

3. The advantages of a real right over a personal right come from its absolute character, which means that it can be exercised *erga omnes*, against all third parties or "the whole world." In contrast, a personal right can be exercised only against the debtor.

4. A legal problem might arise if the landowner wishes to sell the forest independent of the land for a purpose other than harvesting. The Civil Code considers that the forest is adhered to the land, so the two are a single entity. They may be separated only for purposes of selling the wood of the forest. A legal option for a landowner who wants to transfer the right to the forest but not to the land would be to constitute the right of usufruct over the forest. Nevertheless, this option would not transfer the right of ownership of the forest.

5. "Prescription for acquisition of property" is explained by article 2498 of the Civil Code, which establishes that it is possible to acquire property rights (and other real rights) in "immovable or movable goods" when they have been legally or regularly possessed.

6. Prescription can also imply the extinction of a right when it has not been exercised during a certain period of time (Civil Code, article 2492).

7. Because the Chilean administrative legal system is now based strongly on principles such as publicity, transparency, and access to information, the general rule is that bids are public. Exceptionally, they will be private. When the concession is allocated directly, the ministry will have to justify this decision.

8. The right can also include the possibility of planting new forests on the land, as long as this conserves the "original form and substance" of the land.

9. The main difference between usufruct and rent is that the former is a real right (and therefore enforceable against all parties, or *erga omnes*) and the latter generates a personal right (enforceable only against the debtor). Furthermore, usufruct allows the use and enjoyment of the thing whereas renting allows only limited use of the thing that is rented.

10. The legal reform did not imply that CONAMA is now a Ministry of the Environment. Instead the new minister must elaborate a proposal to discuss the creation of such a ministry, determining its functions and attributes, a proposal that will have to be submitted to the Chilean Congress.

11. Article 2(k) of the EFL defines an "environmental impact" as "any alteration to the environment provoked directly or indirectly by a project or activity in a particular area."

12. Article 2(i) defines an environmental impact study as "the document that describes in detail the characteristics of a project or activity intended to be performed or its modification. It shall give reasoned background for the prediction, identification, and interpretation of its environmental impact and describe the action or actions to be executed to prevent or reduce its adverse negative effects." Article 2(f) defines an environmental impact declaration as "a descriptive document of an activity or project intended to be carried out, or the modifications that will be introduced to it, submitted under oath by the corresponding Head, whose content allows the competent organ to assess whether its environmental impacts conform to the current environmental legislation."

13. The obligation to submit an EIS is determined by article 11 of the EFL, which identifies the effects, characteristics, and circumstances of the project that give rise to this duty.

14. A determinate air basin corresponds to an area with generally similar meteorological and geographic conditions throughout, in which air quality is influenced by human and natural emission sources inside the area and in certain cases by the transport of pollutants from other air basins.

15. The two options argued, both nationally and internationally, as defining the legal nature of this kind of right are "property rights" and "administrative grants." It is also argued that these rights have characteristics of both categories. See M. J. Mace, "The Legal Nature of Emission Reductions and EU Allowances: Issues Addressed in an International Workshop," *Journal of European Environmental and Planning Law* 2 (2005): 123–34.

16. If the right to a CER is considered a property right, then it is enforceable even against the government. Indeed, this consequence is one reason environmentalists should consider that this kind of right might not best be considered a property right. Intangible or movable goods are those that consist of mere rights, such as credits and active easements (Civil Code, article 565). Movable goods are those that can be transported from one place to another, either by themselves, such as animals, or by an external force, such as inanimate things (Civil Code, article 567).

17. An "area under official protection" for the purpose of the EIAS is defined by article 2(a) of EFL bylaw SD 95. This definition determines that such areas are those that are

geographically delimited, that are established by a formal act of the corresponding authority, and whose objective is the conservation of biodiversity and environmental heritage or the preservation of nature. Indigenous lands are not included in this concept, because they have not been established for these purposes.

18. Corporación Chilena de la Madera (CORMA), "Sector forestal chileno: Una visión del sector privado," 2003 (www.corma.cl).

19. On this issue, see www.inviertaenaysen.com.

20. Besides, too, the definitions and conditions of the Kyoto Protocol, which has been ratified by Chile.

12

Legal Issues and Contractual Solutions for LULUCF Projects under the Clean Development Mechanism

MONIQUE MILLER, MARTIJN WILDER, AND ERIC KNIGHT

Over the past decade, substantial scientific work has been done to enable the relatively accurate modeling of and accounting for greenhouse gases sequestered by land use, land-use change, and forestry (LULUCF) activities. Recognizing that the sequestration of carbon by forestry activities provides a valuable environmental service toward mitigating climate change, organizations have developed voluntary regulatory schemes to assign values to emission reductions arising from LULUCF projects and enable trading in such "LULUCF credits." As an example, the Kyoto Protocol recognizes that Annex I Parties to the protocol can use emission reductions from carbon sequestered by certain domestic LULUCF activities when complying with their commitment to limit or reduce their emissions.[1] The Clean Development Mechanism (CDM) also allows Annex I Parties to procure and use emission reductions arising from certain LULUCF projects in developing, non–Annex I countries.

Although it is possible to generate and trade LULUCF credits under the Joint Implementation framework and even outside of Kyoto Protocol rules,[2] the focus of this chapter is the CDM framework. The CDM enables participants in eligible LULUCF projects to have certain LULUCF credits recognized and issued in the form of Certified Emission Reductions (CERs).

The CDM rules provide special treatment for CERs generated from LULUCF projects and enable the creation of two special types of CERs from reforestation, or the human-induced generation of forest on land that was clear of forest on January 1, 1990, and from afforestation, the human-induced generation of forest on land that has been devoid of forest for at least fifty years. The two types of CERs that can be created from LULUCF projects under the CDM are temporary CERs, commonly referred to as tCERs, and long-term CERs, commonly referred to as lCERs.

LULUCF credits are differentiated from other types of emission reductions in CDM rules because, under the current accounting rules of the Kyoto Protocol, carbon sequestered in a forest is deemed to be released into the atmosphere upon the destruction or removal of the forest. This inherent nonpermanence of sequestered carbon and its vulnerability to events such as bushfires, pests, and uncontrolled harvesting mean that LULUCF credits differ from permanent emission reductions such as those generated by renewable energy projects.

In this chapter we look at the legal issues and risks arising from LULUCF activities and the transaction of tCERs and lCERs generated under the CDM.[3] Specifically, we address whether and how it is possible to establish which entity has legal title to create and sell LULUCF credits, in light of other ownership issues such as legal title to the land and trees, the unique legal risks surrounding the delivery and permanence of LULUCF credits, and the types of legal and contractual solutions that can be employed to clarify ambiguities and minimize risks. We consider these issues in the light of both practical experience and case studies.

Establishing Legal Title to Sequestered Carbon

When one seeks to develop a LULUCF CDM project or to transact LULUCF credits (specifically tCERs or lCERs) from that project, it is crucial to identify the entity that has legal title to undertake such development and transaction. Unless the entity purporting to sell the LULUCF credits has the right to do so under the laws of the country hosting the CDM project, it will not be authorized to sell the carbon associated with the biomass stored in the project area, and disputes may arise with the rightful owner of the LULUCF credits during the course of the CDM project.

Under the CDM rules, if the carbon underlying a tCER or lCER is released into the atmosphere during the validity of that LULUCF credit, then the country that used the credit for compliance purposes is obligated to replace the credit. It could be exposed to significant market price risk if it is required to purchase replacement credits on the market. Although the Kyoto Protocol obligation is on the *country* that used the LULUCF credit for compliance purposes, each country is likely to pass this obligation through to the individual entities that used the relevant credit for compliance under any national greenhouse policy framework. Purchasers of tCERs and lCERs should be careful to ensure that the initial project developer has the right and title to develop the CDM project and sell the resultant LULUCF credits. If the project developer does not have such legal title, then the risk profile of the project increases substantially. There is clearly a greater risk that the carbon underlying the tCER or lCER will be released into the atmosphere, through, for example, harvesting of the forest from which the LULUCF credits were generated.

Legal Title under the International Rules

It is useful to differentiate between the legal title to carbon sequestered on a particular portion of land (referred to for the purposes of this chapter as the carbon property right) and the actual volume of carbon sequestered on this land (measured in tonnes of carbon dioxide equivalent, with each tonne referred to here as a LULUCF credit). It is possible to hold a carbon property right over a piece of land on which no LULUCF activity has actually occurred. Alternatively, it is possible that a LULUCF project may have been implemented on a piece of land but that the project developer does not actually have an entitlement to sell LULUCF credits arising from that activity (that is, the project developer does not hold the carbon property right).

Despite specific rules and guidelines within the CDM framework for LULUCF projects,[4] the parties to the United Nations Framework Convention on Climate Change (UNFCCC), when negotiating the modalities for the CDM, did not assume any particular land right or usage regime for land on which LULUCF activities were to be undertaken. That is, the CDM rules do not specify who holds the carbon property right in respect of a CDM project. Moreover, there is no stipulation about the particular relationship between the right to land and the right to the carbon sequestered on that land through afforestation-reforestation (AR) activities.

Instead, the international rules allow the domestic law of the host country to address this legal vacuum, and they require the proponents of a CDM activity to include in their project design document "a description of legal title to the land, right of access to the sequestered carbon, current land tenure and land use."[5] The host country is required under the international rules to approve the project in accordance with its own criteria.

Carbon Property Rights under the Host Country Law

Because host countries' domestic laws may not adequately address the requirements of CDM investors in establishing carbon property rights and therefore may not ensure that project participants have the legal right (and obligation) to maintain the sequestered carbon for a long period, contractual arrangements must be used to mitigate investment risk. These issues are best addressed at the project design and early contract negotiation stages so that the project operates smoothly and disputes do not arise after the sale of CERs.

When one structures the sale of CERs from a LULUCF or any other CDM project, it is important first to identify the entity that holds the initial legal title to the underlying greenhouse gas emission removals and reductions—that is, the carbon property right. In the case of LULUCF projects, identifying which entity holds carbon property rights is particularly complex, for a number of reasons. First, a range of entities may be responsible for sequestering the carbon. For example, contractual arrangements might provide that one person plants and harvests the trees, another has legal title to the land, and a third person owns the timber. These arrangements may not clarify who holds the carbon property rights. Second, the way in which ownership is allocated among entities within the project will depend to a large extent on local property laws. These property laws may be uncertain on the specific issue of legal title to sequestered carbon, particularly in developing countries. And third, there may be countervailing property issues over the land where the forestry project has been developed, such as indigenous land rights.

It is crucial for the party selling the credits arising from the LULUCF project to be able to establish carbon property rights sufficient to exercise long-term control over the land on which the LULUCF activity occurs. Such control will ensure that the forest sinks are maintained for the period required under long-term supply agreements for LULUCF credits, commonly referred to as Emission Reductions Purchase Agreements, or ERPAs.

Resolving the uncertainties surrounding legal title to the sequestered carbon is critical to securing its market value in a CDM transaction. Buyers are wary of purchasing credits from LULUCF projects when there is a risk of a domestic law challenge over legal title. Legal solutions to managing this risk will depend on the legislative framework in the country where the LULUCF project is being developed. Ideally, entitlement to carbon property rights would be clear from the legal system of each host country.

As of the date of writing, few host countries have passed legislation that expressly or implicitly allocates carbon property rights and resolves the relevant legal issues.[6] However, under many host country legal systems it is possible under general legal principles to establish the prima facie holder of carbon property rights (for example, on the basis of the existing property law regime). We consider in turn legislative and contractual solutions to address these scenarios.

Specific Legislation for Carbon Property Rights. If a host country implements clear legislation or issues guidance on the creation and sale of carbon property rights, this should resolve any ambiguity over the appropriate entity to create and sell LULUCF credits in that country. To date, however, few jurisdictions have enacted explicit legislation on this issue. In the few examples in which countries have specified the holder of legal title, either of two approaches has been taken: title is vested in the government or title is vested in ownership of the land.

In some countries the government may choose to assume prima facie ownership of all sink credits. In this case, transfer of legal title for the sequestered carbon is within the host country government's discretion. For example, in New Zealand the government initially chose to retain all rights and obligations arising out of forest sinks in New Zealand. This position has been modified slightly under the Permanent Forest Sink Initiative, whereby the New Zealand government is prepared to provide credits to certain private sector sink project developers who are investing in permanent reforestation. The details of New Zealand's scheme are set out in chapter 18.

If the host country of a CDM project asserts that carbon property rights (and the resultant LULUCF credits, tCERs and lCERs) are initially vested in the host country government, then a private sector entity wishing to develop a CDM project would need to obtain a concession arrangement or assignment of carbon property rights from the government before it could validly create and sell LULUCF credits. In host countries where more than one level of government is responsible for property and forest management. such as Brazil, national and state governments may disagree over which level of government is vested with carbon property rights.

An alternative approach may be for the host country to pass specific legislation under which carbon property rights can be vested in landholders. For example, this approach has been adopted by the six states in Australia. Each state has enacted legislation developing a basic legal framework for the recognition of carbon property rights. This makes it possible to determine the owner of the sequestered carbon (by a land title search) and to transact this carbon separately from the timber and the land to which it relates. Interestingly, however, although all Australian states recognize carbon property rights, only one state, New South Wales, has to date developed a legislatively recognized LULUCF credit that can be transacted under an emissions trading regime (see chapter 18). With the recent change of government in Australia, it is likely that an emissions trading regime will be introduced on a national level and will include the ability to trade LULUCF credits.

Establishing the Holder of a Carbon Property Right under Host Country Legal Principles. If it is impossible to establish the holder of a carbon property right under general host country legal principles, this creates a lack of security with respect to the ownership of the credits generated, which may be reflected in lack of investor confidence in LULUCF project development.

There are a number of ways in which carbon property rights may be seen to fit within existing host country legal regimes. The analysis of legal title in a specific LULUCF project needs to be performed on a case-by-case basis. For one example, where a host country vests ownership of all land in the state—as is the case in the Philippines—the carbon property right may be regarded as a state-owned natural resource (in the absence of any concession agreements) by virtue of its connection with land. For another example, the existing property law system might enable the registration of a carbon property right as an interest in land. In Ghana, the right to sequestered carbon is potentially registrable in the land title registry and therefore runs with the land. In Indonesia, LULUCF CDM project developers must satisfy the Indonesian government that any forested land from which they are seeking to generate LULUCF credits holds either an environmental service permit or a wood forest product permit from the regional or national government department responsible for forestry.

In some host countries, particularly those with legal systems based on civil codes, a carbon property right might be characterized as a civil or industrial "fruit" and as such would belong to the owner of the source of the fruit in the absence of other contractual arrangements. This is the prevailing interpretation in Argentina.[7] In other jurisdictions it may be possible to characterize sequestered carbon as a "forest resource" and therefore to consider the right to sequestered carbon the right to "immovable property." Entitlement to sequestered carbon as a forest resource might be freely tradable, and as immovable property the right may be more securely protected under host country law. This is a prevailing interpretation in the Philippines.

If the host country's existing legal framework does not provide a definitive answer regarding entitlement to carbon property rights, then a buyer and seller may seek to put in place legal arrangements in addition to the LULUCF credits sale contract in an effort to establish a carbon property right and secure the long-term use of the relevant land. One alternative for doing so is to register a *profit a prendre,* or a right to take from the land. Under various legal systems, a *profit a prendre* in respect of sequestered carbon might be granted by a landowner to a third person. This grants the third person a right to enter the land and take something that naturally belongs to the land. Landowners historically granted such property rights to people who wanted to hunt or harvest crops or fruit. Because a carbon property right is intangible, it is difficult to see it as a *profit a prendre* in itself. However, it may be possible to couple the carbon property right with the right to harvest trees on the land and for this to become the *profit a prendre.* In some jurisdictions a *profit a prendre* is a legal interest registrable on title. Several Australian states have specifically deemed a carbon property right to be a *profit a prendre.*

Another alternative for an entity seeking to develop a LULUCF CDM project is to enter into a long-term lease with the landowner, allowing the entity the exclusive right to harvest the trees (and conversely, the legal ability to prevent the trees from being harvested). Depending on the host country's legal framework, it may be possible to register such a lease on legal title, granting exclusive possession to the lessee (the CDM project developer).

It may also be possible to grant, by way of a concession agreement, a right to use a certain plot of land in order to generate and sell LULUCF credits. In many legal systems, governments use concession agreements to authorize the private sector to use a natural resource (such as water). In these legal systems it may be possible to adapt such concession agreements to authorize the use of sequestered carbon.

PRACTICAL RECOMMENDATIONS FOR THE LEGAL TITLE ISSUE. For any LULUCF project it is important to complete thorough legal due diligence to resolve a number of issues. Any prudent project participant will seek to ensure that legal title to the sequestered carbon (the carbon property right) is clearly allocated in the contractual arrangements between the parties. These contractual arrangements, however, must work within the domestic law arrangements of the host country that are flexible enough to accommodate carbon property rights and the right to sell LULUCF credits in a range of ways.

Within the varying legal landscapes, a range of contractual solutions may be appropriate in addressing the issue of legal title to tCERs and lCERs under LULUCF projects. We are able to consider contractual and property law issues only in a general manner here, because they are likely to be highly dependent on the specific laws and customs of the particular host country.

It is foreseeable that the following legal interests might exist over a carbon sequestration project: (1) landownership (by either an individual or a community); (2) tenancy of the land (for example, leasehold and traditional or indigenous land rights); (3) the right to take from the land (for example, to harvest the forest)—a *profit a prendre* in common law jurisdictions and a usufructuary right in other jurisdictions; (4) ownership of the timber itself (for example, by a forestry company that does not own or lease the land); (5) rights to natural resources (for example, minerals, but potentially also sequestered carbon, depending on the legal regime of the country in question); (6) concession agreements from the government to undertake forestry; (7) the interest of a bank that holds a mortgage on the land; and (8) rights to land (for example, native title rights or government entitlements) under constitutional law.[8]

In practice, in many host countries the two crucial entities to consider are the landowner and any person entitled to plant or harvest the forest on the land on which the LULUCF activity will occur. This is because the LULUCF credits arising from a planting activity are valuable for compliance purposes only so long as the trees that are sequestering the carbon continue to exist. Because the landowner and the harvester exercise physical control over the trees, it is important to determine between them who owns the carbon property right and therefore is entitled to generate and sell LULUCF credits.

It may be that the relevant legal interests just described have been expressly granted between the parties through contract or have been registered on the land title through one of the registrable entitlements discussed earlier. Alternatively, legal interests may have been granted implicitly through a long pattern of allowing others to harvest trees or crops from the land. In the absence of any legal or contractual arrangements (including leases), the general assumption in many countries in the case of sink projects has been that the owner of the land owns the trees and the carbon sequestered by them. It is therefore important to engage in thorough due diligence and stakeholder consultation to clarify the underlying legal title.

Such issues—among others, such as the environmental and social effects of the project—could be a focus of the formal stakeholder consultation required by designated operational entities (DOEs) for CDM projects. The best case scenario is that any potential disputes will be revealed at this consultation stage rather than once the project has been registered and the buyer has purchased the CERs. The community consultation process should also be used to explain to any entities that might use the land that the trees will need to remain there permanently. Importantly, these issues should be clearly addressed contractually by all the potentially relevant entities.

Other Interests in the Land

Another legal issue that should be addressed through the contract is that of competing rights and interests in the land on which the sink project is to be located. Potentially relevant interests in the land might be both formally registered legal rights, encumbrances, and interests and customary or traditional uses of the land. For instance, when a buyer is looking at purchasing carbon credits from land over which indigenous communities exercise certain customary or legal rights, it may be appropriate to consider entering into some type of agreement with the indigenous community for the co-management of carbon credits.

The possibility that indigenous communities may have certain entitlements to carbon credits reflects the growing international recognition of the contribution indigenous communities make to biodiversity. It is reflected in article 8(j) of the 1992 Convention on Biological Diversity, under which the commercial application of traditional knowledge and practice by third parties requires the approval of the holders of this knowledge (usually the indigenous community) and must be tied to equitable sharing of the benefits arising from its use. The convention states that each member state shall, "subject to its national legislation, respect, preserve and maintain knowledge, innovations and practices of indigenous and local communities embodying traditional lifestyles relevant for the conservation and sustainable use of biological diversity and promote their wider application with the approval and involvement of the holders of such knowledge, innovations and practices and encourage the equitable sharing of the benefits arising from the utilization of such knowledge, innovations and practices." The practicality of such an agreement will depend to a large degree on the particular circumstances of the proposed project.

In developing countries where CDM sink projects are being developed, land property systems can be extremely complex or nonexistent. Often there are no formal records of landownership and no legislative arrangements under which nonlandowners use land. This can create considerable difficulty when multiple claims to carbon property rights or other relevant land rights are made, and it highlights the importance of watertight due diligence and comprehensive contractual provisions. Commonly, developing country property systems rely heavily on long-term leases of land, and therefore it will be imperative to ensure that the lease does not expire within the time frame of the project. Because sink projects rely on maintaining the carbon sequestered, it will also be critical for the project developer (and primary seller) to have continued access and monitoring rights in relation to the maintenance of any forestry rights.

Delivery Risk, Permanence, and Liability

Article 12.1 of the Kyoto Protocol clearly states that an emission reduction can be certified as a CER for the purposes of the CDM only if it is generated by a project that generates emission reductions that are, among other things, "long-term beneficial to the mitigation of climate change."

LULUCF CERs differ from other types of units under the Kyoto Protocol because much more uncertainty surrounds the "permanence" of LULUCF credits. *Permanence* here refers to the extent to which a relevant carbon sequestration project is able to achieve an absolute and irreversible reduction in the volume of carbon dioxide in the atmosphere.

Delivery risk—the risk that an insufficient volume of emission reductions will be generated to meet contractual obligations—is greater in LULUCF projects than in other CDM projects because of the nature of carbon sequestration and the inherent possibility that trees may not grow as quickly as modeled, for natural reasons such as weather, fire, diseases, and insect plagues. In addition, LULUCF projects manifest a unique permanence risk that does not occur in other types of emission reduction projects. It relates to the fact that the carbon stored in forest biomass can be released into the atmosphere at any time after the LULUCF credits have been issued, thereby reversing the sequestration cycle. Buyers and sellers of LULUCF credits, known as "project participants," must allocate the liability for these risks through contractual arrangements as well as prudent project management.

Identifying the Risks

In carbon sequestration projects, risks exist that the carbon dioxide sequestered in the forest sink may be released back into the atmosphere at any time. For example, part of the forest might be destroyed by fire, pests, or disease, releasing the sequestered carbon back into the atmosphere. Alternatively, trees might be illegally logged, so that all the carbon sequestered in the above-ground biomass is deemed to have been emitted into the atmosphere under Kyoto Protocol carbon accounting rules. This is a particular risk in developing countries that may have no strict regulations addressing these issues. These risks may be referred to generally as permanence risks.

In addition, risks are associated with the extent to which a carbon sequestration project successfully delivers the volume of CERs that were initially predicted. For example, in some circumstances it has turned out that trees planted for a carbon sequestration project were unsuited to the soil or were planted in the wrong season, resulting in lower volumes of carbon sequestration than predicted. Alternatively, a forest's growth in a particular year may be stunted by natural occurrences such as drought. These risks may be referred to as delivery risks.

In both instances it is necessary to determine which entity will bear the delivery risk (that is, the risk of underperformance of forest growth) and which entity will be liable for replacing the sequestered carbon or purchasing equivalent LULUCF credits if the carbon is released back into the atmosphere. This can be addressed through the parties' contractual arrangements. The issue of permanence risk has been addressed under the CDM through the LULUCF decision reached at the ninth session of the Conference of the Parties to the UNFCCC, which adopted the two types of credits, tCERs and lCERs. It also established a maximum duration for LULUCF CDM projects, the crediting period.

At the registration of a LULUCF CDM project, the project proponent must select the length of the crediting period for the project and an accounting approach to address the nonpermanence of an afforestation or reforestation project under the CDM. Specifically, the proponent must decide whether the project will create tCERs or lCERs. Both tCERs and lCERs can be used only to meet the obligations of Kyoto Protocol Annex I Parties for the commitment period in which they were initially issued.[9]

Temporary CERs are initially valid only until the end of the commitment period after the one in which they were first issued (the first commitment period runs from 2008 through 2012). Before their expiry, tCERs must be replaced with another credit (either an Assigned Amount Unit [AAU], an Emission Reduction Unit [ERU], a CER, or another tCER). In the meantime, every five years a verification and certification report must be produced containing a request to the

CDM Executive Board to issue new tCERs equivalent to the total amount of carbon sequestered since the start of the project.[10] As a result, the prior tCERs are still valid at the time the new tCERs (for the same sequestered carbon) are issued.[11] The new tCERs will then be valid until the end of the subsequent commitment period, and further tCERs can be issued at certification every five years, until the end of the project crediting period (sixty years if the project proponent elected the maximum project term including renewals).[12]

Long-term CERs are valid until the end of the project's crediting period, although a verification and certification process must take place every five years to confirm that the carbon is still sequestered by the project.[13] Upon expiration of an lCER, the Kyoto Protocol party that used the lCER for compliance purposes must replace it. Replacement can be made only with an AAU, a CER, an ERU, an RMU, or an lCER from the same project activity, and not with another lCER from a different project activity or a tCER.[14] Replacement of an lCER may be required earlier than the end of the crediting period if the certification report that must be provided every five years indicates that the amount of carbon sequestered by the project has decreased since the last certification report or if no certification report is provided at the five-year verification date.

The rules surrounding the creation and use of tCERs and lCERs place the permanence risk with the entity that wishes to use the tCERs or lCERs for compliance purposes. However, this risk can to some extent be passed contractually back to sink project developers by requiring that they maintain the forest in perpetuity and provide replacement credits upon release of the carbon back into the atmosphere.

The replacement obligation means that tCERs and lCERs carry with them an inherent risk for their ultimate recipient (the Kyoto Protocol party that uses them for compliance purposes). Specifically, assuming that the protocol continues in its current form with future commitment periods, then upon expiry of the relevant tCERs or lCERs or release of the carbon from the forest underlying those units, the Kyoto Protocol party will need to obtain replacement units. This may involve purchasing those units at a substantially higher market price than the price initially paid for the tCERs and lCERs. Parties are likely to pass this risk through, by way of regulation, to the public or private sector entities that initially acquired the tCERs or lCERs or used them for compliance purposes under a domestic scheme.[15] The purchaser of tCERs or lCERs may therefore attempt to pass on this risk by way of contract with the project developer, who ultimately has greater control over retaining the carbon sequestered in the forests.

In contrast, forestry-based credits created under the New South Wales Greenhouse Gas Abatement Scheme, an emissions trading scheme in Australia, do not expire, but the scheme requires the entity that is responsible for creating the credits (that is, the seller) to satisfy the scheme administrator that an amount of abatement created and sold under the scheme will be sequestered for 100 years. If the carbon is released back into the atmosphere during this 100-year period, then the project developer must provide equivalent replacement credits to the scheme administrator. This was the first functioning emissions trading regime to recognize credits from carbon sequestration (see chapter 18).

Contractually Managing the Risks

There are three approaches that the parties in a LULUCF project contract may consider in order to address the delivery and permanence risks surrounding tCERs and lCERs.

The first strategy is to use a force majeure clause to encompass the permanence risks. This will result in the termination of the contract when an event occurs that is beyond the reasonable control of either party—for example, storms, floods, droughts, frosts, fires, and insect plagues. It is obviously advantageous for the party bearing the ultimate liability for delivering the volumes of sequestered carbon under the contract (ordinarily the seller) to define force majeure events as broadly as possible. But if the seller is absolved by force majeure under the contract, this does not mean that the buyer gets exemption under international rules. The buyer is still required to replace the credit. Therefore, purchasers may seek to define force majeure narrowly and to exclude events that are common in forestry projects, such as forest fires, droughts, and pests.

A second strategy for negotiating permanence risk may be to impose detailed conditions in the contract for forest management. For example, the parties might agree to adequate firebreaks and pest control strategies that would not otherwise be required under the domestic law of the host country. This might mitigate the risk that fire or pests will destroy a substantial portion of the plantation.

The third and most significant contractual solution is to adequately negotiate liability between the parties. Under the CDM rules, the general requirement is that the Kyoto Protocol party that used the lCER or tCER for compliance purposes bears the liability of replacing it in the event that the credit expires or a permanence or delivery risk materializes. Kyoto Protocol parties are likely to seek to pass this risk through to private sector entities that surrender tCERs or lCERs under a domestic scheme that recognizes these units.

Providing replacement tCERs or lCERs might be relatively easy when the holder of the tCER or lCER has a direct contractual relationship with the project developer or owner of the land where the carbon property right originated. In instances where the tCER or lCER has been sold to multiple buyers, however, the ultimate holder of that right will not have a direct contractual relationship with the landowner or project developer and may therefore be unable to directly enforce any obligations to replace the lost carbon or pay the cash equivalent.

A forward sale agreement for lCERs may seek to impose obligations on the project developer to replace lCERs in the event that the underlying sequestered carbon is lost. Replacement may be satisfied in various ways, such as by replanting the trees, purchasing replacement lCERs, or paying "mark-to-market" damages, which compensate the counterparty for the loss it has suffered because of movements in market price since the forward sale contract was entered into.

In these cases, purchasers of tCERs or lCERs may need to put in place mitigation and hedging strategies that would not otherwise be required for permanent CERs created from other CDM projects such as renewable energy generation. These strategies might include paying a substantially lower price to account for the increased risk. Alternatively, the purchaser might pool a reserve of tCERs or lCERs generated in other projects ("carbon pooling") that can be drawn on if unexpected losses arise—for example, from disputes over legal ownership or from any of the events affecting delivery or permanence risk discussed earlier. This will allow the purchaser to avoid default of its other contractual obligations in emergency situations without needing to go to the market to purchase permanent credits at a potentially higher price.

Conclusion

Although the varied jurisdictions in which LULUCF projects are conducted have varying levels of clarity in their domestic legal systems, it is possible for parties to overcome uncertainty in part through strong contractual arrangements. In practice this requires comprehensive legal due diligence and proper consultation with the relevant parties.

Although it is difficult to predict the future course of the emerging market for LULUCF credits, it is likely that growing interest from landowners and investors will encourage local host country lawyers to analyze their existing legal frameworks to determine whether and how rights and benefits from sequestered carbon (that is, carbon property rights) can be assigned. As legal analysis of this issue evolves in host countries, it is likely that contracts for the sale of LULUCF credits will be modified to suit the legal property regimes of individual host countries, depending on how carbon property rights are dealt with in that regime.

Where it is difficult to obtain clarity of entitlement to carbon property rights, host country governments—in order to generate sufficient investor certainty in afforestation and reforestation projects, which can bring many local environmental and social benefits—may well clarify and strengthen their legal regimes around legal title in carbon sequestration activities.

Notes

1. See, for example, article 3.3 of the Kyoto Protocol.

2. Examples are the voluntary carbon market and particular emissions trading schemes such as the New South Wales Greenhouse Gas Abatement Scheme in Australia.

3. The rules relating to the treatment of LULUCF projects under the Joint Implementation framework are different and beyond the scope of this chapter.

4. LULUCF Decision, ninth Conference of Parties to the UNFCCC, Milan.

5. Decision 19/CP.9, Appendix B, "Project Design Document," paragraph 2(c).

6. The Forestry Rights Registration and Timber Harvest Guarantee Act (FRRTHG Act) of Vanuatu defines a "forestry right" in relation to land to include "a carbon sequestration right in respect of the land."

7. See International Union for the Conservation of Nature, "Legal Aspects in the Implementation of CDM Forestry Projects," IUCN Environmental Policy and Law Papers 59, 2005, p. 50.

8. For further examples, see Australian Greenhouse Office, *Planning Forest Sink Projects: A Guide to Legal, Tax and Accounting Issues* 2005 (www.greenhouse.gov.au/nrm/publications/forestsinks.html).

9. Paragraphs 23, 38, 41, and 45 of the LULUCF decision.

10. Paragraphs 32, 42, and 44 of the LULUCF Decision.

11. Because the validity of an initial lot of tCERs can extend beyond the five-year certification and issuance of new tCERs (that is, the end of a subsequent crediting period is often more than five years away), it is possible that two lots of tCERs are created from the same stock of sequestered carbon and are valid at the same time, even though no additional sequestration has taken place and no tCERs have yet been replaced. In effect this allows LULUCF CDM projects to create additional carbon credits in the market despite the fact that no additional emissions have been reduced. This seems to be an unusual result. However, ultimately all tCERs issued must be replaced by other emission reductions and therefore are accounted for despite the "gap" between the expiry and replacement of tCERs and the five-year issuance of new tCERs.

12. Paragraph 36 of the LULUCF decision.

13. Paragraphs 32, 36, and 46 of the LULUCF decision.

14. Paragraph 49(d) of the LULUCF decision. If an lCER needs to be replaced before the end of the crediting period because there has been a net reversal, then the lCER can be replaced by another lCER "from the same project activity."

15. This was proposed by Charlotte Streck and Robert O'Sullivan in "Briefing Note: LULUCF Amendment to the EU ETS" at the BioCarbon Fund's "Technical Workshop on the Role of Forests in the Carbon Market, Focusing on the EU ETS," Brussels, March 2006.

PART FOUR

Outlook: Avoided Deforestation and the Post-Kyoto Agenda

13

Reducing Emissions from Deforestation in Developing Countries: An Introduction

ROBERT O'SULLIVAN

Deforestation is one of the underlying causes of current levels of atmospheric CO_2 concentration. It has been estimated that about 40 percent of CO_2 emissions over the last 200 years have been from changes in land use and land management, most of which have been deforestation.[1] The remaining forest ecosystems still store an estimated 638 gigatonnes (Gt) of carbon (measured to a soil depth of 30 cm), 283 Gt of which is in the forest biomass alone. This is a significant amount of carbon—more than there is carbon in the atmosphere.[2]

Although rates of net deforestation are decreasing because of increased reforestation, gross deforestation (that is, deforestation only, not taking into account reforestation) averages 13 million hectares per year and does not appear to be decreasing. Rates of deforestation and modification of primary forests are also alarmingly high, at 6 million hectares per year, and do not appear to be decreasing either.[3] If these rates continue, approximately 15 percent of remaining forests will be lost by 2050, and approximately 70 percent will be lost within the next 200 years.[4] This rate is higher for primary forests, with 19 percent expected to be lost or modified by 2050 and 84 percent expected to be lost or modified within the next 200 years if current deforestation rates continue.[5] However, this loss is not expected to be uniform across all countries. According to the Millennium Ecosystem Assessment scenarios, forest area in industrialized regions will increase between 2000 and 2050 by 60 million to 230 million hectares.[6] At the same time, the forest area in developing countries is expected to decrease by 200 million to 490 million hectares.[7] This means that between 10 and 22 percent of forests in developing countries can be expected to be lost by 2050.[8] Some of these areas are already under threat from deforestation, but others may have seen little or no deforestation historically.

What does this deforestation mean for the climate? With up to 25 percent of global greenhouse gas (GHG) emissions coming from tropical deforestation, this sector is a significant source of GHG emissions and one of the underlying drivers of climate change.[9] If these emissions are not reduced, they have the potential to undercut reductions in energy-related and industrial GHG emissions and frustrate the objectives of the United Nations Framework Convention on Climate Change (UNFCCC).[10] GHG emissions currently produced by deforestation in Brazil and Indonesia alone are estimated to equal four-fifths of the emission reductions generated under the Kyoto Protocol's first commitment period.[11] If the global community is serious about trying to prevent significant climate change, then emissions from deforestation need to be addressed as a priority.

The effects of deforestation should not be measured only in terms of global climate change. Deforestation has significant negative effects on soil quality, biodiversity, local livelihoods, and indigenous communities. It destabilizes local climate and weather by disrupting historical hydrological cycles, albedo, and large-scale circulation patterns. On the other hand, forest conservation, sustainable management, planting, and rehabilitation of forests can mitigate CO_2 emissions though carbon sequestration, protect biodiversity, and deliver a range of socioeconomic benefits.[12]

Deforestation under the UNFCCC and Kyoto Protocol

The UNFCCC and the Kyoto Protocol are principally concerned with addressing GHG emissions in order to avoid serious anthropogenic interference in global climate. The focus of their instruments is on accounting for GHG *emissions* (including emission reductions and removals), and any consideration of forests and deforestation must take place within this framework. Although emissions from fossil fuels and industrial gases can be accurately monitored, accounted for, and reduced by undertaking certain activities, dealing with deforestation within this framework creates significant complexity, because the sources of emissions are not completely analogous. For example, monitoring and accounting for reduced emissions from a coal-fired power plant through an activity such as switching to a fuel with a lower emission factor (such as natural gas) is simple. The quantity of emissions generated under a coal-fired baseline is easily calculated, as are the emissions generated under the new baseline of an alternative fuel. If the switch to the new fuel was not part of the power plant's business-as-usual plans, then the difference between the two baselines is the amount of the emission reduction. The focus is on monitoring the amount of electricity produced from the activity of fuel switching and calculating any resulting reductions. The activity of electricity generation is considered the source of the emissions, and the underlying source of the avoided emissions—the coal that is not burned by the power plant—is not monitored or tracked in any way.

Applying to forests this approach toward monitoring emissions produces significantly different results. The focus shifts to undertaking certain activities to protect a forest and then determining emission reductions through ongoing monitoring of the underlying source of potential emissions—the forest.[13] The focus on monitoring and accounting for emissions from forests creates significant complexity for this sector that does not arise in other sectors. This increased complexity is the main reason addressing emissions from deforestation has been, and continues to be, a difficult, complex, and often controversial issue.

The Kyoto Protocol contains incentive mechanisms to reduce emissions from deforestation (RED)—and, as a result, to protect forests—in industrialized countries only. RED was excluded from the Clean Development Mechanism (CDM) for a number of reasons, including concerns over "leakage," permanence, potential for market flooding, and general opposition from a number of nongovernmental organizations (NGOs) to the use of forest-based credits to offset emissions from fossil fuels.[14] The creation of incentives to reduce emissions from deforestation in developing countries is currently being discussed under the auspices of the UNFCCC. The discussions were initiated by a proposal put forward in 2005, at the eleventh session of the Conference of the Parties (COP) to the UNFCCC, by the governments of Papua New Guinea and Costa Rica.[15] The formation of the Coalition for Rainforest Nations created additional momentum to address the issue, and it has since become a highly visible topic discussed at all levels, from the UN Security Council to "talk back" radio.[16]

The 2005 submission started a two-year learning process for the UNFCCC's Subsidiary Body for Scientific and Technical Advice (SBSTA), during which it was to focus on "relevant scientific, technical and methodological issues, and the exchange of relevant information and experiences, including policy approaches and positive incentives."[17] Among other things, the SBSTA held two technical workshops, one from August 30 to September 1, 2006, in Rome, Italy, and the other on March 7–9, 2007, in Cairns, Australia. A report summarizing the two-year process and recommendations was submitted to the thirteenth COP in Bali in December 2007. The official UNFCCC events have been augmented by additional initiatives, conferences, workshops, and meetings organized by governments, NGOs, the World Bank, research institutions, and, more recently, the private sector. The objectives of these events have largely been to inform and to try to influence the outcomes of the UNFCCC process and any decisions taken in Bali. These events, both official and unofficial, have produced a plethora of views on how to address deforestation in developing countries.

Key Policy Issues

In the following sections I outline some key issues that were initially raised and discussed during the Kyoto Protocol negotiations and that have resurfaced and been discussed in greater detail during the recent two-year process to address RED. In line with the theme of this book, the list is focused on policy issues and implications rather than some of the more scientific and technical issues.

Monitoring and Accounting: National and Subnational

Determining the area in which emissions from deforestation are to be monitored and accounted for is simple on its face but quickly becomes one of the most critical—and complex—policy issues under discussion. The issue is inextricably linked to other key issues, often in significant yet subtle ways that are easy to overlook.

Different boundaries within which to monitor and account for RED have been proposed, from global and country to project based (see chapters 14, 16, 17). The existence of such a broad continuum reflects different priorities and the interplay between sometimes conflicting considerations. Some policymakers see global monitoring and accounting as the most methodologically sound way to check that deforestation is in fact reduced globally. This is because it is the only way to accurately take into account "leakage." Leakage occurs when reduced deforestation in one location causes increased deforestation in another. It can be both international (from one country to another) and domestic (within a country). For example, if illegal logging is stopped in one area, loggers may simply relocate to another and continue logging there. No net reduction in logging (or emissions) takes place. The reduction in emissions in the first area should therefore not be credited as a real reduction. This is true regardless of whether the area being protected is a discrete project area, a county or province within a country, or an entire country.

To better understand some of the options for dealing with boundaries and leakage in RED, it is useful to look at the way leakage is addressed in existing Kyoto Protocol mechanisms. Domestic leakage under the CDM (for all projects) is accounted for on the project level through strict rules and regulations. Concern over how to address project-level leakage was one of the prominent reasons for originally excluding RED from the CDM. Joint Implementation, on the other hand, does not require the same strict project-level accounting, because domestic leakage will be reflected in a country's national inventory and overall emissions accounting. International leakage is not taken into consideration at all in any sector or mechanism under the Kyoto Protocol.[18] An industrialized country can therefore meet part of its Kyoto Protocol commitment if, for example, heavy industry is shut down and relocated offshore to a developing country. It is often argued that if this sort of leakage is not accounted for in other sectors, then the land-use sector should not be unfairly burdened by being required to account for it.

To tease out some of the subtler differences and issues associated with global, national, and subnational accounting, factors other than monitoring, accounting, and leakage need to be explored. These factors include political considerations as well as implementation of any policies and measures to reduce deforestation.

National monitoring and accounting have been the favored approaches since deforestation was first reintroduced into the UNFCCC agenda in 2005. They were seen as significant steps toward addressing domestic leakage. "National monitoring and accounting" can be understood in a number of different ways. The most common view to date involves national governments' (1) determining and negotiating national deforestation reference levels, (2) implementing national policies and measures to ensure that deforestation is reduced nationally, (3) monitoring to check actual rates of deforestation against the reference level, and (4) receiving credits for national reductions. This type of national monitoring and accounting is favored by some because it centralizes management and control over land-use choices in the government. It is argued that a top-down approach in which the central government is responsible for reforming land and forest policy and laws, corruption, and law enforcement is the most appropriate approach to reducing deforestation. This also centralizes income generated from the sale of credits with the federal government, which is appealing to some governments. Another advantage seen in national accounting is that it is an initial step toward developing countries' assumption of voluntary targets under a post-Kyoto regime. Criticisms of national-level monitoring and accounting include the reduced accuracy and increased uncertainty associated with national monitoring, increased sovereign responsibilities and potential for associated corruption, and uncertainty over how the private sector might participate in this type of system.

Subnational and project-based approaches focus primarily on monitoring and accounting for emissions from discrete areas within a country. Such approaches are preferred if the obligations of the national accounting method are viewed as potential risks rather than advantages. A lack of capacity or challenges with governance in some developing countries may make it difficult to reliably and accurately account for emissions and effectively reform current practices nationally to reduce deforestation. The private sector has also expressed concern that the national approach creates too much risk to warrant up-front or on-the-ground private sector participation. This is because if the private sector invests in a project in a country that has national accounting and crediting, then investors will earn credits only if the project performs as expected *and* the government effectively protects other forests outside the project boundary. This is true even if an increase in deforestation outside the project boundary is not due to leakage. Possible solutions, such as sovereign guarantees to provide credits to projects that perform irrespective of national performance, are viewed as either unlikely (because some countries may not want to assume this liability) or too risky to stimulate much private sector interest (because such guarantees are subject to the [often poor] credit ratings of the issuing governments). A number of private sector investors have indicated a preference for project-based accounting and crediting because it allows private sector investment in individual projects without exposure to sovereign performance risk.[19] For this reason, some view a project-based approach that takes leakage into account to be a more effective mechanism for reducing deforestation than the national approach.

Project-based accounting may also enable funding to flow more readily to local communities and landowners, because it promotes the direct participation of these stakeholders. It also allows for more accurate monitoring of emissions and emission reductions in a distinct area. As noted earlier, issues of leakage have traditionally been the main stumbling blocks of project-level accounting.

The national and project approaches represent two ends of a possible continuum. An approach that sits between the two could, for example, require a country only to determine and negotiate a national reference level and to undertake periodic national monitoring against this reference level. Implementation could be left either to the government or to private entities. Under this approach the difficulty lies in issuing credits and reconciling any such crediting against national performance. As soon as projects are issued credits—even if it is done under a national accounting approach—issues such as accounting for domestic leakage and balancing project-level success against national failure must be addressed.[20] These issues could be addressed by deducting credits issued to projects from emission reductions monitored nationally. Any remaining national-level reductions would then be issued as credits to the government, and any national-level deficiency could be carried over to subsequent accounting periods or taken into account in future renegotiations of the national reference scenario.

A significant push by a number of countries and other interested groups leading up to COP 13 in Bali to include the possibility of project-level accounting for RED was successful, with the decision on RED referring to the possibility of undertaking "subnational demonstration activities" and "subnational approaches."[21]

Monitoring and Accounting: Degradation

Reducing emissions from forest degradation was not explicitly included in the original UNFCCC process, which refers to "reducing emissions from deforestation" only. After much debate, the COP 13 round of negotiations opened up the possibility of accounting for degradation. Degradation can be important, because depending on how (or whether) deforestation is defined, significant amounts of carbon can be released from a forest before deforestation occurs (see chapter 14). Forest degradation also results in a loss of biodiversity, can lead to the establishment of invasive species, and decreases a forest's ability to adapt to changing climatic conditions.[22] This is particularly important in small island states that are already particularly susceptible to the effects of climate change and will have the greatest difficulty in adapting. However, including degradation in RED (to become REDD, for "reducing emissions from deforestation and degradation") can increase monitoring costs significantly.

Monitoring and Accounting: When to Assess Emissions

Countries have different rates of deforestation that change over time. This makes deciding when to measure baseline or reference emissions in order to calculate any emission reductions important. Most countries argue that historic rates of emission should be used, because these can be accurately assessed, whereas future rates are difficult to quantify. But historic baselines or reference scenarios also have problems. They do not take into account current or future rates of deforestation, which may increase or decrease under a business-as-usual scenario, regardless of a country's efforts to reduce deforestation. Looking at purely historical rates will also make it harder for countries with low levels of historical deforestation to participate in an incentive mechanism designed to reduce rates of deforestation. To address this, it is expected that any (national) reference level will be based on historic emissions but will inevitably be adjusted through negotiation to take into account countries' individual circumstances and development objectives.[23]

Carbon Markets and Market Flooding

RED activities can be funded through traditional forms of governmental funding complemented by official development assistance. Alternatively, funding can be mobilized by harnessing the forces of the carbon market. The Stern Review, produced for the British government in October 2006, determined that reducing emissions from deforestation was one of the most cost-effective ways of reducing GHG emissions globally and estimated that the opportunity cost of stopping deforestation in the eight countries responsible for 70 percent of global deforestation would be between U.S.$5 billion and $10 billion a year. Other estimates put the cost of reducing 50 percent of CO_2 emissions from deforestation over the next twenty years at U.S.$33 billion a year.[24]

Total official development assistance for the entire forestry industry was estimated at approximately U.S.$1.1 billion in 2004—and only a portion of this has been dedicated to forest protection.[25] Most groups recognize that there is a significant shortfall in public funding to address the problem, and the private sector will be expected to play a role through the carbon market.

Opening existing markets to credits generated by avoiding further deforestation, however, has caused concerns of "market flooding" related to the potentially large number of credits generated by RED activities. Two approaches exist to deal with this issue: creating a single market in which deforestation credits are used to offset other emissions, and developing a second, parallel market explicitly for RED credits, which are not fungible with assigned allowances and offsets under a future climate treaty. A number of governments and some NGO commentators advocate capping the number of RED credits that may enter a single market. This is because of concern that the potentially large volume of RED credits might undermine efforts to reduce emissions from industrialized countries and that cheaper RED credits might undercut project-based credits from the CDM, making it harder for these projects to become financially attractive. A ceiling on the number of RED credits that can be used for compliance has been suggested to overcome this. However, it is also argued that for developing countries to accept a ceiling on the number of credits they can generate, the ceiling should be balanced by a commitment from industrialized countries to purchase a minimum number of credits to ensure demand.

A parallel market for RED is seen as another possible solution to the problem of flooding, because RED credits would not compete with other sectors if they are traded in a parallel market. Additional commitments from industrialized countries will be needed to create demand for credits in this sort of dual market. To engage the private sector, these commitments would need to be passed down through domestic legislation or other domestic policy to encourage demand from the private sector.

A variation of this type of parallel market has been proposed by the Center for Clean Air Policy in its "dual market" approach.[26] Under this approach a second, parallel market would be created for REDD credits that is dominated by sovereign-level commitments from industrialized countries to purchase credits from specific developing countries. If a developing country fails to generate sufficient REDD credits to meet the industrialized country's commitment, then the country could buy replacement credits from the regular carbon market or carry over this purchase commitment to subsequent periods.

Permanence

"Permanence" refers to the risk that credits are issued for a forest that is lost in the future. The majority of commentators consider permanence to be a real issue for RED, although this is not universally agreed.[27] There are two basic options to address permanence, which are linked to whether or not any RED credits are used to offset industrial emissions: whether they are traded in a parallel market, and whether they are issued under national reference scenarios or under project baselines.

The first option involves the issuance of permanent credits for any emission reductions generated in developing countries. This approach is typically associated with national-level accounting and crediting, whereby any subsequent increase in deforestation (that is, loss of previous gains) is addressed at the national level. A number of ways to address permanence at the national level have been proposed and can be grouped loosely into the following general categories: (1) tapping into a "reserve pool" of emission credits that were generated in the past but banked as insurance against future reversals; (2) converting the loss into a sovereign liability that may be satisfied by the country's buying replacement credits from the carbon market; and (3) a renegotiation of the country's reference level in the next commitment period. Under all these approaches the country selling the credit ultimately assumes any risk that the forest will be lost in the future.

The second option involves the use of "temporary crediting." This approach is currently used for afforestation and reforestation projects under the CDM whereby a credit either is reissued every five years (that is, the credit lasts for only a short time before it expires and must be replaced) or is reverified every five years and canceled if a verification finds that the underlying carbon stock had not been maintained. Under this approach the buyer of the credits assumes any risk that the forest will be lost in the future.

Another approach is proposed by the Voluntary Carbon Standard (VCS), which uses a collateralization mechanism based on the "buffer approach" to secure the long-term benefits of RED projects. Under the VCS, each project undergoes an independent risk assessment by two verifiers against predefined risk criteria. On the basis of this assessment the project is classified according to risk (low, intermediate, or high) and assigned a buffer withholding percentage (5–30 percent). The project is then required to set aside this portion of the total number of credits generated in a pooled buffer account, which is shared by all VCS agriculture, forestry, and other land use (AFOLU) projects. Over time, as the project proves at subsequent verifications that it has mitigated certain risks, a portion of the project's buffer credits will be released and made available for trading. To maintain overall atmospheric integrity, every few years the entire system is "trued-up" on the basis of actual project performance across the VCS portfolio. Accordingly, adjustments will be made to the risk criteria and buffer withholding percentages to ensure that there is always a net surplus of actual carbon benefits generated relative to those issued and traded.[28]

Conclusion

Deforestation is one of the underlying drivers of climate change. Addressing emissions from deforestation must therefore be part of any solution to reduce emissions globally. Doing so is possible because reducing emissions from deforestation is highly cost effective and has the potential to offer significant reductions fairly quickly.[29] Realizing this potential, however, is notoriously complex and increasingly political. Any workable solution will need to address a number of key issues to produce sound international policy. The treatment of afforestation and reforestation under the CDM and the EU Emissions Trading Scheme is an example of the way overcomplex and discriminatory rules have led to a market failure in this sector.[30] Of the more than 1,000 CDM projects registered to date, only one is from a reforestation project. If international policy to address RED is not developed in a way that effectively stimulates action, a second market failure in the forest sector can be expected, and the dual opportunity to save the remaining forests and reduce global emissions in the most cost-effective manner will be lost.

Notes

1. Millennium Ecosystem Assessment, *Ecosystems and Human Well-Being: Synthesis* (Washington: Island Press, 2005). The primary source of the increased atmospheric concentration of CO_2 since the preindustrial period has been fossil fuel use. See Intergovernmental Panel on Climate Change, "Summary for Policymakers," in *Climate Change 2007: The Physical Science Basis,* edited by S. Solomon and others (Cambridge University Press, 2007), p. 2.

2. Food and Agriculture Organization (FAO), "Global Forest Resources Assessment 2005: Progress towards Sustainable Forest Management," Forestry Paper 147 (Rome: FAO, 2006), p. 14.

3. Ibid.

4. This projection is based on a global average gross deforestation rate of 13 million hectares per year and a total global forest cover of 3,952 million hectares, taken from FAO, "Global Forest Resources Assessment 2005." The calculation is based on a linear loss of forest over time, which is an oversimplification. The 2050 estimates, however, fit within the range of deforestation estimates produced by the Millennium Ecosystem Assessment.

5. These figures are based on FAO 2005 estimates that primary forests make up 36 percent of total forests and are being lost at a rate of 6 million hectares per year.

6. The Millennium Ecosystem Assessment was carried out from 2001 to 2005 by more than 1,360 experts worldwide. The assessment was requested by the UN secretary general in 2000, and the objective of the study was to assess the condition and trends of the world's ecosystems and the consequences of ecosystem change for human well-being. See www.millenniumassessment.org.

7. B. Metz and others, eds., *Climate Change 2007: Mitigation. Contribution of Working Group III to the Fourth Assessment Report of the Intergovernmental Panel on Climate Change* (Cambridge University Press, 2007), p. 545.

8. The FAO has estimated that there were approximately 2,101 million hectares of forest in Africa, South America, Central America, the Caribbean, South and Southeast Asia, and Oceania (excluding Australia and New Zealand) in 2005, and 1,851 million hectares in Europe, North America, Australia, and New Zealand (809 million of which were in Russia). FAO, "Global Forest Resources Assessment 2005."

9. The figure of 25 percent was derived by using 1990 as the baseline and taking into account emissions of carbon dioxide, methane, nitrous oxide, and other chemically reactive gases that result from deforestation and subsequent uses of the land. See R. A. Houghton, "Tropical Deforestation as a Source of Greenhouse Gas Emissions," in *Tropical Deforestation and Climate Change,* edited by P. Moutinho and S. Schwartzman (Washington: Amazon Institute for Environmental Research and Environmental Defense, 2005), p. 13.

10. E. Trines and others, *Integrating Agriculture, Forestry and Other Land Use in Future Climate Regimes: Methodological Issues and Policy Options,* Climate Change Scientific Assessment and Policy Analysis report 500102002 (Netherlands Environmental Protection Agency, 2006).

11. M. Santilli and others, "Tropical Deforestation and Climate Change," *Climate Change* 71 (2005): 267–76.

12. G. Marland and others, "The Climatic Impacts of Land Surface Change and Carbon Management, and the Implications for Climate-Change Mitigation Policy," *Climate Policy* 3 (2003): 149–57; FAO, "Global Forest Resources Assessment 2005."

13. This examination of accounting for emission reductions from an activity should also be extended to other gases, such as CH_4, N_2O, SF_6, and HFC_{23}. In all these instances the analogy with avoided deforestation is less clear. For these gases the focus is on monitoring and accounting for the gases either through a change in practice to prevent their generation in the first place (N_2O and SF_6) or through destroying the gas after it has been created (CH_4, HCF_{23}, N_2O, and SF_6). In all these instances emission reductions are not associated, even indirectly, with storing the gases (or their sources) anywhere, as is the case with fossil-fuel-based CO_2 credits or RED credits.

14. Some of these issues do not arise for industrialized countries that have accepted caps under the Kyoto Protocol, because these emissions can be accounted for as part of the country's overall emissions accounting. The CDM, created by article 12 of the Kyoto Protocol, allows eligible projects in developing countries to generate Certified Emission Reductions, which can be used to help industrialized countries meet their emission reduction commitments.

15. See UNFCCC, "Reducing Emissions from Deforestation in Developing Countries: Approaches to Stimulate Action," FCCC/CP/2005/MISC.1 (http://unfccc.int/resource/docs/2005/cop11/eng/misc01.pdf). Official support for including the issue on the COP's agenda was sent by Bolivia, the Central African Republic, Chile, Congo, the Democratic Republic of Congo, the Dominican Republic, and Nicaragua.

16. The Coalition for Rainforest Nations is a loosely knit group of developing countries with forests. Some of its members participate in joint submissions of views to the UNFCCC.

17. The SBSTA is one of two permanent subsidiary bodies established by the UNFCCC. It provides advice to the COP.

18. The exception is cases in which a group of parties agrees to meet its commitments jointly under article 4. This was done by the European Union, which has adopted an EU-wide target.

19. This was expressed at the workshop "International Roundtable on Reducing Emissions from Deforestation and Forest Degradation in Developing Countries," organized by Climate Focus, the Tropical Agricultural Research and Higher Education Center (CATIE), and Avoided Deforestation Partners, held in Brussels, October 24–25, 2007.

20. Domestic leakage is relevant for project-level crediting under national accounting because failure to account for leakage at the project level will result in a deficit's appearing on a country's account rather than being deducted from a project's credits.

21. See the Annex to the decision from COP 13 titled "Reducing emissions from deforestation in developing countries: approaches to stimulate action" (the UNFCCC document number was unavailable at the time of publication).

22. For example, in Vanuatu the invasive vine *Merremia peltata* becomes established if forest degradation by selective logging opens up too much of the forest canopy. The vine rapidly covers the trees surrounding the break, killing parts of the forest, which leads to further degradation and emissions from loss of biomass.

23. This has been reflected in the COP 13 decision titled "Reducing emissions from deforestation in developing countries: Approaches to stimulate action," which states in paragraph 6 of the Annex, "Reductions in emissions or increases resulting from the demonstration activity should be based on historical emissions, taking into account national circumstances."

24. N. Stern and others, *Stern Review on the Economics of Climate Change* (London, 2006), p. 217, citing a study carried out for the report by M. Grieg-Gran, "The Cost of Avoiding Deforestation," International Institute for Environment and Development, 2006; M. Obersteiner and others, "Economics of Avoided Deforestation," paper presented at the conference "Climate Change Mitigation Measures in the Agro-Forestry Sector and Biodiversity Futures," International Centre for Theoretical Physics, Trieste, Italy, October 16–17, 2006.

25. I. Tomaselli, "Brief Study on Funding and Finance for Forestry and Forest-Based Sector," report prepared for the United Nations Forum on Forests, January 2006.

26. M. Ogonowski and others, *Reducing Emissions from Deforestation and Degradation: The Dual Markets Approach* (Center for Clean Air Policy, 2007).

27. Brazil did not consider permanence in its submission to the Cairns workshop on RED in early 2007. However, this submission also called for public funding to pay for RED credits rather than allowing these credits to be used in the carbon market. This Brazilian example aside, a number of commentators argue that permanence should not be an issue based on the following argument: Reducing emissions from deforestation is similar to reducing emissions by, for example, building a wind farm instead of a coal-fired power station. There is a risk that the coal "saved" by building a wind farm instead of a power station will still be dug up and burnt in the future. There is no difference in the risk that a forest protected now will not be destroyed in the future. Therefore, because the wind farm's emission reductions are considered permanent, emission reductions from protecting forests should also be considered permanent.

28. See www.v-c-s.org.

29. Stern and others, *Stern Review*, p. 537.

30. Credits from CDM afforestation and reforestation projects are excluded from the EU Emissions Trading Scheme.

14

An Accounting Mechanism for Reducing Emissions from Deforestation and Degradation of Forests in Developing Countries

DANILO MOLLICONE, SANDRO FEDERICI, FRÉDÉRIC ACHARD, GIACOMO GRASSI, HUGH D. EVA, EDWARD NIR, ERNST-DETLEF SCHULZE, AND HANS-JÜRGEN STIBIG

Tropical deforestation is an important issue in the debate over the global carbon cycle and climate change. The release of CO_2 due to tropical deforestation can be estimated from three main parameters: the level of tropical deforestation and degradation, the spatial distribution of forest types, and the amount of biomass and soil carbon for different forest types. Our knowledge of the rates of change of tropical forests and the distribution of forest types has greatly improved in the last few years through the use of earth observation technology. At the same time, more information has become available about carbon stocks for different forest types.[1] Using recent figures on rates of net change for the world's tropical forests and refereed data on biomass, the source of atmospheric carbon from tropical deforestation is estimated to have been between 1.1 ± 0.3 gigatonnes of carbon per year (Gt/C yr^{-1}) and 1.6 ± 0.6 Gt/C yr^{-1} for the 1990s.[2] This estimate includes emissions from conversion of forests and loss of soil carbon after deforestation and emissions from forest degradation. It can be compared with CO_2 emissions due to fossil fuel burning, which are estimated to have averaged 6.4 ± 0.4 Gt/C yr^{-1} in the 1990s. Reducing emissions from deforestation is therefore crucial in any effort to combat climate change. Reducing deforestation has many other positive aspects, such as preserving biodiversity, maintaining indigenous rights, and potentially bringing resources to local populations.

The issue is even more important in the light of predicted future increases in deforestation rates. Between 1990 and 2000, the world's total area under agricultural or forest use decreased at a rate of 6.9 million hectares a year, dropping from 41.9 percent to 41.3 percent because of conversion to settlements or abandonment of agricultural or forest use due to soil degradation or desertification. This global pattern is the sum of two opposite trends: land area under agricultural use is increasing, and land area under forest use is decreasing. Furthermore, these trends are linked to development, with developed countries decreasing their agricultural land and increasing their forest area, and developing countries doing the opposite.[3]

Here we propose an accounting mechanism that includes options for determining global and national baselines of forest conversion. The accounting mechanism builds on recent scientific achievements related to the satellite-observation-based estimation of tropical deforestation rates and their consequences for carbon emissions and the assessment of intact forests.[4] We analyze these scientific and technical achievements in the context of one item in the UN Framework Convention on Climate Change (UNFCCC), "reducing emissions from deforestation in developing countries."[5]

Guiding Principles in Accounting for Reduced Emissions from Deforestation

The accounting mechanism we present is based on the principle that any method for calculating the amount of carbon preserved by reducing deforestation must meet the following criteria:

—It neither competes with nor contradicts current and future provisions for mitigation. This means that the mechanism should not consist of or be more profitable than the whole national forest carbon stock—an issue currently addressed under articles 3.3 and 3.4 of the Kyoto Protocol (KP). Instead, such a mechanism should be an additional instrument applicable to countries without KP mitigation commitments, that is, current non–Annex I Parties.

—It correctly documents the amount of carbon that has been preserved at a national level. In tropical forests, conversion of land to other uses is often preceded by forest exploitation, with significant losses of carbon stocks.[6] According to the definition adopted under the UNFCCC, a forest can contain from 10 to 100 percent tree cover; only when cover falls below 10 percent can land be classified as nonforest.[7] Forest exploitation results in carbon pools ranging from fully stocked (100 percent of the original forest biomass remains) to highly degraded (only 10 percent tree canopy is left), even though the land remains classified as forest. For this reason the degradation of fully stocked forests (leading to the reduction of tree canopy cover to as little as 10 percent) could cause a greater loss of carbon than the conversion of already degraded forests (which may be only just above the 10 percent threshold) to other land use. Consequently, conversion processes within the forest category need to be accounted for in addition to forest-to-nonforest conversions.

—It considers only avoided loss of carbon stocks in existing national forests—that is, avoidance of gross deforestation—while excluding forest area expansion (afforestation, reforestation) and forest regrowth (forest management).

—It accounts both for reduction of national forest conversion rates in countries where they are high and for prevention of increases in rates in countries where they are low or have not yet taken hold. This is in order to recognize past efforts made by countries to halt deforestation, and it could counteract the displacement of deforestation between countries (cross-border leakage) and the creation of perverse incentives.[8]

—It can be applied only if emission reductions are achieved by participating countries at the aggregate (worldwide) or semi-aggregate (regional) level. In other words, in order to guarantee environmental integrity, the proposed mechanism would result in incentives to participating countries only if a net reduction in emissions from deforestation were reached at a regional scale or worldwide. This would avoid the displacement of deforestation between countries (cross-border leakage).

—It never results in debts for participating countries that fail to reduce their emissions from forest conversion (that is, it has "no-lose" targets). This is in order to encourage voluntary participation as much as possible while avoiding the possibility that a least developed country (LDC) is asked to pay because its essential income needs have caused deforestation to increase.

—It can be easily and widely applied at national and global scales.

A Measurement System to Account for Reduced Forest Conversion

The forest definition in the UNFCCC framework does not discriminate between forests with different carbon stock levels.[9] Consequently, any accounting mechanism based on this definition will be ineffective in addressing anthropogenic emissions from tropical forests, because it will miss all the carbon lost from forest degradation. Policymakers have tried to reach agreement on a common definition of forest degradation, but no consensus yet exists, and some proposed definitions, such as temporary loss of biomass or canopy cover, could turn out to be impossible to measure or to track over time.

We define forest degradation as the conversion of a forest from one of two subcategories to the other, with a different carbon content. Both subcategories remain within the framework of the UNFCCC forest definition. The first subcategory is that of *intact* forest. These are fully stocked, in that tree cover can range from 10 to 100 percent but must be undisturbed—for example, no timber extraction has taken place. The second subcategory, *non-intact* forest, is not fully stocked. That is, even though tree cover is greater than 10 percent, qualifying as a forest under existing UNFCCC rules, the forest may have undergone some timber exploitation.

The distinction between intact and non-intact forest is important to make, given the current limitation in knowledge about the spatial distribution of biomass. In the future this proxy parameter of biomass (intactness) could be replaced by accurate, spatially explicit estimates of biomass when available. For now the distinction allows us to account for carbon losses from forest degradation—that is, from the conversion of intact forest to non-intact forest—without introducing a definition of forest degradation that has not yet been achieved under the UN's Intergovernmental Panel on Climate Change (IPCC).[10]

Adoption of the "intactness" criterion is also driven by technical and practical reasons. In compliance with current UNFCCC practice, the identification of forests according to the established 10–100 percent cover rule falls under the parties' responsibilities. When the conditions of forests are assessed using satellite remote sensing, the "negative approach" can be used to discriminate between intact and non-intact forests: disturbances such as the development of roads can be easily detected, whereas the absence of such visual evidence of disturbance can be taken as evidence that what is left is intact.[11] Intact forests were originally defined for boreal ecosystems according to six criteria,[12] but these can easily be adapted for our purpose to tropical ecosystems. That is, intact forests meet the current UNFCCC definition of forest; are larger than 1,000 hectares and no narrower than 2 kilometers; contain a contiguous mosaic of natural ecosystems; are not fragmented by infrastructure; show no sign of significant human transformation (the minimum size of isolated deforested or degraded patches, to be identified from satellite imagery, is 5 hectares); and exclude burned land and forest regrowth.

Disturbance is easier to identify unequivocally from satellite imagery than are the forest ecosystem characteristics that would have to be determined if we followed the "positive approach," that is, identifying intact forest and then determining that the rest is non-intact. With the negative approach, conversions from intact forest to non-intact forest and other land uses can be easily measured worldwide through earth-observation satellite imagery. In contrast, forests according to other definitions, such as pristine, virgin, primary, and secondary, are not always measurable.

Forest Conversion

The reduction of deforestation is implemented under a *reduced forest conversion* mechanism. We present the technical options for the accounting system for such a mechanism in the context of the following working definitions:

—*Forest conversion* is defined as one of three potential changes: from intact forest to non-intact forest; from intact forest to other land use; and from non-intact forest to other land use. In particular, once an intact forest has been exploited, it is considered non-intact forest unless it is converted to other land use.

—For each type of potential change, *reduced forest conversion* is defined as the reduction of the conversion rate below a global or national baseline.

—*Reduced deforestation* is then defined as the sum of the preserved forest carbon stocks arising from reductions of the three forest conversion processes.

In order to counteract the displacement of deforestation from countries with high forest conversion rates toward countries with low rates (cross-border leakage), remuneration for avoiding forest conversion should be applicable to both situations. With this aim in mind, we propose to use a global baseline rate to discriminate between countries with low forest conversion rates and those with high rates.[13]

If a hypothetical remuneration mechanism were based only on national baselines, then countries with low forest conversion rates would see little or no benefit in making further reductions—if indeed such reductions were possible. A country with no forest conversion under way could not gain credits from reduced forest conversion, because there is no conversion to avoid. In these kinds of countries, deforestation could easily start, and the mechanism would fail in reducing deforestation worldwide.

Furthermore, in order to apply such a mechanism equally to all countries, data should be compared relative to each country's forest area. This would be possible by reporting forest conversion measures as rates of change. Measuring annual changes in areas of intact, non-intact, and nonforest land relative to total intact and non-intact forest areas allows direct comparisons to be made between countries.

Determining Forest Conversion Rates

Tropical deforestation rates have been high since the 1960s, although with considerable interannual and geographical variability.[14] Establishing a reference value for a baseline deforestation rate is therefore to some extent an open question, but it should certainly be obtained from a representative period of at least a few years' duration. How many years? The monitoring of tropical deforestation through satellite imagery has been carried out systematically since the early 1990s. This experience has highlighted methodological and technical constraints. Analysis of existing online archives for suitable satellite imagery suggests that an optimal solution for setting baselines would be to use the period from 1990 to 2005.[15] A period of this length offers the opportunity to use a large part of the existing satellite image archives, thus allowing the collection of representative and independent historical data on forest conversion and taking into account regional variations in interannual rates, not just regional variability in overall rates of deforestation. Using satellite imagery to fix a baseline has the added advantages of global consistency (the same measurement protocols and basic observation data can be used everywhere), completeness (forest inventories and change statistics pertaining to different KP parties vary in geographical completeness and time span, and the satellite image archives can be used to harmonize this), and neutrality (the same measurements applied to the same data could avoid variations arising from regional differences in inventory and the like).

Average forest conversion rates can be calculated with only two satellite imagery surveys, one at the start and one at the end of the selected period, because earth observation techniques allow forest conversions to be tracked spatially and accurately.[16] This makes the method easily applicable everywhere across the tropics. Other solutions, such as identifying baselines on the basis of temporal trends, require more data (that is, long time series) and appropriate models, making them difficult to implement everywhere.

An additional consideration when the baseline is established is the need to counter in-country leakage. If measures for reduced emissions from deforestation are implemented in only part of a country, it is likely that greater pressure will be exerted elsewhere, resulting in the displacement of deforestation rather than in reduction. In order to avoid such leakage within a country, we propose to set baselines at the national level.

Once a country's baseline conversion rate is set, it can be used as a benchmark against which reduced forest conversions can be measured for any subsequently defined accounting period. In addition to their use for establishing baseline rates of change, rates of conversion between intact forest, non-intact forest, and other land use can be measured through earth observation techniques for any such accounting period.

Assigning Carbon Stocks to Forest Classes

Whereas capacities for monitoring forest area changes are mature, can the same be said for researchers' capacity to measure changes in stand biomasses for different forests? Figures for total carbon stocks in different types of intact forests could be taken from the literature.[17] Considering that changes in forest carbon stocks are driven largely by changes in tree density, that direct measures of carbon stocks are often impossible or very difficult on a large scale, and that there are currently no published figures for forests falling into our non-intact forest subcategory, for the purposes of this chapter we propose to set the carbon stocks of non-intact forests at half the carbon stocks of intact forests. Obviously, countries may use figures for carbon stocks based on forest inventory methodologies or other country-specific data when available. For example, in Papua New Guinea a set of 135 ground sample sites exists in which timber volume has been monitored over fourteen years. These point data can be used at the national level. Carbon stock figures would have to be determined before the start of the accounting period.

Once established, carbon stock figures are used to calculate three carbon preserving factors (CPFs), expressed in tonnes of carbon per hectare per year (t C/ha^{-1}), one for each of the three forest conversion categories (intact to non-intact, intact to nonforest, non-intact to nonforest). CPFs will vary according to main forest type, for which carbon stock figures would be predetermined—for example, humid tropical and dry tropical.

Determining Accountable Preserved Carbon Quantities

The backbone of the accounting mechanism we propose is the *reduced conversion rate* (RCR, in percentage per year), a quantitative expression of a country's efforts to reduce deforestation rates where they are high or to maintain low rates where applicable. RCRs are calculated for each of our three conversion categories. To calculate the RCR, four parameters must be assessed for each type of forest conversion:

—The *global conversion baseline* (GCB) rate is the average annual rate (in percentage per year) of forest area conversion measured during the baseline period (1990–2005) at a global scale. The GCB rate is held constant into and throughout any subsequent accounting period.

—The *national conversion baseline* (NCB) rate is the average annual rate (in percentage per year) of forest area conversion measured during the baseline period (1990–2005) at the country level. Again we consider this to be constant into and during the accounting period.

—The *global conversion accounting* (GCA) rate is the average annual rate (in percentage per year) of forest area conversion measured during the accounting period at a global scale.

—The *national conversion accounting* (NCA) rate is the average annual rate (in percentage per year) of forest area conversion during the accounting period at the country level.

The proposed accounting mechanism introduces first a condition related to the mechanism's global positive contribution at the end of the accounting period. The mechanism would provide incentives only if it leads to a global reduction of carbon emissions due to forest conversion. The term *global* could refer to the sum of all countries participating in the mechanism at either a regional or a world level. The *global accounted preserved carbon* (GAPC), expressed in tCO_2, can be expressed as:

$$\text{GAPC} = n_y \times \Sigma_{ft} \Sigma_{fct} [[\text{GCB}(fct) - \text{GCA}(fct)] \times \text{CPF}(fct) \times \text{Area}(fc)],$$

where n_y is the number of years in the accounting period, *ft* is the forest type (for example, humid tropical, dry tropical), *fc* is the forest subcategory (intact, non-intact), and *fct* is the forest conversion type (intact to non-intact, intact to non-forest, non-intact to nonforest) (figure 14-1).

If the global accounted preserved carbon is positive—that is, GAPC > 0—then two schemes for the accounting of preserved carbon are considered: one for countries with high conversion rates, where the desired outcome is that they reduce their rates, and another for countries with low conversion rates, which do not need to reduce their rates. We discriminate between these two conditions on the basis of the relationship between NCB and GCB/2 (half the global conversion rate during the baseline period). If NCB ≥ GCB/2, then the country has to reduce its conversion rate in order to be able to account for preserved carbon. If NCB < GCB/2, then the country is to keep its conversion rate below half the global rate in order to be able to account for preserved carbon. The reduced conversion rate (RCR) is then calculated as follows (figure 14-2):

$$\text{If NCB} \geq \text{GCB}/2, \text{ then RCR} = \text{NCB} - \text{NCA}.$$
$$\text{If NCB} < \text{GCB}/2, \text{ then RCR} = \text{GCB}/2 - \text{NCA}.$$

Once the RCRs of a country have been determined for each of the three forest conversion types, they can be used to calculate the area of intact and non-intact forest "preserved" by multiplying the RCRs by the area of the corresponding forest subcategory. The preserved forest area for each category can then be converted to preserved carbon by multiplying by the appropriate carbon preserving factors (CPFs). The sum of the resulting values (preserved carbon per conversion type) is then multiplied by the number of years (n_y) in the accounting period to represent the total amount of accountable preserved carbon (APC_{ft}) expressed in tonnes of CO_2:

$$APC_{ft} = n_y \times \Sigma_{fct} [RCR(fct) \times CPF(fct) \times Area(fc)].$$

This calculation procedure is applied for each forest type present in the country, and the total amount of national accountable preserved carbon (NAPC) of the country results from the sum of every APC_{ft}. Finally, the NAPC may be used to distribute the global accounted preserved carbon to the participating countries proportionally to their NAPC. That is, the benefits resulting from reducing emissions from deforestation would be shared between the high-rate countries, which reduced deforestation, and the low-rate countries, which did not increase their deforestation over GCB/2.

Hypothetical Application of the Accounting Mechanism

Table 14-1 provides, for Brazil, Congo, and Papua New Guinea, examples of estimates of the potential for preserved carbon related to avoided forest conversion,

Figure 14-1. *Procedure for the Accounting of Preserved Carbon*

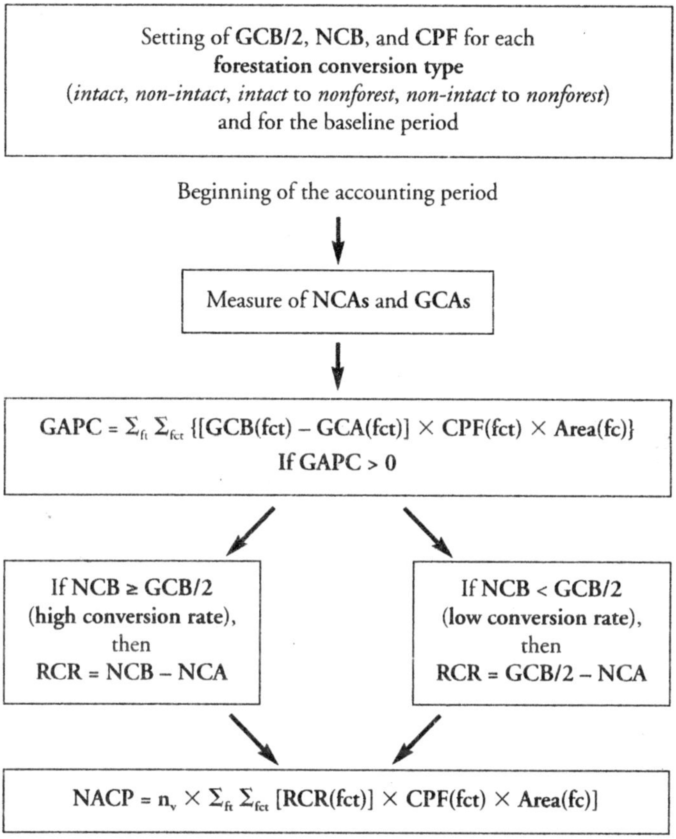

Where
GCB = global conversion rate during baseline period
NCB = national conversion rate during baseline period
CPF = carbon preserving factors
NCA = national conversion rate measured during accounting period
GCA = global conversion rate measured during accounting period
GAPC = global accounted preserved carbon
RCR = reduced conversion rate
NAPC = national accountable preserved carbon
ft = forest type (humid tropic, dry tropic, etc.)
fc = forest subcategory (intact, non-intact)
fct = forest conversion type (intact to non-intact, intact to nonforest, non-intact to nonforest)
ny = number of years in accounting period

Figure 14-2. *Examples of High and Low Conversion Rates during Baseline Period*

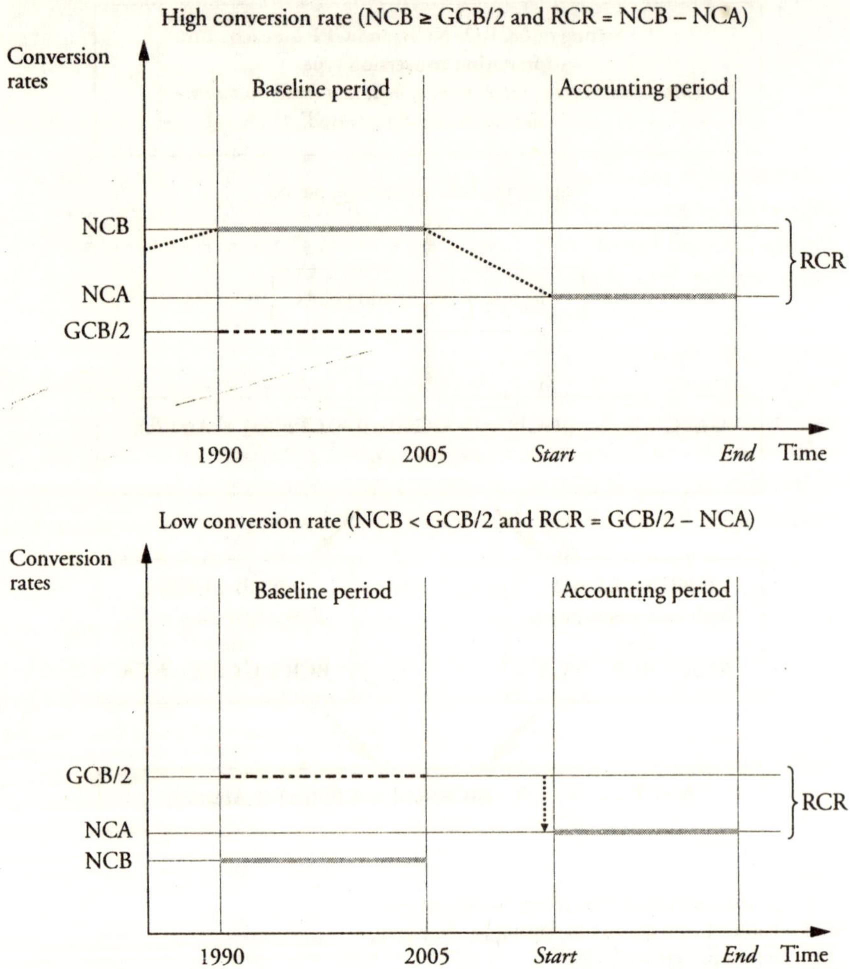

calculated by means of the preceding equations. All input parameters (forest areas, forest conversion rates, carbon preserving factors) will have to be assessed by combining earth observation techniques with field surveys for the baseline and accounting periods for all forest types (for example, humid, dry) and subcategories (intact, non-intact). Hypotheses of a 10 percent reduction for all forest conversion rates for Brazil, of a 10 percent reduction in conversion from intact forest to other types for PNG, of no change for all conversion rates for Congo, and of no change for the non-intact to nonforest conversion rate for PNG lead to the following estimates of national accountable preserved carbon (NAPC) at the end of the accounting period: 720 million and 52 million t CO_2 for Brazil and PNG,

Table 14-1. *Examples of Calculations of Preserved Carbon for Brazil, Congo, and Papua New Guinea (PNG)*[a]

	Brazil					
Item	*Humid*	*Cerrado*	*Total*	*Congo*	*PNG*	*Global*[b]
Forest area in 2005 (10^6 ha)[c]						
Intact (FAO primary)	301	44	416	7.5	25.2	765
Non-intact (FAO secondary)	41	16	56	15.0	4.1	475
Total	341	131	472	22.4	29.3	1,303
Carbon preserving factor (t C/ha^{-1})						
Intact to nonforest	186	47	—	143	151	—
Non-intact to nonforest	93	23.5	—	71.5	75.5	—
Intact to non-intact	93	23.5	—	71.5	75.5	—
Annual forest area change during baseline period, 1990–2005 (10^3 ha/yr^{-1})[d]						
Intact to nonforest	–980	–520	–1,500	0	–130	–2,800
Non-intact to nonforest	–880	–470	–1,350	–10	–10	–5,000
Intact to non-intact	–970	–510	–1,480	0	–140	–2,900
Annual conversion rate during baseline period 1990–2005: NCB (%)						*½ GCB*
Intact to nonforest	–0.31	–0.43	–0.35	–0.04	–0.48	–0.19
Non-intact to nonforest	–2.16	–2.98	–2.39	–0.10	0.00	–0.51
Intact to non-intact	–0.32	–0.44	–0.34	–0.04	–0.52	–0.12
Annual conversion rate during accounting period 2013–17: NCA (%)[e]						
Intact to nonforest	–0.28	–0.39	—	–0.04	–0.44	—
Non-intact to nonforest	–1.94	–2.68	—	–0.10	0.00	—
Intact to non-intact	–0.29	–0.40	—	–0.04	–0.47	—

(continued)

respectively, related to the reduction of their high conversion rates, and 120 million and 12 million t CO_2 for Congo and PNG, respectively, related to the preservation of their low conversion rates (non-intact to nonforest for PNG).

The Proposed Mechanism in Relation to Reporting Requirements under the UNFCCC

Existing rules (IPCC guidelines and guidance) pertinent to Annex B Parties to the Kyoto Protocol for reporting on deforestation may be consistently applied to non–Annex I Parties if we consider "gross deforestation" the main indicator of carbon losses, because the same definitions and principles have been adopted for this proposed mechanism. In fact, remote sensing techniques could be widely applied to non–Annex I Parties, providing data on land representation at Tier 3, while data on carbon stocks could be provided at Tier 1–2 (depending on the availability and quality of related data at the national level).

Table 14-1. *Examples of Calculations of Preserved Carbon for Brazil, Congo, and Papua New Guinea (PNG)*[a] *(continued)*

Item	Brazil			Congo	PNG	Global[b]
	Humid	*Cerrado*	*Total*			
Reduced forest conversion rate: RCR1 (NCA – NCB) or RCR_2 (GCB/2 – NCA) (%)						
Intact to nonforest	0.03	0.04	—	0.15	0.05	—
Non-intact to nonforest	0.22	0.30	—	0.43	0.52	—
Intact to non-intact	0.03	0.04	—	0.08	0.05	—
Annual accountable preserved carbon (10^3 t CO_2/yr^{-1})[f]						
Intact to nonforest	64,000	8,600	—	5,900	6,800	285,000
Non-intact to nonforest	30,000	4,000	—	16,700	2,400	191,000
Intact to non-intact	32,900	4,400	—	1,500	3,700	100,000
NAPC for five-year commitment period (Mt CO_2)		720		121	64	2,900

a. Estimates are extrapolated from FAO net change estimates and should be considered indicative.

b. In Food and Agriculture Organization (FAO), *Global Forest Resources Assessment 2005* (Rome, 2006), primary and secondary forest areas are not reported for a number of countries, such as Venezuela and India. Consequently, global estimates of areas, changed areas, and rates correspond to only part of the tropical forest domain—1,303 million hectares of forests—relative to the total tropical forest domain, 1,810 million hectares in 2005.

c. National forest areas and change rates for the period 1990–2005 are derived from FAO, *Global Forest Resources Assessment 2005.*

d. Intact and non-intact forest areas are taken as FAO's primary and secondary forest areas (FAO, *Global Forest Resources Assessment 2005*). The annual change areas from intact forests to non-intact forests are approximated at 0.52 using the 2002 ratio between gross deforestation (INPE, *Monitoramento da Floresta Amazônica Brasileira por Satelite: Projeto PRODES*, 2006 [www.obt.inpe.br/prodes/index.html]) and the logging rate estimate for Brazilian Amazonia (G. P. Asner and others, "Selective Logging in the Brazilian Amazon," *Science* 310 [2005]: 480–82).

e. These change rates for the period 2013–17 are hypothetical.

f. Global APC figures are estimated from all countries with available data (that is, for a 1,303 million ha forest domain) and correspond to a hypothesis of a 10 percent reduction for high rates and the preservation of low rates.

In our proposal, lack of accuracy is partially addressed by assumptions about changes in carbon stocks due to forest conversion from intact to non-intact forest (that is, carbon stocks are reduced to half) and from non-intact forest to other use (that is, carbon stocks are zeroed) and by the exclusion of any removals from the accounting. Further conservative assumptions could be added, such as the adoption of a discount factor for forest management under article 3.4, in order to help address the (ultimately insoluble) problem of accuracy of carbon stock data for developing countries.

Moreover, regarding carbon stocks, we believe the lack of "accurate" data from most tropical countries should not necessarily be considered a major obstacle to a mechanism for reducing emissions from deforestation. Although we are far from having accurate estimates of total carbon stock changes from deforestation (for most tropical countries, data availability is "relatively good" for above-ground biomass but poor for other pools, especially soil), we think it is feasible to develop conservative estimates that are realistically measurable and useful. Our reasoning is that during any forest conversion, all the pools (biomass, dead organic matter, and soil organic matter) can be assumed to be sources of carbon, and if data from a pool are unavailable, then total emissions will be underestimated. Consequently, if the deforested area decreases, then the total reduction of emissions (national accountable preserved carbon in our proposed mechanism) will also be underestimated. Thus the resulting estimate of total reduction of emissions, although imperfect, will be conservative. If this conservative approach is accepted (as it is already for current accounting under the Kyoto Protocol), then the quantity of data necessary to participate in a mechanism for reducing emissions from deforestation could be considerably less than that required for "accurate" estimates (for example, a country could participate even with no reliable data for some pools).

Conclusion

Our proposed mechanism represents one of the first attempts to create an operational, scientifically grounded tool that is suitable to account for the amount of carbon that would be preserved from reducing deforestation in countries with high forest conversion rates and maintaining low forest conversion rates in the other countries. The mechanism has been developed according to the main principles of the Rio Conventions (for example, global baselines to pursue the equity principle; accounting for conversion from intact to non-intact forest to follow the effectiveness principle) and of the UNFCCC (for example, use of a conservative technical option to calculate the baselines—based on rates and not on parameters such as area or trends—which guarantees the additionality of actions). Furthermore, our approach responds to issues raised at the eleventh Conference of the Parties by a few tropical countries that wanted to consider proactive moves to reduce deforestation rates in their territories.

The proposed accounting mechanism is not in contrast to or in competition with the current set of provisions for mitigation (for example, the Kyoto Protocol). By distinguishing two forest subcategories, intact and non-intact, by accounting on a national basis, and by applying a global baseline for forest conversion, this proposal realistically accounts for preserved carbon stocks and considers possible carbon losses related to forest degradation and the displacement of deforestation within and among countries.

We developed this proposal on the basis of recent scientific achievements related to the estimation of tropical deforestation rates and their consequences on carbon emissions and to the assessment of intact forests. The current state of earth observation techniques using satellite sensors allows one to measure forest conversion rates rapidly and widely, at national and global scales. It is important to stress, however, that although the data set is certainly global and a single, universally agreed-upon measurement protocol could be adopted, there is no need for centralized analysis once a global baseline conversion rate has been agreed upon. Calculation of the reduced conversion rates and the resulting national accountable preserved carbon would firmly remain the responsibility of individual parties: common data, common methodology, but independent implementation.

The technological basis of the accounting mechanism presented here is a reality. Driven in part by the desire to achieve verified reductions in deforestation rates and in part by the use of earth observation technology, such an approach has been demonstrated recently in Brazil, where a 30 percent reduction in deforestation rates has been measured between 2003–04 and 2004–05, from 2.72 million hectares to 1.89 million hectares per year.[18] As data from ground surveys become increasingly available, the carbon preserving factor will be calculated from more reliable forest-type and country-specific data. Therefore this proposal shows that a mechanism for reducing emissions from deforestation could be implemented operationally in line with decisions governing current reporting under the UNFCCC. The science to support it is mature, and the required technical capabilities already exist.

Notes

1. R. A. Houghton, "Aboveground Forest Biomass and the Global Carbon Balance," *Global Change Biology* 11 (2005): 945–58; M. Santilli and others, "Tropical Deforestation and the Kyoto Protocol: An Editorial Essay," *Climatic Change* 71 (2005): 267–76; P. Mayaux and others, "Tropical Forest Cover Change in the 1990s and Options for Future Monitoring," *Philosophical Transactions of the Royal Society*, Series B, 360 (2005): 373–84; Food and Agriculture Organization, "Global Forest Resources Assessment 2005: Progress towards Sustainable Forest Management," Forestry Paper 147 (Rome: FAO, 2006), p. 322.

2. F. Achard and others, "Determination of Deforestation Rates of the World's Humid Tropical Forests," *Science* 297 (2002): 999–1002; F. Achard and others, "Improved Estimates of Net Carbon Emissions from Land Cover Change in the Tropics for the 1990s," *Global Biogeochemical Cycles* 18 (2004): GB2008; Houghton, "Aboveground Forest Biomass."

3. FAO, *Summary of World Food and Agricultural Statistics* (Rome, 2004); E. Lepers and others, "A Synthesis of Rapid Land-Cover Change Information for the 1981–2000 Period," *BioScience* 55 (2005): 115–24; FAO, "Global Forest Resources Assessment 2005."

4. Mayaux and others, "Tropical Forest Cover Change in the 1990s"; Achard and others, "Improved Estimates of Net Carbon Emissions"; N. Ramankutty and others, "Challenges to Estimating Carbon Emissions from Tropical Deforestation," *Global Change Biology* 13 (2007): 51–66; A. Yaroshenko, P. Potapov, and S. A. Turubanova, *The Last Intact Forest Landscapes of Northern European Russia* (Moscow: Greenpeace Russia, 2001), p. 75.

5. UNFCCC, "Draft Conclusions for Agenda Item 6: Reducing Emissions from Deforestation in Developing Countries," 2005 (http://unfccc.int/resource/docs/2005/cop11/eng/l02.pdf).

6. G. P. Asner and others, "Selective Logging in the Brazilian Amazon," *Science* 310 (2005): 480–82.

7. UNFCCC, "Definitions, Modalities, Rules and Guidelines Relating to LULUCF Activities under the Kyoto Protocol," 2001 (http://unfccc.int/resource/docs/cop6/05a03v04.pdf).

8. E.-D. Schulze and others, "Making Deforestation Pay under the Kyoto Protocol?" *Science* 299 (2003): 1669.

9. UNFCCC, "Definitions, Modalities, Rules."

10. J. Penman and others, eds., *Definitions and Methodological Options to Inventory Emissions from Direct Human-Induced Degradation of Forests and Devegetation of Other Vegetation Types* (Kanagawa, Japan: Institute for Global Environmental Strategies, 2003).

11. Yaroshenko, Potapov, and Turubanova, *Last Intact Forest Landscapes.*

12. Greenpeace, *The World's Last Intact Forest Landscapes* (Moscow: Greenpeace Russia, 2006).

13. D. Mollicone and others, "An Incentive Mechanism for Reducing Emissions from Conversion of Intact and Non-Intact Forests," *Climatic Change* 83 (2007): 477–93.

14. FAO, "Global Forest Resources Assessment 2005"; Achard and others, "Determination of Deforestation Rates."

15. Schulze and others, "Making Deforestation Pay"; D. Mollicone and others, *Land Use Change Monitoring in the Framework of the UNFCCC and Its Kyoto Protocol: Report on Current Capabilities of Satellite Remote Sensing Technology* (Luxembourg: European Communities, 2003), p. 48.

16. Achard and others, "Determination of Deforestation Rates."

17. See, for example, Achard and others, "Improved Estimates of Net Carbon Emissions"; FAO, "Global Forest Resources Assessment 2005"; Houghton, "Aboveground Forest Biomass."

18. Instituto Nacional de Pesquisas Espaciais, *Monitoramento da Floresta Amazônica Brasileira por Satelite: Projeto PRODES*, 2006 (www.obt.inpe.br/prodes/index.html).

CASE STUDY

Creative Financing and Multisector Partners in Madagascar

JEANNICQ RANDRIANARISOA, BEN VITALE, AND SONAL PANDYA

The Ankeneny-Zahamena-Mantadia Biodiversity Conservation Corridor and Restoration Project ("Mantadia") is situated in the eastern portion of Madagascar, touching on five major protected areas and a forest station. These are Zahamena National Park, the Zahamena Strict Nature Reserve, the Mangerivola Special Reserve, Mantadia National Park, the Analamazaotra Special Reserve (commonly known as Andasibe Reserve), and the Vohimana Forest. The government of Madagascar, working through the Ministry of Environment, Water, and Forests and a network of national and international nonprofit organizations, including Conservation International, formed a partnership to design and launch this forest restoration and conservation project.

The project consists of three main components: the restoration of 3,000 hectares of natural forest corridors to restore viable biological connectivity among several isolated forests and protected areas; promotion of sustainable cultivation systems to increase soil fertility, protect watersheds, and stabilize land-use patterns across 2,000 hectares; and protection of 425,000 hectares of native forest by reducing deforestation driven by unsustainable agricultural expansion and fuelwood harvesting. The Kyoto-compliant restoration activities involve planting more than ninety native species in clusters that resemble natural forest succession in order to encourage natural regeneration. Along with generating carbon offsets, these activities will protect against siltation of irrigated rice fields, soil erosion, and flooding.

The sustainable forest and community gardens will provide potential alternative uses of degraded agricultural land, especially hillsides, that is no longer useful to local people and is at risk of further degradation by soil erosion. Besides fruit trees, the forest gardens will be composed largely of local trees and plants, which can be planted to mimic local forest structure and function. These gardens will enable a shift in land use from slash-and-burn (*tavy*) agriculture to more sustainable activities. The resulting valuable products will provide food and income to local communities. The establishment of fuelwood plantations will provide additional fuel sources to local communities that currently use wood from existing natural forests. These activities will help minimize potential leakage associated with forest protection activities.

The forest protection component will prevent the loss of one of the last remaining areas of contiguous forest in eastern Madagascar, namely, the large tract between Mantadia and Zahamena National Parks. As part of the project, in December 2005 the government declared the entire corridor a preliminary multiuse protected area, with the expressed goal of reducing emissions from deforestation by transitioning some areas to local community management while improving ecological monitoring and patrolling by local forest agents.

The project was designed according to the Climate, Community, and Biodiversity Standards to ensure that it would generate significant climate, local community, and biodiversity benefits. Using an approved CDM methodology, the project will sequester more than 1 million tons of CO_2 over a thirty-year period. Overall, the project activities will improve and diversify agricultural productivity on degraded slopes and protect future development options for the region, including eco-tourism, irrigated agriculture, and nontimber and fiber-based forest products. In addition, the project will employ more than 200 members of the local community during the seven-year forest establishment period.

People downriver of the project site may benefit from reduced risk of flooding, reduced siltation of irrigated rice fields, and more consistent hydrological flows. The local economy will benefit from the increased agricultural alternatives provided by people who switch to sustainably growing a wider range of crops with market value. The project will also protect this attractive eco-tourism destination, which holds the most heavily visited sites in Madagascar. The local and regional economies will benefit from increased tourism revenues, sustainable production of fuelwood, sale of a diverse range of fruits, and low-level, sustainable production of high-value timber from native trees. Finally, the project will protect and restore a globally important and threatened high-biodiversity corridor, avoiding the release of tens of millions of tons of CO_2 into the atmosphere and simultaneously enabling the movement of endangered species and seed dispersers over greater ranges.

One lesson learned from the Mantadia project is that a landscape-scale project that designs in multiple benefits can effectively tap a mix of carbon offset financing and philanthropic and development funding. The Mantadia project was designed to include the multiple benefits of carbon sequestration from reforestation, creation of sustainable livelihoods, reduced emissions from deforestation, and biodiversity protection. This has allowed the government of Madagascar to structure the project's financing to leverage multiple revenue streams. The project partners expect that future carbon offset sales will generate more than one-third of project revenues, with additional funding for environmental protection coming from the government of Madagascar, biodiversity conservation organizations, economic development assistance for community activities, and other sources.

Philanthropic and development funding was used to define the project, develop a community outreach program, and provide start-up funding for early project activities. Carbon financiers often limit up-front payments for future carbon credits to less than 20 percent of total payments, and they pay the balance of funds upon certification and delivery of the credits. Therefore it was important to obtain a mix of early project revenue, including revenue from the government of Madagascar, for project activities that were front-loaded, such as community engagement and forest restoration.

The largest source of financing for the project will come from carbon investors such as the World Bank's BioCarbon Fund. These investors are purchasing carbon credits produced by Kyoto-eligible forest restoration activities and non-Kyoto emission reductions generated by forest protection. Additional financing to support the sustainable livelihood components, capacity-building, and biodiversity conservation has been provided by the U.S. Agency for International Development, Conservation International, and other philanthropic donors.

Another lesson learned is that projects that involve significant community-managed land require best-in-class community engagement, monitoring, and enforcement. In order to create and deliver carbon credits to a carbon offset investor, project proponents must ensure that landownership and formal property rights are well defined and documented. This creates a special risk for projects in developing countries that have inaccessible or costly land-titling procedures. For this project, the government of Madagascar is establishing a local registry office near the project site to enable local families to secure land tenure.

Landscape-scale projects in developing countries require agreement and cooperation among local and national government agencies, local communities, individual farmers, and civil society organizations such as nonprofit organizations. It is critical that the project management structure and role definitions be established at the outset, to ensure acceptance by all key project stakeholders during the long time frames required to achieve the climate change mitigation, community, and conservation goals of the project.

15

A Latin American Perspective on Land Use, Land-Use Change, and Forestry Negotiations under the United Nations Framework Convention on Climate Change

MANUEL ESTRADA PORRUA AND ANDREA GARCÍA-GUERRERO

Our goal in this chapter is to provide some basic elements to facilitate understanding of the dynamics and reasons behind the positions of Latin American countries in the land use, land-use change, and forestry (LULUCF) negotiations under the United Nations Framework Convention on Climate Change (UNFCCC). To this end we present an overview of the negotiation groups involving Latin American countries and a summary of the most relevant positions assumed by them, from negotiations on the inclusion of LULUCF activities in the Kyoto Protocol to current discussions of incentives to reduce greenhouse gas (GHG) emissions from deforestation in developing countries.

With the aim of presenting the social, environmental, and economic context of the Latin American negotiation positions, we further provide a summary of the situation of the forest sector in Latin America. Finally, we analyze the importance of carbon stocks in Latin American forests and elaborate on potential opportunities in this sector to contribute to the international GHG emissions mitigation effort while advancing sustainable development goals.

The views expressed in this chapter are solely those of the authors and do not represent the position of the Mexican or Colombian government (or any other Latin American government).

The Role of Latin American Countries in Negotiations on LULUCF Issues under the UNFCCC

Although national circumstances vary widely, in most Latin American countries forests are valued highly for their environmental, social, cultural, and economic benefits. At the same time, many countries of the region are particularly vulnerable to the effects of climate change and therefore depend on the successful and timely implementation of an effective global regime to reduce GHG emissions worldwide.

Historically the region has supported the multilateral negotiation process through constructive participation and creative proposals based on the principles of sovereignty, equity, environmental integrity, sustainable development, common but differentiated responsibilities, and cost-effectiveness.

Negotiating Groups in Latin America

The parties to the UNFCCC and the Kyoto Protocol are divided into regional groups, which serve as constituencies for the purposes of nominating candidates for elections of members of the UNFCCC Bureau and other bodies such as the Clean Development Mechanism (CDM) Executive Board, the Joint Implementation Supervisory Committee, and the Compliance Committee of the Kyoto Protocol. Latin American countries are represented for these purposes by the "Group of Latin American Countries" (GRULAC). But although delegates sometimes use GRULAC meetings to informally discuss issues under negotiation, it does not act as a negotiating group.

With the exception of Mexico, all Latin American countries are part of the Group of 77 (G77) and China.[1] However, common positions of the G77 have proved particularly difficult to reach in negotiations on LULUCF issues under the Kyoto Protocol. Latin American members of the group have usually promoted the wide inclusion of carbon sinks in the international climate regime, particularly in the CDM, facing opposition from G77 members with different interests. Mexico, which left the G77 in 1994—when it joined the Organization for Economic Cooperation and Development (OECD)—and which later became a member of the Environmental Integrity Group,[2] maintains permanent cooperation with most Latin American countries on LULUCF issues in the negotiations and is usually involved in regional processes and joint submissions to the UNFCCC.

The inclusion of sinks in the CDM has been one of the most important negotiation topics under the UNFCCC for Latin American countries. This was reflected by the creation of the Latin American Initiatives Group (GRILA) during the fifth session of the Conference of the Parties (COP 5) in 1999. The creation of GRILA has been the only formally organized attempt by Latin American countries to date to promote a set of particular subjects in the UNFCCC negotiations.[3] Bolivia was one of the main promoters of this group, which included almost all the countries of the region with the exception of Brazil, Peru, and Argentina (although the latter was originally part of the group). GRILA held regular meetings that allowed its members to produce joint submissions and a common negotiation strategy.

The inclusion of sinks was also supported to some extent by Brazil, Peru, and Argentina, but many other G77 countries, such as those included in the Alliance of Small Island States (AOSIS), China, and India, opposed it. As a consequence, GRILA has been considered one of the factors that caused a weakening in the G77 in the final negotiations of COP 6.[4]

After COP 6, GRILA stopped meeting, but many Latin American countries continued working together informally. During the negotiations on modalities and procedures for afforestation and reforestation projects in the CDM, some of these countries—among them Bolivia, Chile, Uruguay, and Mexico—started a process of regional consultations, with the financial, technical, and legal support of various UN bodies (the Food and Agriculture Organization [FAO], the United Nations Environment Program [UNEP]) and the World Conservation Union (IUCN). Additionally, the process facilitated dialogue with African, Asian, and European countries. This regional effort finished after COP 10, but Latin American countries—now supported by the Centro Agronómico Tropical de Investigación y Enseñanza (CATIE), the Swiss Cooperation, and the French Organisation Nationale des Forêts—have continued holding workshops regularly, mostly on issues related to the implementation and methodological aspects of CDM LULUCF projects and, after COP 11, on international mechanisms to reduce emissions from deforestation in developing countries.

Position of Latin American Countries in LULUCF Negotiations

The inclusion of an ample variety of carbon sequestration projects in the CDM was a constant demand of many Latin American countries during the negotiations over the principles, modalities, rules, and guidelines for the flexible mechanisms under the Kyoto Protocol.

At the twelfth session of the UNFCCC Subsidiary Body for Implementation (SBI), held in 2000, Costa Rica presented a submission on behalf of Argentina, Bolivia, Chile, Colombia, the Dominican Republic, Ecuador, Guatemala, Honduras, Mexico, Nicaragua, Panama, Paraguay, and Uruguay. It stated that "the scope of eligible LULUCF projects should correspond to the activities established under Article 3.3 and those to be established under Article 3.4." The submission underscored that "projects that effectively and credibly avoid, slow, or reduce deforestation are covered under Article 3.3, whether the project includes total protection or forest management."[5]

Later that year, at the Subsidiary Bodies' thirteenth general meeting (SB 13), Honduras, on behalf of Bolivia, Chile, Colombia, Costa Rica, Ecuador, Guatemala, Mexico, Nicaragua, Paraguay, and Uruguay, proposed the inclusion of projects for the prevention of deforestation in the consolidated text on principles, modalities, rules, and guidelines pursuant to articles 6, 12, and 17 of the Kyoto Protocol.[6] Moreover, during a special session on the subject at SB 13, Brazil, Peru, Chile, Colombia, Costa Rica, Bolivia, and Uruguay, supported by the Environmental Integrity Group, New Zealand, Australia, the United States, Japan, Canada, and Norway—which often shared views on this issue—argued in favor of sinks in the CDM.

These demands were supported by creative proposals to address the main concerns associated with LULUCF projects in the CDM, namely, permanence, leakage, additionality, and environmental and social effects. Eventually the technical expertise of some countries of the region proved to be fundamental for the "environmentally safe" inclusion of sinks in the CDM. This is best exemplified by the Colombian proposal on expiring Certified Emission Reductions—presented at COP 6—to deal with the nonpermanence of carbon sequestration in LULUCF activities, which served as the basis for the temporary and long-term Certified Emission Reductions approach adopted by COP 9 in 2003.

In November 2005, during COP 11, Costa Rica and Papua New Guinea, with the support of numerous developing countries, submitted a proposal to start the consideration of options and incentives to reduce GHG emissions from deforestation in developing countries, thus reviving a historical interest of Latin American countries in the negotiations.

During the first UNFCCC workshop on this issue, Brazil presented initial thoughts on a scheme to achieve this goal. Later, at COP 12, it introduced a submission on "positive incentives for voluntary action in developing countries to address climate change: Brazilian perspective on reducing emissions from deforestation."[7]

Although some Latin American countries have shown interest in the existing proposals to deal with emissions from deforestation (that is, the Brazilian and the Papuan schemes), most of them have pointed out the need for a variety of options. According to these countries, the arrangements eventually adopted under the UNFCCC should successfully address deforestation while allowing for equitable participation, taking into account the different capacities, needs, interests, and circumstances of developing countries.

This position has been collectively expressed by many countries of the region on numerous occasions, through, among other things, a joint submission supported by more than ten countries.[8] Moreover, many Latin American countries are concerned about the emphasis put on approaches that are national in scope as an exclusive option for providing incentives to reduce emissions from deforestation through the UNFCCC, because it implies enormous institutional and technical capacities and ignores the limitations currently faced by most developing countries.

Overview of the Forest Sector in Latin America

Latin America has a large tropical forest area and houses rich biodiversity as well as vast forest plantations. According to the FAO's *Global Forest Resources Assessment 2005,* the world's forest area is 3,952 million hectares, which corresponds to 30 percent of the planet's land surface. Latin America and the Caribbean account for 924 million hectares, 23.4 percent of the global total. Six of the world's fifteen richest countries in terms of native forest resources are located in the region: Bolivia (1.46 percent), Brazil (14.26 percent), Colombia (1.50 percent), Mexico (1.59 percent), Peru (1.40 percent), and Venezuela (1.18 percent). Eight countries of the region are classified as biologically megadiverse: Bolivia, Brazil, Colombia, Costa Rica, Ecuador, Mexico, Peru, and Venezuela.[9]

From an economic perspective, the forestry sector's contribution to gross domestic product (GDP) in Latin American countries ranges from 0.1 percent to 8.0 percent. Five countries (Argentina, Brazil, Chile, Mexico, and Uruguay) account for 90 percent of forest products generation and exports in the region and are host to more than 90 percent of the region's forest plantations. Latin America and the Caribbean account for between 3 and 4 percent of global exports of wood products—a figure that could be substantially increased.[10]

The effect of rural natural resource activities (primary agriculture, forestry, and fisheries) on national growth and poverty reduction in the region is about twice as large as the sector's GDP share. In average, for each 1 percent of growth in the rural natural resource sector, there is a 0.22 percent increase in national GDP and a 0.28 percent increase in the incomes of the poorest families, which represents more than twice the expected 0.12 percent increase according to the sector's share of GDP.[11]

Protected Areas

Large human populations in the region are heavily dependent on forests for food, particularly in tropical South America. Changes to the forests therefore affect people's quality of life and their social and cultural customs directly and sometimes dramatically. Concern over deforestation and its ecological, social, and economic repercussions has led to increased efforts in conservation and improved forest management.

However, these efforts have been partly offset by national economic needs that are responsible for an increase in forest harvesting. Deforestation and forest degradation and fragmentation cause major losses in biodiversity. Weaknesses in forest management in the region are due to unclear land tenure, lack of information about proper management techniques, limited economic resources, and insufficient institutional capacity and technical knowledge. Where management plans have been drawn up, they are frequently geared exclusively toward the production of timber, neglecting any development of nonwood forest products or forest services. When forest management plans have been approved by government bodies, they may in some places amount to little more than straightforward felling permits, because of inadequate capacity to monitor the plans and ensure that they are implemented properly.[12]

Overall, more than 10 percent of the territory of the region (213.54 million ha) is currently protected. In addition, the attractiveness of the cloud and tropical rainforests and recognition of their economic, social, and cultural value by concerned persons has led to the creation of many private forest reserves, often linked to scientific research programs and eco-tourism ventures. A related trend in the 1990s was the creation of community-managed montane forest reserves.[13]

However, the indicator for protected areas reflects their legal status, not their degree of effective protection. A comparison between maps of protected areas and the locations of human settlements in countries of the region shows considerable overlap, a clear sign of human pressure.[14] Moreover, the average budget per protected area in Latin America is eight times less than that of Europe, for example, even though in the former case protected areas are likely to be far larger. Likewise, the average staff number per protected area in Latin America and the Caribbean is one of the lowest in the world, amounting to only 6 persons per area, versus 29 in Africa and Madagascar, 57 in Asia and the Pacific, and 41 in Europe and the Middle East.[15]

In Latin America, as in most developing countries, international funding for conservation has been important but insufficient. Even under a best-case scenario, in the next ten years total official development aid (ODA), Global Environmental Facility (GEF) funding, and private philanthropic contributions to forestry in developing countries will not exceed some U.S.$2.5 billion a year. Assuming a 50 percent cost of overhead and delivery, this would work out to about U.S.$1.8 per hectare per year when spread across the some 700 million hectares of forest in public protected areas in developing countries. If extended to the entire developing-country forest estate, about 1.7 billion hectares, it would work out to about U.S.$0.7 per hectare per year.

Although these sums are important contributions, they are insufficient to significantly enhance the incentive to safeguard protected areas, much less the forests outside them. Even if government, private philanthropic, and GEF funds double, they will be insufficient to substantially advance conservation in developing countries.[16] Furthermore, donations and ODA are by definition nonbinding, making them unreliable in the long term. Conservation is a long-term goal for which resources are needed for establishment, maintenance, monitoring, and, especially, providing productive alternatives to communities that deforest or degrade forest resources in order to survive.

As we look toward the next ten years, it is clear that ODA and other concessionary sources of support will remain inadequate to the challenge of the global forest estate. The sources with greatest potential to provide financial incentives for forest conservation are the CDM of the Kyoto Protocol and other emerging carbon mechanisms, such as the voluntary carbon market, international, market-based compensation mechanisms, higher prices for sustainable forest management timber, and domestic markets for environmental services. It will also be critical to build on the in-kind contributions of indigenous and other communities and smallholders who own forest resources. Rather than competing or working in parallel with these private markets, a more strategic use of ODA funds would be to leverage these private flows and incentives—transforming these markets and instruments into more positive contributors to forest conservation and poverty alleviation.[17]

Deforestation in Latin America

Currently Latin America experiences high levels of deforestation. During the period 2000–05 the net total of native forest surface was reduced by 4.7 million hectares a year, with an annual rate equivalent to 0.50 percent in South America (about 4.3 million ha per year), and 1.23 percent in Central America (around 350,000 ha per year) (figure 15-1). Conversely, net forest area increased by 0.9 percent per year in the Caribbean.[18]

As a result of habitat conversion and loss, 31 of the 178 eco-regions in Latin America are in a critical state, 51 are endangered, and 55 are vulnerable. Most endangered eco-regions are found in the northern and central Andes, Central America, the steppe and winter rainfall areas of the southern cone, the Cerrado and other dry forests south of the Amazon basin, and the Caribbean.[19] Seven of the world's 25 biodiversity hot spots (where exceptional concentrations of endemic species are undergoing exceptional loss of habitat) are located in Latin America.[20]

Considering only carbon stocks in living biomass, the projected forest loss in Latin America could entail the emission to the atmosphere of around 4,235.7 million tonnes of carbon (15,530.75 million tonnes of CO_2) between 2000 and 2020.[21] The magnitude of these emissions is overwhelming in comparison with the region's contribution to the Clean Development Mechanism. They are more than fifty times larger than the total Certified Emission Reductions (CERs) expected from current projects in the region up to 2012.[22]

On the other hand, completely avoiding those emissions would be similar to mitigating almost twenty years' worth of GHGs from road transportation in the

Figure 15-1. *Annual Net Change in Forest Area, by Region, 1990–2005*

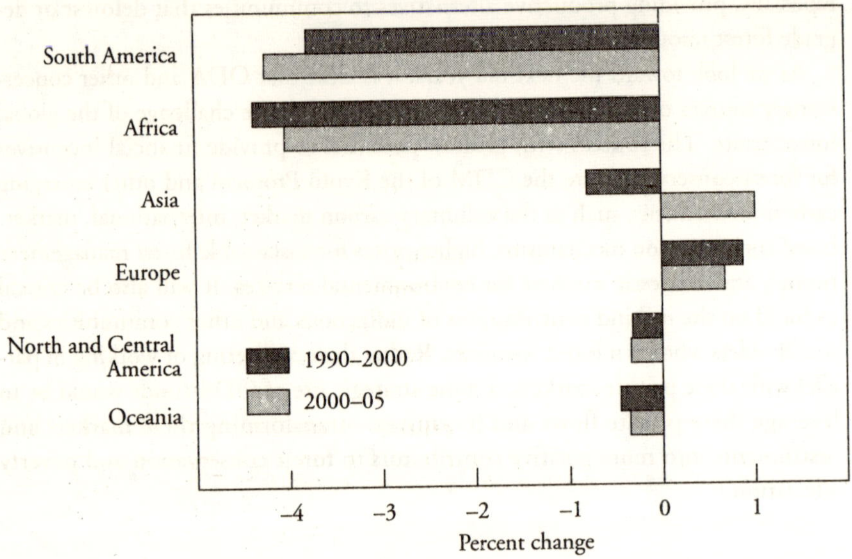

Source: Food and Agriculture Organization, "Global Forest Resources Assessment 2005: Progress towards Sustainable Forest Management," Forestry Paper 147 (Rome: FAO, 2006).

European Union—the EU's second most important source of GHGs—at their 2003 levels (790.73 million tonnes of CO_2 equivalent [t CO_2e]) and to obviating the total GHG emissions of the United States for two years at their 2004 levels (7,122.10 million t CO_2e).[23]

The expansion of the agricultural frontier has been one of the main causes of deforestation in Latin America. Commercial farmers have cleared large areas for soybeans grown for export in Brazil, Bolivia, and Paraguay, for coffee in Brazil, and for bananas in Central America, Colombia, Ecuador, and the Caribbean. Small-scale farmers also cause deforestation by employing slash-and-burn practices to extend their agricultural land into forests. Forest fires, always a natural force in forest ecosystems, have also become a major problem in the region,[24] accentuated by increased vulnerability to fire due to forest fragmentation and increased exposure to ignition sources due to slash-and-burn practices.

Land tenure regulations are also part of the problem. When legal property rights in land are unclear, people tend to clear and build on areas to establish claims to them. Forest cover may also be removed to keep areas accessible when forest communities fear that forests may be declared protected areas, limiting community rights to use the forest. This happened in Costa Rica when the government intended to expand its protected area system.[25]

Deforestation has worsened in some countries because of policies designed to increase economic growth. Livestock expansion and mechanized agriculture account for more loss of forest cover than wood production, which is concentrated in relatively few countries.[26] Subsidies are a contributing factor. Agricultural incentives can result in greater landownership and more mechanized, capital-intensive methods of production that displace farmworkers. Unemployed workers have migrated into the Cerrado of Brazil and forests in the Amazon, the department of Santa Cruz in Bolivia, and parts of Paraguay, causing further forest clearance.

The construction of roads also contributes to loss of forest cover. Some 400 to 2,000 hectares of a forest may be removed for each kilometer of new road built through it. In the Brazilian state of Pará, deforestation due to road construction increased from 0.6 percent to 17.3 percent of the state's area during 1972–85. In Ecuador, Peru, and Venezuela, mining corporations and individual miners clear large areas of forests. Additionally, biological phenomena such as the proliferation of pests are causes of irreversible damage to some forests.[27]

Civil unrest in some countries of the region has had a double effect on forests. On one hand, many times it has increased conservation and forest recovery. Wars, civil wars, and guerrilla wars generally take place in rural areas and can force landowners to cease their activities, be afraid to expand, or even abandon their land. Rural land purchase stops, so land conversion rates decrease dramatically and forest regrowth occurs. On the other hand, civil unrest causes communities to be displaced, forcing them to move to other rural areas where they generally arrive in extreme poverty and have to subsist on whatever resources they find. Many times this causes forest degradation and deforestation.

Illicit crops in Latin America cause deforestation in as many as two-thirds of land conversion cases in some countries.[28] These crops are seldom grown in the same place for long, in order for growers to avoid detection or because of governmental eradication efforts. Therefore they cause frequent land-use changes from primary and secondary forest to crop land and from legal crops and pastures to illicit plantations. Most Latin American countries in which illicit crops are grown are fighting the problem. They continually seek additional sources of funding to do so and consequently to decrease deforestation. However, one strategy used by some Latin American countries involves spraying herbicides such as glyphosphate over vast areas, which has a negative effect on forests as well.

Degradation estimates are scarce because of the difficulty and cost of measuring this type of biomass loss. The rare efforts that are made are reported on a study-site basis and therefore are inappropriate for national, regional, or global estimates. It is believed, however, that loss of biomass by degradation is significant in many Latin American countries, judging from some identified drivers that cause not land conversion but a slow decrease in biomass, such as the collection of small-diameter trees

Figure 15-2. *Total Carbon Stock (C) in Forests, by Region, 2005*

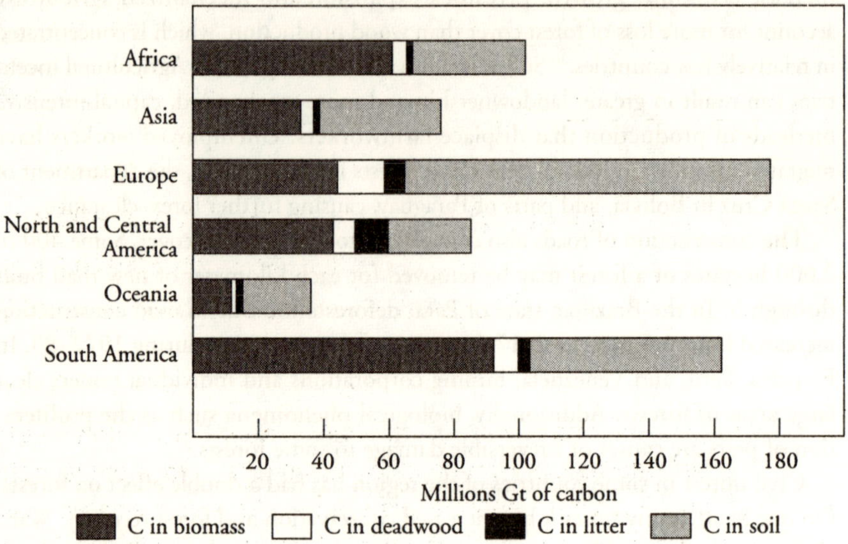

Source: FAO, "Global Forest Resources Assessment 2005."

for energy needs and the extraction of valuable species that grow at very low densities in natural forests.

GHG Emissions from Deforestation in the Region

It has been estimated that Latin America and the Caribbean hold between 18 and 26 percent of the global carbon contained in forest ecosystems, 11 percent of the carbon in pasture land, and 17 percent of the carbon in agro-ecosystems.[29] The total carbon content of forest ecosystems in Central and South America (without taking into account Mexico's forests) for 2005 ranked first at the global level in terms of carbon stock in biomass (figure 15-2). From 1990 to 2005, carbon in biomass decreased in Central America (21 percent) and South America (6 percent) while it increased in North America (5 percent) and the Caribbean (50 percent) (table 15-1). The importance of conserving the forests in the region to protect the global climate becomes evident when one observes that the carbon stock in living biomass in Central and South America (94.2 Gt) was comparable or even superior to the carbon content in China's coal reserves in 2005.[30]

The FAO estimates that if current rates of land-use change continue, Latin American forests are likely to shrink by 9 percent (around 82.8 million ha) between 2000 and 2020. Permanent pasture, permanent crop land, and arable

Table 15-1. *Trends in Carbon Stocks in Forest Biomass, 1990–2005*

	Carbon in living biomass (Gt)			
Region/subregion	*1990*	*2000*	*2005*	*% Change, 1990–2005*
Eastern and southern Africa	15.9	14.8	14.4	–9.43
Northern Africa	3.8	3.5	3.4	–10.53
Western and central Africa	46.0	43.9	43.1	–6.30
Total Africa	65.8	62.2	60.8	–7.60
East Asia	7.2	8.4	9.1	26.39
South and Southeast Asia	32.3	25.5	21.8	–32.51
West and Central Asia	1.6	1.7	1.7	6.25
Total Asia	41.1	35.6	32.6	–20.68
Total Europe	42.0	43.1	43.9	4.52
Caribbean	0.4	0.5	0.6	50.00
Central America	3.4	2.9	2.7	–20.59
North America	37.2	38.5	39.2	5.38
Total North and Central America	41.0	41.9	42.4	3.41
Total Oceania	11.6	11.4	11.4	–1.72
Total South America	97.7	94.2	91.5	–6.35
World	299.2	288.6	282.7	–5.51

Source: FAO, *Global Forest Resources Assessment 2005* (Rome, 2006).

land would increase by 2.0 percent, 24.7 percent, and 13.1 percent, respectively. Countries that are likely to lose at least 20 percent of their forests by 2020 are Mexico, Belize, Ecuador, Guatemala, Nicaragua, and Panama. El Salvador is likely to lose more than 60 percent if current deforestation rates continue. On the other hand, by 2020 El Salvador is expected to expand its pasture land by more than 60 percent.[31]

Conclusion

Latin America has historically supported the multilateral climate change negotiations represented by the United Nations Framework Convention on Climate Change. Particularly, discussions of land use, land-use change, and forestry issues have been a priority for many of the region's countries, which usually act together in the negotiations. Latin American countries are likely to play a major role in discussions of the LULUCF aspects of the post-2012 regime, guided by the principles of sovereignty, equity, environmental integrity, sustainable development, common but differentiated responsibilities, and cost-effectiveness.

Latin America accounts for an important share of the global forest area, and its share contains some of the richest forest and biodiversity resources in the world. These forests provide social, cultural, economic, and environmental benefits at local, national, regional, and global levels. The role of forests in fostering development is particularly relevant at the local level, because some of the poorest people in Latin America—almost a third of them indigenous people—live in rural areas and depend to a large extent on the resources provided by forests for their survival (including the opportunity of using them for other land uses as short-term solutions to cover their most urgent needs).

Deforestation and forest degradation have been occurring at alarming rates in many Latin American countries during the last decades, owing to varied, complex, and interacting drivers. Although the region's governments have made important efforts to stop these processes—efforts that include establishing protected areas, reforestation, and paying for environmental services programs—the complexity of the problem, together with a chronic lack of adequate resources (such as funding, capacities, and technologies) and enforcement capacity, has impeded their success. International financial support has been important, although insufficient, and is considered critical to reducing deforestation rates in the region in the coming years.

Carbon finance is the most important new and additional source of development funding, potentially exceeding U.S.$1 billion a year over the next two to three years and U.S.$50 billion to $120 billion a year in the long term.[32] Finding mechanisms to allow for the participation of the Latin American forest sector in the carbon market, taking into account the development needs and circumstances of the various countries in the region, will be paramount for the global climate change regime. Carbon stocks in living biomass in Central and South America represent a potential contribution to global emissions parallel to that of China's known coal reserves. This comparison is also noteworthy in terms of the expected role of these two natural resources (forested areas and all their potential uses, and coal) to foster the economic development of Latin America and China.

On the other hand, finding adequate international funding options to stop deforestation in Latin America in the next twenty years could potentially be as relevant as stopping GHG emissions from the transportation sector in the European Union for twenty years at their 2003 levels or eliminating total U.S. GHG emissions for two years at their 2004 levels. This would also entail associated benefits, from local environmental improvements (air and water quality, for instance) to the conservation of globally significant species and ecosystems and the sustainable development of rural communities affected by poverty.

Notes

1. The Group of 77 (G77) was established on June 15, 1964, by seventy-seven developing country signatories of the "Joint Declaration of the Seventy-Seven Countries," issued at the end of the first session of the United Nations Conference on Trade and Development (UNCTAD) in Geneva. Although the members of the G77 have increased

to 130 countries, the original name was retained because of its historic significance. The Group of 77 is the largest intergovernmental organization of developing states in the United Nations, which provides the means for the countries of the South to articulate and promote their collective economic interests and enhance their joint negotiating capacity on all major international economic issues within the United Nations system and to promote South-South cooperation for development.

2. The Environmental Integrity Group emerged at the thirteenth session of the Subsidiary Bodies in Lyon, France, in 2000. It includes Switzerland, Mexico, South Korea, Monaco, and Liechtenstein, countries that are not members of the other negotiating groups. Its goal is to emphasize the need to achieve "environmental integrity" in the outcomes of climate change negotiations.

3. GRILA also covered many agenda items relevant to its members, such as other CDM and LULUCF issues, vulnerability and adaptation, technology transfer, capacity-building, and compliance.

4. Centre for Socio-Economic Development, "NSS Bolivia Training Module Project, 2000" (contract 7113150) (Swiss SP-P067482-CSS-TF040428), Appendix 8, "National Negotiation Strategy," by Sergio Jáuregui, Geneva.

5. FCCC/SB/2000/MISC.1/Add.2.

6. FCCC/SB/2000/MISC.4/Add.2/Rev.1.

7. UNFCCC, Dialogue Working Paper 21, 2006.

8. FCCC/SBSTA/2006/MISC.5.

9. UN Economic Commission for Latin America and the Caribbean (ECLAC), *The Millennium Development Goals: A Latin American and Caribbean Perspective* (Santiago, Chile: United Nations, 2005).

10. Food and Agriculture Organization, "Global Forest Resources Assessment 2005: Progress towards Sustainable Forest Management," Forestry Paper 147 (Rome: FAO, 2006).

11. D. Ferranti and others, *Beyond the City: The Rural Contribution to Development* (Washington: World Bank, 2005).

12. FAO, "State of the World's Forests," 1997 (www.fao.org/docrep/W4345E/W4345E00.htm).

13. United Nations Environment Programme–World Conservation Monitoring Centre (UNEP-WCMC), "GEO3 Protected Areas Snapshot," 2001 (http://valhalla.unep-wcmc.org/wdbpa/GEO3.cfm).

14. ECLAC, *Millennium Development Goals.*

15. World Wildlife Fund (WWF), "How Effective Are Protected Areas? A Preliminary Analysis of Forest Protected Areas by WWF," report prepared for the seventh Conference of Parties of the Convention on Biological Diversity, 2004 (http://assets.panda.org/downloads/protectedareamanagementreport.pdf#search=%22How%20effective%20are%20protected%20areas%3F%20A%20preliminary%20analysis%20of%20forest%20protected%20areas%20by%20WWF%22).

16. A. White and others, "To Johannesburg and Beyond: Strategic Options to Advance the Conservation of Natural Forests," discussion paper for the Global Environment Facility (GEF) Forest Roundtable, March 11, 2002, New York (www.forest-trends.org/documents/publications/grf_revisedfinal1.pdf#search=%22To%20Johannesburg%20and%20Beyond%3A%20Strategic%20Options%20to%20Advance%20the%20Conservation%20of%20Natural%20Forests.%20Discussion%20Paper%20for%20the%20Global%20Environment%20Facility%20(GEF)%20%22).

17. Ibid.

18. Latin American and Caribbean Forestry Commission, "Review of the Latin American and Caribbean Forestry Commission (LACFC) and Other FAO Supported Activities, Including Follow-Up to the Recommendations of the 23rd Session of the Commission," Twenty-Fourth Session Overview of Forestry Activities of Interest to the Region, Secretariat Note, Santo Domingo, Dominican Republic, 2006 (FO:LACFC/2006/3.1) (ftp://ftp.fao.org/docrep/fao/meeting/011/ag281e.pdf).

19. The Cerrado is the world's most biologically rich savanna. It has more than 10,000 species of plants, of which 45 percent are exclusive to the Cerrado, and it stretches across nearly 500 million acres of Brazil—an area nearly three times the size of Texas. The Cerrado also feeds three of the major water basins in South America, the basins of the Amazon, Paraguay, and São Francisco Rivers.

20. UNEP, *GEO: Global Environment Outlook 3. Past, Present and Future Perspectives,* 2002 (www.grida.no/geo/geo3/).

21. This assumes carbon stocks in living biomass per hectare for Central and South America of 119.4 and 110 tonnes per hectare, respectively, and 57.8 tonnes per hectare for Mexico. It is also assumed, for the sake of simplicity, that the land use to which the forest area is converted has a carbon content in living biomass equal to zero.

22. According to the CDM pipeline overview prepared by the UNEP Risø Centre, Latin America had, up to March 6, 2006, 415 project activities, which were expected to generate a total of 301.67 Mt CO_2e until 2012.

23. "Annual European Community Greenhouse Gas Inventory, 1990–2003" and "Inventory Report, 2005," submissions to the UNFCCC Secretariat; "Emissions of Greenhouse Gases in the United States, 2004: Executive Summary," U.S. Energy Information Administration.

24. UNEP, *GEO: Global Environment Outlook 3.*

25. Ibid.

26. Ibid.

27. Ibid.

28. United Nations Office for Drugs and Crime (UNODC), "Colombia: Monitoreo de Cultivos de Coca," 2006 (www.acnur.org/index.php?id_pag=5239).

29. ECLAC, *Millennium Development Goals.*

30. China has proven, recoverable coal reserves estimated at about 113,400 million tonnes as of January 2005, according to the Carbon Sequestration Leadership Forum. Assuming a carbon content of 45 to 86 percent, the total carbon stock in China's coal reserves would range from 97.52 Gt to 51.03 Gt.

31. FAO, "Socio-Economic Trends and Outlook in Latin America: Implications for the Forestry Sector to 2020," Latin American Forestry Sector Outlook Study Working Paper, 2006 (www.fao.org/DOCREP/006/J2459E/J2459E00.HTM).

32. World Bank, "G8 Climate Change Action Plan and the Investment Framework: The Technology and Financing Context and Issues," Montreal, COP 11, UK-IEA side event, December 5, 2005.

CASE STUDY

The Noel Kempff Climate Action Project, Bolivia

JÖRG SEIFERT-GRANZIN

The Noel Kempff Climate Action Project (NKCAP) is the second largest greenhouse gas emission reduction project developed under the Activity Implemented Jointly program of the United Nations Framework Convention on Climate Change (UNFCCC). Designed by the Nature Conservancy and Fundación Amigos de la Naturaleza (FAN Bolivia), NKCAP is mitigating climate change by avoiding degradation caused by industrial logging and small-scale deforestation by slash-and-burn practices. Based on the political commitment of the Bolivian government, the project indemnified the timber concessions adjacent to Noel Kempff Mercado National Park with financial support from three corporate partners—American Electric Power Company, BP America, and PacifiCorp—and the Nature Conservancy, extended the park to its natural boundaries, initiated a community support and development program, and launched a rigorous scientific program to measure avoided emissions in the project area (642,458 ha). In addition, NKCAP supported the formation of the Organization of Indigenous Communities in Bajo Paraguá (CIBABA) and land title claims within the indigenous territory. In 2005, SGS UK Ltd. certified the project, validating and verifying its baseline methodologies, monitoring schemes, and environmental and socioeconomic effects. Although the project is ineligible under the CDM, it has been assessed as if it were an eligible activity to make it comparable with other CDM projects. SGS verified that NKCAP achieved net emission reductions of 1,034,107 t CO_2 between 1997 and 2005.[1]

Emission reductions gained by cessation of logging (degradation avoidance) were quantified using a nonlinear dynamic econometric optimization model driven by the dynamics of domestic and international timber markets.[2] The model simulates timber harvests over 125 years. Corresponding emissions are calculated considering collateral damage in vegetation, regrowth, decomposition, and carbon sequestration in harvested wood products.[3] Carbon accounting is being used to assess the effects on all the five carbon pools, using methodologies developed by Winrock International.[4] Avoided degradation accounts for 68 percent (791,444 t CO_2) of the total gross emission reductions. The model was used to calculate leakage considering possible activity shifting of timber production within Bolivian territory. Results indicated that leakage negates 16 percent of the gross degradation emission reductions achieved.

The project avoids deforestation along its border by promoting alternative income programs for the surrounding communities and reducing slash-and-burn practices. GHG effects over the thirty-year life of the project (1997–2026) were quantified using a dedicated modeling approach based on detected deforestation patterns, rates, and drivers. Projected forest cover changes were combined with the results of a comprehensive biomass inventory to establish a spatially explicit baseline and to calculate avoided emissions.

According to the project agreement, 51 percent of the certified emission reductions belong to the corporate partners and 49 percent to the Bolivian government, which agreed to earmark 15 percent for the protection of the park, 5 percent for the national system of protected areas, and 29 percent for other purposes, including improving the livelihoods of the indigenous communities adjacent to the park. The Bolivian government aims to sell part of the voluntary emission reductions (VERs) at the Chicago Climate Exchange.

What lessons have been learned from NKCAP? A number of issues have hampered UNFCCC efforts to include emissions from deforestation, especially the issues of additionality, leakage, permanence, and the feasibility of monitoring. The certification of NKCAP proves that projects reducing emissions from deforestation and degradation (REDD) can produce real, measurable, and additional carbon offsets. Setting aside logging areas to include them under a protection scheme can generate substantial benefits to the climate, community, and biodiversity. To secure these benefits, binding rules on how to distribute revenues from carbon finance should be defined in advance. Community benefits in particular depend on fair, transparent, and effective benefit-allocation schemes. Otherwise these benefits may vanish when public budgets face cutbacks.

Leakage remains an issue for project-based REDD mechanisms. Under the specific conditions of NKCAP it is possible to calculate and monitor leakage in both components of REDD, including national degradation baseline tracking associated with the timber-producing sector and the monitoring of regional leakage caused by shifted small-scale deforestation. However, other projects might have to consider timber piracy and activity shifting of cash crop production caused by large-scale deforestation avoidance, which would be challenging to quantify without additional sectoral economic modeling efforts. A comprehensive national accounting mechanism for REDD complying with agreed-upon rules under the international climate regime would reduce these challenges substantially.

Any national-level REDD mechanism will have to be based on either a scientifically credible projection of a business-as-usual emissions scenario or a political consensus about allowed emission levels in a given period. Considering the high variability of biomass across Bolivian forest types, spatially explicit projections are needed to determine business-as-usual emissions. However, spatiotemporal modeling of land-use change at high resolution (approximately 30 meters) and on an annual basis is still challenging, even though spatially explicit land-use change modeling can be validated using accepted statistical methods.[5]

By detecting deforestation and degradation patterns, remote sensing plays an essential role in projecting likely future emissions. However, further inputs and modeling approaches are needed to establish credible projections, taking into account the complex relation between proximate, underlying causes and other factors driving deforestation. Econometric modeling proved to be highly suitable to capturing timber market behavior and determining sectoral business as usual.

Baseline projections have to be revised periodically, because deforestation and degradation drivers change over time. New roads tend to change deforestation patterns; changes in settlement or timber harvesting regimes and changes in prices of agricultural goods can have strong effects on land-use decisions. Furthermore, methodologies should be continually revised to anticipate technological changes and state-of-the-art science. In the case of NKCAP, the dynamic baseline approaches have been certified and are revised every five years, and monitoring of the key parameters is conducted annually.

Notes

1. SGS UK Ltd., "Validation and Verification Report, Programa Nacional de Cambio Climático, Noel Kempff Climate Action Project. Summary Only. Project No. Vol 0001 Date: 27 November 2005" (www.fan-bo.org/serviciosambientales/proyectos/redd/certificacion.htm).

2. B. Sohngen and S. Brown, "Measuring Leakage from Carbon Projects in Open Economies: A Stop Timber Harvesting Project in Bolivia as a Case Study," *Canadian Journal of Forest Research* 34 (2004): 829–39.

3. FAN Bolivia, "Project Design Document of Noel Kempff Mercado Climate Action Project," 2005 (www.fan-bo.org/serviciosambientales/proyectos/redd/monitoreo.htm).

4. S. Brown, M. Delaney, and T. Pearson, "Carbon Monitoring and Verification Protocols for the Noel Kempff Climate Action Project," Winrock International, Arlington, Virginia, 2003.

5. R. G. Pontius, D. Huffaker, and K. Denman, "Useful Techniques of Validation for Spatially Explicit Land-Change Models," *Ecological Modelling* 179 (2004): 445–61; R. G. Pontius and others, "Comparing Input, Output, and Validation Maps for Several Models of Land Change," *Annals of Regional Science*, in press (www.clarku.edu/~rpontius/pontius_etal_2007_ars.doc.)

16

Compensated Reductions: Rewarding Developing Countries for Protecting Forest Carbon

STEPHAN SCHWARTZMAN AND PAULO MOUTINHO

Tropical forests store 200 billion tonnes of carbon (200 petagrams [Pg] C) globally.[1] Deforestation is releasing these stocks into the atmosphere, and both regional and global feedbacks could cause massive carbon emissions, contributing to global warming and the collapse of the ecological equilibrium of tropical forest ecosystems.[2] In Amazonia, for example, one-third of the forest could be transformed into savanna.[3] Although greenhouse gas (GHG) emissions from the burning of fossil fuels are the principal cause of global warming, tropical deforestation causes approximately 18 percent of annual global emissions of carbon dioxide (CO_2).[4]

A consensus now exists in the international community that to avoid "dangerous interference" in the global climate system—the primary objective of the United Nations Framework Convention on Climate Change (UNFCCC, article 2)—tropical deforestation should be greatly reduced.[5] The Kyoto Protocol, although an important step for reducing GHG emissions, has no means of addressing tropical deforestation. In order to ensure that atmospheric CO_2 concentrations remain below 450 parts per million by volume (ppmv) by 2100 and to avoid "dangerous interference," annual global emission reductions must be greater than 2 percent per year starting in 2010.[6] Given the inertia of global power consumption and costs of changing the energy matrix, it is likely for both developed and developing countries that large emission reductions (>2 percent per year) from fossil fuels will be unrealistic in the short term. Reductions in tropical deforestation, however, may be a bridge to technological transformation, offering a viable, cost-effective means by which to begin reducing GHG emissions before the technology needed to transform the energy and transportation sectors globally is developed.

In this chapter we present a brief analysis of GHG emissions from deforestation in Amazonia and describe "compensated reduction," a mechanism for addressing deforestation in the context of the UNFCCC. Through this mechanism, developing countries, in which most standing forests are located, could be compensated through international emissions trading for emission reductions from avoided deforestation below historical baselines.[7]

Amazon Deforestation and GHG Emissions

During the 1980s, deforestation in Latin America alone produced a net carbon flux into the atmosphere on the order of 0.3 Pg C a year. This amount rose to 0.4 Pg C a year through the 1990s as a result of additional deforestation of more than 4 million hectares a year. Of this total, 0.2 Pg C resulted from deforestation in Brazilian Amazonia. Real emissions were certainly higher, because emissions from forest fires and logging were not included in the calculation.[8]

The estimated contribution of forest fires to GHG emissions from the Brazilian Amazon during the 1998 El Niño was 0.2 Pg of carbon. In that year, 30 percent of the forests in the region were at high risk from fire; 1.3 million hectares of standing forest burned in the state of Roraima, and another 2.5 million hectares were affected by fire in southern Pará and northern Mato Grosso. More recently, Asner and colleagues estimated that 0.1 Pg C a year is released by logging activities in the Brazilian Amazon, although there is still no clear idea of carbon volumes absorbed by regeneration of vegetation after selective logging.[9] In addition, deforestation may be eliminating part of the carbon sink function of the forest.

Deforestation emissions in regions with large areas of remaining tropical forest are expected to continue at current rates or to increase in coming decades. It should not be assumed that once currently accessible forest regions are deforested, emissions will decline, because currently inaccessible regions will become accessible with frontier expansion. Over the next 100 years, between 80 and 130 Pg C will have been released into the atmosphere by tropical deforestation, an amount equal to or higher than, for example, the entire carbon stock stored in the Amazon forest.[10]

In the Brazilian Amazon, annual deforestation rates increased 30 percent from 2001 (18,165 km^2) to 2002 (23,266 km^2) and 2004 (23,750 ± 950 km^2), generating net emissions on the order of 0.2 Pg C a year (3 percent of the global total) or more. Despite recent reduction (2005–07) of deforestation rate, Brazil is among the top five or six GHG emitters globally, and 75 percent of the country's emissions come from deforestation.[11]

Recent studies show that an additional 32 Pg C will be emitted into the atmosphere by 2050 if deforestation in the Amazon Basin follows the trend of the last two decades. Scenarios for other tropical countries appear to follow the same trend. In Indonesia, 17,000 square kilometers were cut down annually between 1987 and 1997, and 21,000 square kilometers in 2003, resulting in average emissions of 0.2 Pg C a year.[12]

Compensated Reduction: Decreasing Deforestation to Reduce GHG Emissions

During the ninth Conference of the Parties (COP 9) of the UNFCCC, in Milan, an international group of scientists proposed a new concept for addressing emissions from tropical deforestation.[13] The compensated reduction concept proposes that developing countries that elect to reduce their national deforestation rates below a historical baseline could receive compensation through trading in international carbon markets. Reductions below agreed-upon baselines could be issued in Assigned Amount Units (AAUs) or certificates similar to Certified Emission Reductions (CERs) and sold to Annex I countries, companies, or traders. Reductions before 2012 (the end of the first Kyoto commitment period), once verified, would be tradable starting in 2013. Reductions would be calculated against national (or potentially regional, such as pan-Amazon or Central African) deforestation baselines, so this would not be a project-based mechanism like the Clean Development Mechanism (CDM). Reductions would be credited post facto, after robust verification—for example, using remote sensing technology. Various technologies adequate to this end are already in use and are improving rapidly.[14]

Figure 16-1 illustrates the way the compensated reduction concept might work in Brazilian Amazonia. The deforestation baseline would be set at 20,000 km^2, the average for annual deforestation in the 1980s. In this hypothetical case, deforestation decreases annually by 4 to 6 percent below that baseline during a five-year commitment period. Carbon emissions are correspondingly reduced by a total of 64 million tonnes over the five years, calculated as annual rate of deforestation (km^2) x 12,000 tonnes of C/km^2 (120 t/ha). The annual mean for avoided emissions per year is 12,000 (± 2) million tonnes of carbon.

The risk that countries having received compensation for reducing their emissions might subsequently increase them could be addressed in several ways. Disallowing access to the market for countries that had received compensation and subsequently exceeded their baseline deforestation rate until net deforestation was reduced below the baseline would limit crediting of impermanent reductions while maintaining the voluntary character of the mechanism.

Figure 16-1. *The Compensation Reduction Concepts as It Might Play Out in Brazilian Amazonia*

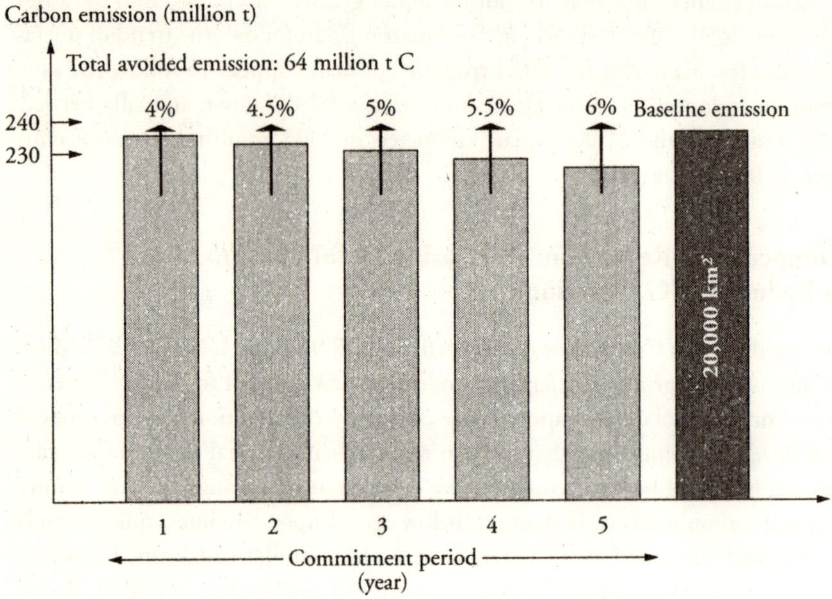

A mechanism such as compensated reductions not only would provide an incentive for developing countries to take immediate action to mitigate climate change but also could allow more stringent developed-country emissions targets to be set. If tropical nations were to offer Annex I countries a certain volume of emission offsets for the second commitment period, a proportional or even greater increase in the GHG reduction targets of these countries could be required.

More Aggressive Goals for Developed Countries and Greater Benefits for the Atmosphere

Including voluntary emission reductions from tropical deforestation can and should result in substantial increases in obligatory goals for developed countries while providing greater benefits to the atmosphere. Two scenarios illustrate how this could be handled. In scenario A, Annex I countries decide simply to stabilize emissions at the first commitment period levels (repetition of Kyoto goals). No compensation for deforestation reduction would be viable in such cases. In scenario B, Annex I countries agree to triple their reduction goals for the second commitment period (15 percent below 1990). Under this scenario, parties might authorize some quantity or proportion of the target for which Annex I countries could use deforestation offsets. Post-2012 targets should be set according to science-based measures of what is needed to avoid dangerous climate change and evaluation of all feasible means of achieving reductions.

Under scenario B, tropical nations would obtain significant rewards and developed countries would be stimulated to establish higher goals than previously contemplated. Fears of flooding the market with cheap carbon from avoided deforestation are often voiced. Obviously some limit could be placed on allowable offsets, but even in the absence of a cap, flooding is unlikely to occur. Even if carbon prices were to exceed those of alternative land uses, compensated reductions, to be effective, must reward past conservation as well as give the beneficiaries of deforestation incentives to reduce. Moreover, in all large remaining or future tropical forest frontiers, governments must substantially invest in long-term governance frameworks (monitoring and enforcement capability, organization of land tenure, and allocation of property rights) before carbon compensation can become a direct economic alternative for individuals and companies. Consequently, sudden wholesale replacement of logging in Indonesia or cattle ranching in the Amazon by carbon offsets is unlikely. Gradual reductions are the most that can realistically be expected in most tropical regions.

Funds obtained from compensation for deforestation reduction could be invested in public programs and policies aimed at enforcing environmental legislation, providing support for economic alternatives to felling of the forest. This would strengthen institutional capacity in remote forest regions along the lines recently demonstrated in parts of Brazilian Amazonia through environmental licensing in Amazonian states and the Deforestation Control Program.[15] Substantial forest areas can be protected through the securitization of conservation units if adequate funding is available.[16]

Design of the Compensated Reductions Mechanism

The issues of additionality, base period, leakage, and nonpermanence arise frequently in discussions about reducing deforestation in the UNFCCC context. Some claim that crediting forest conservation will simply allow industrialized countries to continue emitting, generating no net benefit to the atmosphere.[17] Forests can be cut, burned, and logged, leading some to doubt that reduced deforestation would represent "permanent" emission reductions. The issue of "leakage"—the possibility that forest could be conserved in one area but cleared in another—was prominent in the polemics surrounding the project-based CDM under the Kyoto Protocol. However, CDM discussions of forests under the Kyoto Protocol focused largely on carbon sinks and sequestration in forest plantations and not on the essential role of existing natural tropical forests in the stability of the global climate system. We discuss each of four issues in turn, as they relate to the concept of compensated reductions of deforestation.

Additionality and Base Periods

Base periods are established according to historical deforestation rates (for example, the average rate for the 1980s).[18] Because there is no evidence that tropical deforestation will decrease significantly in the short term (several decades) without appropriate incentives (such as monetizing the standing forests' carbon value), additionality is not an issue. In all remaining large tropical forests, deforestation tends to continue at current rates or to increase until the forests are eliminated.[19] Therefore, any reduction below recent deforestation rates represents a real and additional gain for the atmosphere.

The procedure for selecting historical base periods (or reduction goals) must take into account the different regional dynamics of deforestation in the tropics. In Amazonia, for example, with approximately 80 percent of its original forest cover and high current deforestation rates, a base period for average annual deforestation rates set in the 1980s (since 1990 is the reference year for the Kyoto goals) would be adequate. Countries with substantial tropical forests and relatively little deforestation to date (for example, Peru and Bolivia) should be able to adopt higher baselines than their recent deforestation rates to provide an incentive to participate and to avoid future increases. Regions that have been heavily degraded, such as Kalimantan, Sumatra, and Sulawesi, where 70 to 80 percent of the lowland *Dipterocarpaceae* forest cover has been removed in areas deforested and where conversion to oil-producing palm species is under way, would have to be handled differently. In such cases, a baseline could be expressed in terms of carbon stocks at some time in the past, with credit for any increase above this between, for example, 2008 and 2012, making reforestation or regeneration an alternative to palm plantations. A historical baseline could be revised downward in, for example, twenty years to provide an incentive for countries to continue reducing deforestation rates. This is a plausible length of time for a nation such as Brazil to transform its land-use practices. Some researchers have proposed a global forest baseline that would provide incentives for both low and high deforestation tropical countries to protect forests, as well as potentially avoiding leakage between countries.[20]

Leakage

Calculating emission reductions from deforestation against a national baseline addresses the problem of leakage that affects CDM projects.[21] It is less likely that deforestation would "leak" from one country to another than from one project to another. In addition, monitoring national deforestation rates would help to identify potential leakage between countries.

Deforestation can be measured from the beginning to the end of the commitment period, precisely as is done in the case of national emissions from Annex I countries. There is a risk of so-called international market leakage for timber exports, in which a participating country stops exporting timber in order to obtain carbon investments and a nonparticipating country correspondingly increases exports. Such market leakage is only one case of a broader issue arising from an emissions control regime that includes only some countries within the global economy. Already firms with emission limitations in Annex I countries can choose to move high-emission activities to non–Annex I countries.

With respect to forestry, the current rules under the Kyoto Protocol create greater imbalances in global carbon accounting than would compensated reductions. That is, sinks and activities that increase carbon stocks in Annex I countries are credited, but forest destruction in developing countries is not debited.[22] By bringing more large-scale emitters into the international emissions control regime, compensated reductions would help to address this issue. Leakage of deforestation per se from one country to another (for example, Brazilian soy planters who move to Bolivia) might occur, although participation by several countries in a geographical region (Amazonian countries, for example) in a compensated reduction mechanism would help address this problem. Furthermore, economic modeling of international timber trade might be able to identify international export leakage. Ultimately, international leakage—in all sectors, not just forestry—will be resolved only when all major emitters participate in an international emissions control regime.

Nonpermanence

Protected forests can be lost in the future because of natural disturbances (from fire, pests, and so forth) or through direct human action. Carbon offsets from reducing deforestation are therefore potentially nonpermanent reductions. Permanence can be ensured in a compensated reduction mechanism by requiring participating countries that increase deforestation (emissions) above the levels of their base periods subsequently to assume the surplus emitted as an obligatory reduction goal. Security of emissions offsets might be enhanced by an insurance system in which only a portion of the credits from reductions achieved in the first five-year commitment period would be made available for emission offsets starting in 2013. Another portion could be held in reserve as insurance against potential future loss of the forest used for compensation. It is also important to keep in mind that permanence of reductions is an issue not only for forestry but potentially for all sectors. A country that fulfills its emission limitation commitment in the first period might decide subsequently to accept no limitations and to increase its emissions.

In a sufficiently developed market, insurers could offer investors carbon insurance based in AAUs or CERs acquired for that purpose. With regard to tropical forests in particular, reducing emissions from deforestation will have other climate stabilizing benefits in addition to reducing CO_2 emissions, such as preserving the forests' critical water cycling function. Large-scale deforestation tends to reduce rainfall regionally, so preventing deforestation can reduce forests' vulnerability to fires. Therefore, large-scale reductions in deforestation will probably be necessary to avert the collapse of surrounding ecosystems as predicted in various climate models.

Conclusion

Emissions from tropical deforestation, although not as important a cause of the greenhouse effect as those resulting from the burning of fossil fuels, are of sufficient scale to aggravate it and may compromise a good part of the international efforts to reduce emissions during the first Kyoto commitment period as well as after 2012. Moreover, if the earth's temperature increase provokes drastic reductions in tropical forest stocks, substantial additional quantities of greenhouse gases will be injected into the atmosphere, potentially creating a disastrous positive feedback loop with regard to climate change. Regional deforestation dynamics alone, independent of global warming, may be enough to provoke large-scale ecosystem collapse.[23] Given the inertia of global energy consumption and the costs of transforming the global energy matrix, for both developed and developing countries, rapid emission reductions (> 2 percent per year) under current circumstances are unlikely. Reductions in tropical deforestation may be a bridge to a low-carbon future, a relatively quick and inexpensive means to reduce emissions while new technologies to cut fossil fuel emissions are developed.

Halting or decreasing deforestation can contribute to the continuity and strengthening of a robust, inclusive international emission reduction regime after 2012—or vice versa. Furthermore, nothing could be more useful to the preservation of biological diversity on the planet. More hazardous to the global climate system than any issue of leakage or permanence in carbon offsets from reducing deforestation is the prospect of failing to sustain and increase developing-country participation in a mandatory international emission reduction system. Finally, during the last COP (14) in Bali, the UNFCCC opened its agenda for a new climate change framework agreement (to be implemented in 2012) that includes the emissions from tropical deforestation (identified as Reducing Emission from Deforestation and Degradation, or REDD). REDD regimes could be based on the compensated reduction concept and offer substantial incentives for large developing countries to reduce emissions through means of their own choice, averting the global climate crisis while we still have time.

Notes

1. IPCC, "Fourth Assessment Report: Climate Change 2007" (www.ipcc.ch/ipcc reports/assessments-reports.htm).

2. *Deforestation* is defined here as the complete removal of forest cover, or clear-cutting. This is distinct from logging, which is characterized by selective removal of trees and therefore of only part of the forest cover.

3. M. D. Oyama and C. A. Nobre, "Climatic Consequences of a Large-Scale Desertification in Northeast Brazil: A GCM Simulation Study," *Journal of Climate* 17, no. 16 (2004): 3203–13; P. M. Cox and others, "Amazonian Forest Dieback under Climate–Carbon Cycle Projections for the 21st Century," *Theoretical and Applied Climatology* 78, no. 1–3 (2004): 137–56.

4. R. A. Houghton, "Revised Estimates of the Annual Net Flux of Carbon to the Atmosphere from Changes in Land Use and Land Management, 1850–2000," *Tellus* 55B (2003): 378–90; F. Achard and others, "Determination of Deforestation Rates of the World's Humid Tropical Forests," *Science* 297 (2002): 999–1002; R. S. DeFries and others, "Carbon Emissions from Tropical Deforestation and Regrowth Based on Satellite Observations for the 1980s and 1990s," *Proceedings of the National Academy of Sciences of the United States of America* 99 (2002): 14256–61.

5. "Dangerous interference" can be defined in various ways, such as interference sufficient to shut down the ocean's thermohaline circulation or to melt the West Antarctic ice sheet. B. C. O'Neill and M. Oppenheimer, "Dangerous Climate Impacts and the Kyoto Protocol," *Science* 296 (2002): 1971–72.

6. Ibid.

7. M. P. Santilli and others, "Tropical Deforestation and the Kyoto Protocol: An Editorial Essay," *Climatic Change* 71 (2005): 267–76; P. Moutinho and S. Schwartzman, eds., *Tropical Deforestation and Climate Change* (Belém, Brazil: Instituto de Pesquisa Ambiental da Amazônia and Environmental Defense, 2005).

8. DeFries and others, "Carbon Emissions from Tropical Deforestation"; R. A. Houghton, "Tropical Deforestation as a Source of Greenhouse Gas Emission," in *Tropical Deforestation and Climate Change*, edited by Moutinho and Schwartzman, pp. 13–22; Houghton, "Revised Estimates of the Annual Net Flux of Carbon"; A. Alencar, D. Nepstad, and P. Moutinho, "Carbon Emission Associated with Forest Fires in Brazil," in *Tropical Deforestation and Climate Change*, edited by Moutinho and Schwartzman, pp. 13–21; G. P. Asner and others, "Selective Logging in the Brazilian Amazon," *Science* 310 (2005): 480–82.

9. M. J. C de Mendonça and others, "The Economic Cost of the Use of Fire in Amazon," *Ecological Economics* 49 (2004): 89–105; D. Nepstad and others, "Amazon Drought and Its Implications for Forest Flammability and Tree Growth: A Basin-Wide Analysis," *Global Change Biology* 10 (2004): 1–14; A. Alencar, L. A. Solorzano, and D. C. Nepstad, "Modeling Forest Understory Fires in an Eastern Amazonian Landscape," *Ecological Applications* 14 (2004): 139–49; A. Alencar and others, "Deforestation in the Amazon: Getting beyond the 'Chronic Emergency,'" Amazon Institute for Environmental Research (IPAM), Belém, 2004 (in Portuguese); Asner and others, "Selective Logging in the Brazilian Amazon."

10. R. A. Houghton, "Aboveground Forest Biomass and the Global Carbon Balance," *Global Change Biology* 11 (2005): 945–58; Instituto Nacional de Pesquisas Especiais, "Monitoramento da Floresta Amazônica Brasileira," São Paulo, 2005 (www.obt.inpe.br/prodes/index.html); Houghton, "Tropical Deforestation as a Source of Greenhouse Gas Emission."

11. Houghton, "Tropical Deforestation as a Source of Greenhouse Gas Emission"; Santilli and others, "Tropical Deforestation and the Kyoto Protocol."

12. B. S. Soares-Filho and others, "Modelling Conservation in the Amazon Basin," *Nature* 440 (2006): 520–23; Houghton, "Revised Estimates of the Annual Net Flux of Carbon."

13. Santilli and others, "Tropical Deforestation and the Kyoto Protocol"; Moutinho and Schwartzman, *Tropical Deforestation and Climate Change.* The participating scientists represented the Amazon Institute for Environmental Research (IPAM), the Instituto Socioambiental (ISA), Environmental Defense, the Instituto Nacional de Pesquisas Espaciais (INPE), and Woods Hole Research Center.

14. R. DeFries and others, "Reducing Greenhouse Gas Emissions from Deforestation in Developing Countries: Considerations for Monitoring and Measuring," Report 26, Global Observation of Forest and Land Cover Dynamics (GOFC-GOLD), 2006 (http://nofc.cfs.nrcan.gc.ca/gofc-gold/Reportpercent20Series/GOLD_26.pdf).

15. Fundação Estadual do Meio Ambiente, "Environmental Control System on Rural Properties in Mato Grosso" (Cuiabá, Mato Grosso, 2001); D. Nepstad and others, "Frontier Governance in Amazonia," *Science* 295 (2002): 629–30; P. M. Fearnside and W. F. Laurance, "Comment on 'Determination of Deforestation Rates of the World's Humid Tropical Forests,'" *Science* 299 (2003): 1015.

16. A. G. Bruner and others, "Effectiveness of Parks in Protecting Tropical Biodiversity," *Science* 291 (2001): 125–28; S. L. Pimm and others, "Can We Defy Nature's End?" *Science* 293 (2001): 2207–08; D. C. Nepstad and others, "Inhibition of Amazon Deforestation and Fire by Parks and Indigenous Reserves," *Conservation Biology* 20, no. 1 (2006): 65–73.

17. P. Fearnside, "Environmentalists Split over Kyoto and Amazonian Deforestation," *Environmental Conservation* 28, no. 4 (2001): 295–99.

18. Base periods are different from the concept of baseline, which is defined as a basis for extrapolations in the future for emission reductions that occur because of a change in human activity.

19. B. S. Soares-Filho and others, "Modelling Conservation in the Amazon Basin."

20. D. Mollicone and others, "An Incentive Mechanism for Reducing Emissions from Conversion of Intact and Non-Intact Forests," *Climatic Change* 83 (2007): 477–93.

21. Santilli and others, "Tropical Deforestation and the Kyoto Protocol"; Moutinho and Schwartzman, *Tropical Deforestation and Climate Change*; B. Schlamadinger and others, "Should We Include Avoidance of Deforestation in the International Response to Climate Change?" in *Tropical Deforestation and Climate Change,* edited by Moutinho and Schwartzman, pp. 13–22.

22. E. Niesten and others, "Designing a Carbon Market that Protects Forests in Developing Countries," *Philosophical Transactions: Mathematical, Physical and Engineering Sciences* 360 (2002): 1875–88.

23. Oyama and Nobre, "Climatic Consequences of a Large-Scale Desertification in Northeast Brazil."

17

Creating Incentives for Avoiding Further Deforestation: The Nested Approach

CHARLOTTE STRECK, LUCIO PEDRONI,
MANUEL ESTRADA PORRUA, AND MICHAEL DUTSCHKE

Forests are the world's most important terrestrial storehouses of carbon and play an important role in controlling the climate. Yet in many parts of the world forests are degraded and destroyed to expand agricultural land, gain timber, or clear space for infrastructure or mining. Tropical deforestation has severe consequences for biodiversity. It affects water quality and storage, exacerbates flooding, landslides, and soil erosion, and threatens the livelihoods and cultural integrity of 1.2 billion forest-dependent people. It is also a major contributor to global climate change. About 36 percent of the carbon that was added to the atmosphere between 1850 and 2000 came from forests that were eliminated, and about 18 percent of the carbon added in the 1990s came from land-use change. At a worldwide scale, gross deforestation is about 12.3 million hectares per year.[1] Effectively reducing deforestation is therefore a strategic issue on the climate change and development agendas for the period after 2012.

Despite the negative effects of deforestation, mitigation of emissions of greenhouse gases (GHGs) from deforestation and forest degradation in developing

For their valuable contributions, we would like to thank Bernhard Schlamadinger, Axel Michelowa, Robert O'Sullivan, and the many other colleagues and friends who provided feedback on previous versions of this chapter and who encouraged us to present our ideas in many workshops and panel discussions around the globe.

countries was not adequately addressed in the Kyoto Protocol. During the negotiations of the protocol the most important goal was to create a framework in which industrialized countries would reduce their industrial emissions through binding commitments. Negotiators considered emissions stemming from agriculture, forestry, and other land uses (AFOLU) an issue to be subordinated to any agreement on the reduction of industry- and energy-related emissions. Nevertheless, the topic came up various times during the negotiations, and controversies spun around the ability to devise a practical means to include the complex accounting of sinks in a manner that adequately maintained the environmental integrity of the overall agreement.[2] The solution found in Kyoto was based on a partial accounting framework that probably was the best that could have been achieved in 1997 but remains unsatisfactory in the longer run. Despite its inconsistencies, the system devised in Kyoto rewards carbon emission reductions and removals in industrial countries and creates incentives for forest conservation and afforestation in those countries. On the other hand, the decision not to permit "avoided deforestation" as a project class under the Clean Development Mechanism (CDM) leaves the largest source of GHG emissions in many developing countries unaddressed.

It is essential that a post-Kyoto agreement address this gap and include policy and economic incentives to reduce further emissions from deforestation. Since 2005, when Papua New Guinea and Costa Rica put forward their proposal to consider whether and how incentives to reduce tropical deforestation could be included in the future climate regime under the United Nations Framework Convention on Climate Change (UNFCCC) or the Kyoto Protocol,[3] international negotiators have discussed how an international mechanism that would trigger emission reductions by avoiding deforestation could be shaped.

Recognizing that reducing emissions from deforestation will require a continuous international effort to build the required capacities and sustain adequate levels of funding, we propose in this chapter an integrated approach to rewarding emission reductions from avoided deforestation and, eventually, forest degradation (REDD). The objective of what has become known as the "nested approach" is to devise a framework aimed at achieving meaningful reductions in GHG emissions from improved forest governance and management in developing countries. The nested approach is based on a set of mechanisms allowing for immediate and broad participation by developing countries while facilitating the integration of private investors in such efforts.[4]

The Financing Gap

Long-term protection of tropical forests will depend on the mobilization of sufficient human and financial resources and on the capacity of public institutions to promote efficient and sustainable forest protection and management. To achieve the required level of resource mobilization, public financing needs to be matched with private capital.

Research carried out for the Stern Review, released in 2006, indicated that "the opportunity cost of forest protection in 8 countries responsible for 70 per cent of emissions from land use could be around U.S.$5 billion annually, initially, although over time marginal costs would rise."[5] A review of relevant data shows that only market instruments can mobilize this level of investment and induce GHG emission reduction activities at a scale adequate for pursuing the ultimate objective of the UNFCCC.

The total volume transacted in the regulated carbon markets in 2007 was an estimated 2.7 billion tonnes of CO_2 equivalent (t CO_2e), worth a financial value of approximately EUR 40 billion.[6] In comparison, the voluntary carbon market traded in 2006 a volume of 24 million t CO_2e, worth EUR 67 million.[7] Although the lion's share of this market has to be attributed to transactions under the EU's Emissions Trading Scheme, the CDM, with trades representing a value of more than EUR 12 billion, is the second largest segment of the global carbon market. About three quarters of the CDM market, and more than 90 percent of the carbon market as a whole, is made up of private carbon purchases and trading.

Through the CDM, the Kyoto Protocol has established a financial tool that is of unprecedented success in international environmental law. In 2006 the primary market of CDM transactions alone triggered about double the funds of the fourth replenishment of the Global Environment Facility (GEF), the single largest environmental trust fund and the financial mechanism for four international environmental conventions. Whereas the CDM mobilized U.S.$5 billion in 2006, the GEF received U.S.$3.13 billion in August 2006 from thirty-two donor governments for its operations between 2006 and 2010.[8]

Private investments in avoiding further deforestation are thus essential to mobilizing the required level of funding. It is also clear that such a level of investment in climate change mitigation cannot be mobilized by developing countries alone. Relative to more traditional commodity markets, the carbon market remains small, yet it still dwarfs financing made available through official development assistance, and its continuous growth evinces market participants' trust that credits with vintages beyond 2012 will come into play. It is therefore probable that the carbon markets triggered by the Kyoto Protocol will outlive the protocol itself and will remain a force behind attempts toward a more carbon- and emission-conscientious economy. In order to channel finance made available by the demand for carbon credits into activities that reduce emissions from deforestation, negotiators are called upon to create a robust framework and stable incentives for investments in sustainable forestry and forest protection.

Markets depend on a reliable regulatory framework that enables participants to count on certain returns on their investments. The level of investment reflects the level of risk exposure market participants attach to certain activities, countries, and mechanisms. Private investors as well as communities, local governments, and other private and public entities wish to control the risks associated with their investments and to be rewarded for their efforts. They will deploy capital only if they can manage the risks associated with the activities they finance. Whereas it is essential to integrate subnational REDD activities into broader public programs, the rewarding of such subnational activities should be unlinked from the risk of broader policy failure. The "nested approach" proposed here defines a mechanism that is informed by experience in existing markets and seeks to match the need to scale up emission reduction activities with the capacity of different stakeholders within an environmentally robust and effective regulatory framework. The nested approach is based on the principle that every entity should be held liable only for activities under its responsibility. It also provides a framework to enhance the contribution of developing countries to global emission reductions that is consistent with the principle of common but differentiated responsibilities and capabilities.

The Governance Gap

Reducing emissions from deforestation at a national level implies that countries are able to successfully implement effective policy, legal, and institutional reforms nationwide and are in the position to formulate and enforce appropriate social and economic safeguards. A review of selected governance indicators for the eight countries that contribute the 70 percent of total GHG emissions from land use referred to in the Stern Review provides a sobering view of the ability of some of the governments and public sectors of these countries to provide and enforce robust policies in their territories (table 17-1).

Poor forest governance, characterized by illegal logging, corruption, and land speculation, is a common phenomenon among rain-forest-rich countries. Public sector performance thus often aggravates the problem of deforestation rather than contributing to a country's attempt to protect forest resources. A 2006 World Bank report on forest governance emphasized that failures of law and enforcement must be addressed to improve forest sector governance and ensure that the forest-dependent poor are not unfairly punished.[9] Abusive central government intervention that awards lucrative forestry concessions to political allies and foreign corporations contributes to the general problem. Improved forest management, land reform, and a fight against corruption and forest crime are essential to enable countries to effectively protect their forests. However, the central insight

Table 17-1. *Governance Indicators for Eight Countries, 2002 and 2005*

	Governance indicator							
	Government effectiveness[a]		*Regulatory quality*[b]		*Rule of law*[c]		*Control of corruption*[d]	
Country	*2002*	*2005*	*2002*	*2005*	*2002*	*2005*	*2002*	*2005*
Bolivia	35.4	23.9	47.8	32.7	29.8	27.1	22.5	23.6
Brazil	53.6	55.0	61.1	55.0	43.3	43.0	54.4	48.3
Cameroon	25.8	21.5	21.7	23.3	10.1	15.5	10.8	8.4
Congo, D.R. of	1.4	1.0	4.4	4.5	1.0	1.0	2.0	3.0
Ghana	54.5	53.6	44.3	49.5	49.0	48.3	44.6	45.3
Indonesia	34.0	37.3	23.6	36.6	18.3	20.3	6.9	21.2
Malaysia	80.9	80.4	67.5	66.8	64.4	66.2	66.7	64.5
PNG	21.5	16.7	35.5	19.8	14.9	18.8	25.0	12.8
Average	38.39	36.18	38.24	36.03	28.85	30.03	29.11	28.39

Source: World Bank, "Worldwide Governance Indicators Country Snapshot," with special thanks to Manuel Estrada Porrua.

a. Measures the quality of public services, the quality of the civil service and its degree of independence from political pressures, the quality of policy formulation and implementation, and the credibility of the government's commitment to such policies.

b. Measures the ability of the government to formulate and implement sound policies and regulations that permit and promote private sector development.

c. Measures the extent to which agents have confidence in and abide by the rules of society, in particular the quality of contract enforcement, the police, and the courts, as well as the likelihood of crime and violence.

d. Measures the extent to which public power is exercised for private gain, including petty and grand forms of corruption as well as "capture" of the state by elites and private interests.

of the community-based forestry effort is that the law enforced by a centralized state can be only one component of a broader forestry solution.[10]

In the context of a REDD-based system, the foregoing means that significantly reducing the national rate of deforestation in most developing countries will take time and will carry important socioeconomic and political costs. It will have to rely on the joint efforts of national and local, public and private actors.

Challenges of a REDD Mechanism Based Exclusively on National-Level Accounting

A REDD mechanism built exclusively on national accounting would require the definition of a national reference emission level and a nationwide monitoring system. Credits would be issued once emission reductions have been achieved and GHG emissions from deforestation have dropped below the reference emission level. This means that in principle the cost of the additional effort required to change historic deforestation patterns would have to be covered up front by developing country governments. Moreover, under such a scheme countries would require the capacity and willingness to adopt and enforce forest management and protection policies as well as to accurately account for forest carbon and be accountable for subsequent carbon losses.

Consequently, a national approach to REDD is based on the assumption that countries are able to fund and successfully implement effective policy, legal, and institutional reforms nationwide and that they are in the position to formulate and enforce appropriate social and economic safeguards—something that is a challenge for many developing countries. History shows that deforestation has rarely been successfully limited at the national level through specific policy interventions. Indeed, as noted in the report from a workshop organized by the UNFCCC in Rome in 2006, all but one of the reported policies for reducing deforestation applied so far show variable, moderate, low, or even negative success, with success often dependent on local circumstances.[11] From this it seems that policies are likely to fail if countries lack adequate resources to identify, develop, and implement appropriate policies.[12] The massive capacity-building effort and long-term funding required to effectively reduce emissions from deforestation are currently not a policy priority of most developing countries, and even if they were, the effort would be far beyond the possibilities of the public resources available in most countries. A consequence of mechanisms based on national performance would thus be that countries with little capacity to implement forest protection measures and therefore most in need of international support would be unable to benefit from incentives for the reduction of emissions from deforestation and would be subject to (most likely unaccounted) international leakage from the few successful countries.

Given the level of resources required to reduce emissions from deforestation, the participation of private funds in any mechanism adopted by the UNFCCC toward this end will be critical. However, the private sector has expressed reluctance to invest directly in developing country governments or in projects for which performance is linked to government performance in reducing emissions from deforestation and degradation nationally.[13] In a system in which the allocation of funds and potential carbon credits takes place through host country governments, the political and legal risk of the mechanism is considered too high to attract private finance. The experience of the Kyoto Protocol's Joint Implementation (JI) mechanism shows that the mere fact that credits will be issued by national governments rather than an international body creates a significant barrier to the market.[14]

In sum, in most of the relevant countries a REDD system based *exclusively* on a national emission reduction crediting system is likely to take significant time to be established, and the risk of failure will remain high.

Cornerstones of an Inclusive REDD Mechanism

Decades of international cooperation for the protection and development of forest resources in developing countries, as well as the recent establishment and rapid growth of the international carbon market, provide the foundation for identifying some basic elements crucial for the success of any mechanism aimed at reducing emissions from deforestation in developing countries. These elements are the following:

—Incentives to undertake REDD measures under the UNFCCC framework should accommodate different national circumstances and levels of capacity, so that countries are able to participate immediately and to increase their participation as they enhance their capacities, thus allowing for growing involvement in global emission reduction efforts.

—Incentives to reduce emissions from deforestation should be complemented by instruments to allow countries to build capacity and enhance the availability and quality of data, as well as to propose and implement effective policy measures.

—In order to mobilize the necessary investment flows into developing countries, a private-sector-driven mechanism allowing for the commercialization of carbon credits is essential.

—Any mechanism must be embedded in wider participation and deeper GHG emission reduction commitments by industrialized countries and enhanced responsibility on the part of developing countries.

—Scientific and methodological uncertainties have decreased, so the proposed mechanisms should be consistent with the principles of the carbon market and should rely on the technical and institutional infrastructure already in place.

—Effective policies to address technical issues such as baselines and leakage should be formulated so that these issues do not become barriers to initiating REDD activities.

On the basis of these cornerstones, we propose a phased, double baseline-and-credit mechanism that rewards government as well as public and private entities for lowering deforestation rates.

Proposal for a REDD Mechanism

The success of a REDD mechanism will depend on the adoption of a flexible policy approach to address deforestation drivers at both the national and subnational levels while creating an enabling environment for forest carbon conservation and management. Such an enabling framework should put in place the conditions for private and public activities at the national, subnational, local, and project levels.

The National Approach

Developing countries should be encouraged to account for and control their emissions from deforestation at the national level as soon as possible. However, owing to different national circumstances, some countries will be unable to do this in the short run. We therefore propose a double baseline approach whereby both subnational activities (that is, projects, programs, and activities at the municipal and state levels) and national ones can be started immediately. Developing countries would be able to decide on their initial level of participation in this mechanism according to their circumstances and interests. In the case of implementation of activities at the subnational level, once the affected area of the participating country reached a determined percentage of its total forest territory, or once more than an agreed-upon number of years had elapsed since the start of the first subnational activity, the country would have to adopt a voluntary goal to reduce its national emissions from deforestation. We consider such an obligation acceptable given that the implementation of subnational activities and developing projects will create capacities in the countries while the proposed "area trigger" provides an incentive for Annex I countries to invest in REDD activities in developing countries, to buy the ensuing emissions reductions, or both. Participating developing countries could, at any time before reaching the proposed limit, decide voluntarily to adopt a national emission reduction target.

In addition, the approach should provide incentives for participation in national-level initiatives. One incentive could be that national-level accounting might use IPCC Tier 2 or Tier 3 methods while project-level accounting uses only Tier 3.[15] Higher-tier methods are more data intensive and require monitoring of local variables and are therefore more accurate, precise, and expensive. Another incentive is that national-level initiatives would not have to assess leakage (provided that there is continued international agreement to ignore international leakage), whereas project level accounting would have to assess, verify, and subtract leakage, thus enhancing the atmospheric integrity of project activities relative to national accounting schemes.

National target GHG emission levels may be above or below the empirically known deforestation level of the base year or base reference period and may be reviewed periodically (that is, for each crediting period) to account for structural and other relevant changes. Because exact deforestation levels and future land-use trends are uncertain, developing countries should be given sufficient time (and assistance) to assess these issues. Moreover, in order to be realistic and achievable, the emission reduction target to be pursued by each country should first be discussed and agreed upon internally by each host country, taking into account institutional barriers, agents and drivers of land-use change, growth projections, contrasting interests of different economic agents, and multiple views on national sustainable development. The success of reducing emissions from deforestation depends on the ability of developing countries to conduct this process with sufficient technical and financial assistance. This takes time. We therefore envision a road map with clearly defined milestones to reach the goal of establishing a national target level of GHG emissions from deforestation.

REDD credits would be issued for any emission reduction below the agreed national target emission level. An established percentage of REDD credits issued from a country would not be transferred to industrialized countries but would be held by the host country in a mandatory reserve account to guarantee compliance with the agreed target in future verification periods. Credits in this reserve account would represent net contributions to global emission reductions because they would be unavailable to offset emissions in industrialized countries. Issuance of REDD credits would be overseen by a UNFCCC body according to the following principles:

—A target emission level would be defined for each relevant crediting period.

—If emissions from deforestation remain above the target emission level during the initial verification period, no credits would be issued and no emission debits accounted for.

—If emissions from deforestation remain below the target emission level during a verification period and if REDD credits were issued for that period, then the implementing country would have to compensate for any potential future overemissions in subsequent verification periods.

—Consequently, in case of future emissions above the target emission level, the implementing country could choose to (1) offset the excess emissions by canceling REDD credits from its reserve account or by acquiring REDD credits from other implementing countries' reserve accounts; (2) overcomply in the subsequent verification period by a quantity of emission reductions equivalent to the excess deforestation emissions of the previous verification period; or (3) request an adjustment of its target emission level for the subsequent verification period, arguing justifiable reasons of force majeure (such as large-scale forest destruction due to extreme climatic events and their consequences, war, terrorism, and so forth) or improvements in the availability of data and methods. Any adjustment of the target emission level would be subject to either review and approval by the parties or independent validation and certification following transparent procedures, agreed upon by the parties.

The Subnational Mechanism

In order to mobilize emission reductions from the broadest possible range of activities, it is necessary to allow activities at the subnational level and to encourage the participation of the private sector. Subnational activities could start without further delay and in all countries, independent of emission reduction targets at the national level and of international support. Successful REDD activities would further encourage governments to take action and would bend the learning curve upward, because the private sector brings not only finance but also human resources. Some countries might be able to speed up their national process and announce internationally a voluntary national emission target for the post-2012 period. Any country could authorize private or public entities to develop and implement REDD activities at the subnational level.

Such REDD activities would have to either adopt already existing regional emission reference levels (baselines) or establish their own emission reference levels. REDD activities would have to be authorized by the host country and implemented in accordance with its sustainable development policies. REDD credits issued for such projects would be backed by real, measurable, and additional emission reductions. They would be issued directly to the authorized project participants by the competent UNFCCC body, even in the case of excess deforestation emissions at the national level. Issuance of REDD credits for project activities would require that the activities be subject to a validation, verification, and certification procedure by an independent, accredited body.

In order to avoid leakage, one of the main environmental concerns associated with subnational REDD activities, leakage-prone types of projects would be avoided under strict criteria for eligibility. In addition, leakage would need to be monitored carefully, and if detected it would have to be independently verified and subtracted in the calculation of emission reductions attributable to the REDD activity. Good policies and methodological guidance would have to be developed to accomplish the following:

—Promote project designs that address the drivers of deforestation rather than displace them elsewhere—for example, by providing alternative livelihood options to the agents of deforestation.

—Attribute leakage to deforestation and forest degradation agents rather than to agents of conservation and sustainable forest management. For instance, if illegal logging increases in a country after the implementation of a REDD project, but the project has actually provided sustainable livelihood alternatives to the communities that were previously deforesting the project area, then leakage should be attributed to poor law enforcement rather than to the project.

—Design and enforce policies and regulations that minimize the risk of leakage and complement conservation activities such as displacement of illegal logging.

—Add independently verified leakage to the national target emission level once this level has been negotiated and registered, in order to avoid increasing the burden to governments. In countries that have not adopted a national emission target, leakage should be determined using an approved methodology, and the amount of leakage detected should be deducted from the project credits. Host countries may decide to request projects to transfer a certain amount of credits to a national reserve account to address leakage internally.

To further enhance the contribution of developing countries to global emission reductions, credits issued rewarding emission reductions from specific activities could be either temporary, with no project and no host-country liability (similar to tCERs), or permanent, with a mandatory reserve of credits to be transferred to the national reserve account. A third solution would be to follow the example of the Voluntary Carbon Standard and require the establishment of buffers of credits at the project level or a collective buffer at the global level.[16] Once a country has adopted a national emission target, only permanent credits would be issued.

Supporting Instruments

We further encourage the establishment of a multilateral fund that would finance activities aimed at creating enabling conditions, including institutional and technical capacities. The fund might include an enabling window (readiness) and an activity window. The enabling window of the fund would be disbursed on a grant basis. Part of its tasks would be to develop reliable carbon monitoring systems. An activity window of the fund might enable early action activities, implemented before 2012, and pilot activities designed to test the effectiveness of capacities and measures for reducing emissions from deforestation.

A new fund, or the replenishment of an existing one, has also been proposed to finance the government costs of REDD. To achieve climatically meaningful emission reductions will require identifying sources of sufficient, continued, and predictable replenishments from industrial countries, especially if these are seen by the parties as an alternative to market instruments. Therefore, in addition to voluntary contributions to kick-start capacity-building and early action activities in developing countries, any new fund should be fed by institutionalized mechanisms such as, among others, a levy on Assigned Amount Units first traded in the carbon market, similar to the one imposed on CERs; fees on carbon-intensive commodities and services in industrialized countries; a levy on international transport emissions; revenues from the auctioning of credits in emission trading systems; and, where emission trading systems have price caps, revenues from selling credits at the price-cap level.

Conclusion

Deforestation is a symptom of a multicausal disease for which a proven cure does not yet exist. Governments and multilateral agencies have been trying to promote forest conservation and halt illegal logging for almost three decades. Success has been modest. Taking into account the multiple challenges developing countries face, the risk of governments' failure to reduce emissions from deforestation is real and undeniable. The world community does not have the time and resources to take this risk. The consequences of global climate change are real and imminent, and they will hit developing countries hardest. According to the IPCC's Working Group III, the world has fifteen years to meet the climate target of limiting global mean temperature rise to 2°C above the preindustrial level. This means we have no time to lose, and relying exclusively on governments to act would be negligent. To maximize chances to reduce emissions from deforestation, negotiators are called upon to put in place a system of incentives that encourages all layers of society and all sectors to contribute to the enormous effort required to decrease tropical deforestation and move toward a fair and robust land management system.

In our view, the REDD crediting system we have described would be able to attract private capital into REDD activities because successful project-based activities would be credited even in the case of excess deforestation emissions at the national level. At the same time, the nested approach encourages governments to adopt internationally negotiated and agreed-upon target levels for reductions of emissions from deforestation, which rewards the lowering of national deforestation emission levels.

Although we think this approach will facilitate national as well as subnational activities to reduce emissions from deforestation, we are aware that the system in itself remains incomplete. It can be only an element that encourages broader conservation, including conservation of forests that are not under actual threat of deforestation. Once such a threat can be measured, in many cases it will be too late to effectively protect the forest and the treasures it harbors.

Notes

1. R. A. Houghton, "Role of Forests, in Particular Tropical Forests, in the Global Carbon Cycle," presentation at UNFCCC workshop, Rome, 2006 (http://unfccc.int/files/methods_and_science/lulucf/application/pdf/060830_houghton.pdf); Intergovernmental Panel on Climate Change, "Fourth Assessment Report," 2007; Food and Agriculture Organization, "Global Forest Resources Assessment 2005: Progress towards Sustainable Forest Management," Forestry Paper 147 (Rome: FAO, 2006). Because of the rapid expansion of emissions from fossil fuel consumption in the last decades, the *relative* contribution of deforestation to global GHG emissions is decreasing, but the *absolute* emission level is not.

2. Murray Ward, "Where to with LULUCF? First, How Did We Get There?" Global Climate Change Consultancy (http://homepages.paradise.net.nz/murrayw3/documents/pdf/Where percent20to percent20with percent20LULUCF.pdf).

3. FCCC/CP/2005/MISC.1.

4. The "nested approach" was first presented by the Tropical Agricultural Research and Higher Education Center (CATIE) and the German Emissions Trading Association (BVEK) in a submission to the UNFCCC Secretariat in February 2007. Our proposal is

intended to contribute to the ongoing discussions in the context of the UNFCCC by incorporating the different views and concerns expressed by parties during the negotiation process started at the thirteenth Conference of the Parties in Montreal in 2005.

5. N. Stern and others, *Stern Review on the Economics of Climate Change* (London, 2006).

6. Market data from Point Carbon, "Carbon 2008—Post 2012 Is now—." Oslo, March 11, 2008.

7. "Carbon 2007: A New Climate for Carbon Trading," Point Carbon, 2007 (www.pointcarbon.com/getfile.php/fileelement_105366/Carbon_2007_final.pdf); K. Hamilton and others, "State of the Voluntary Carbon Markets 2007: Picking Up Steam," Ecosystem Marketplace and New Carbon Finance, Washington and London, 2007.

8. See Capoor and Ambrosi, "State and Trends of the Carbon Market 2007." Documents relevant to the GEF fourth replenishment are at www.gefweb.org/interior.aspx?id=48.

9. World Bank, "Strengthening Forest Law Enforcement and Governance: Addressing a Systemic Constraint to Sustainable Development," 2006 (http://siteresources.worldbank.org/INTFORESTS/Resources/ForestLawFINAL_HI_RES_9_27_06_FINAL_web.pdf?).

10. Craig Segall, "The Forestry Crisis as a Crisis of the Rule of Law," *Stanford Law Review* 58, no. 5 (2006): 1539–41, 1543.

11. See http://unfccc.int/methods_and_science/lulucf/items/4123.php.

12. "Views on the range of topics and other relevant information relating to reducing emissions from deforestation in developing countries," submission by the Republic of Vanuatu, February 2007 (FCCC/SBSTA/2007/MISC.2).

13. Report of the public dialogue on mechanisms to reduce emissions from deforestation (RED) in developing countries organized by Avoided Deforestation Partners, Brussels, October 24–25, 2007. This dialogue was supported by the governments of Costa Rica, Peru, Colombia, Paraguay, Panama, and Ecuador, as well as the UN Economic Commission on Latin America and the Caribbean and GTZ. The event attracted more than seventy attendees from governments and the private, for-profit, and nonprofit sectors.

14. The slow progress of countries such as those of the Russian Federation in issuing approvals for JI activities creates additional risks. By April 2008 only a handful of JI projects had achieved international registration, in comparison with more than 1,000 CDM projects.

15. On IPCC methodologies and reporting principles, see IPCC, "Good Practice Guidance for Land Use, Land Use Change and Forestry" (Institute for Global Environmental Strategies, 2003). See also A. Ravin and T. Raine, "Best Practices for Including Carbon Sinks and Greenhouse Gas Inventories" (www.epa.gov/ttn/chief/conference/ei16/session3/ravin.pdf, accessed January 31, 2008).

16. "Voluntary Carbon Standard, Guidance for Agriculture, Forestry and Other Land Use Projects" (www.v-c-s.org/docs/AFOLUpercent20Guidancepercent20Document.pdf).

PART FIVE

National Systems and Voluntary Carbon Offsets

18

Legislative Approaches to Forest Sinks in Australia and New Zealand: Working Models for Other Jurisdictions?

KAREN GOULD, MONIQUE MILLER, AND MARTIJN WILDER

In this chapter we summarize the approaches taken by Australia and New Zealand, two neighboring countries with historically different views on participation in the Kyoto Protocol, toward regulating and encouraging the creation and trading of carbon commodities from forest sinks.

Australia and New Zealand have both expended significant thought and effort on the regulation of forest sink activities and have put in place a range of legislation and regulatory frameworks to assist in identifying the legal ownership of sequestered carbon, encourage additional investment in forest sinks, and regulate the trading of carbon commodities from carbon sinks. Their experiences provide a useful starting point for jurisdictions that are considering implementing policies or legislation to support forest sink projects.

Australia

Australia has a strong forestry industry and a major environmental legacy associated with the clearing of extensive regions of native forest for agricultural purposes. It is a world leader in forest accounting practices, and the forestry industry is a strong lobby group at both the state and Commonwealth government levels. Despite its former stance on the Kyoto Protocol, Australia has developed some domestic policies and legislation to facilitate security in carbon sequestration projects. These may provide some valuable lessons to jurisdictions that are currently grappling with the regulation and promotion of carbon sequestration projects in the context of emissions trading.

On November 24, 2007, the Australian Labor Party was elected to government following eleven years of Australian Liberal Party rule. The Liberal Party had refused to ratify the Kyoto Protocol in its current form, but the newly elected Commonwealth government ratified the Kyoto Protocol on December 3, 2007, as its first official act. Ratification enters into force ninety days after the Instrument of Ratification is received by the United Nations, so Australia became a full member of the Kyoto Protocol at the end of March 2008. Although short- and medium-term targets will not be set until late 2008, the Commonwealth government has committed to reducing Australia's greenhouse gas emissions to 60 percent below 2000 levels by 2050.

A national emissions trading scheme including provision for forest sinks is firmly on the agenda for Australia. Two alternatives have been proposed in detail. The first was proposed as a joint venture between the Australian state and territory (Labor) governments. The second, more recent proposal was made by the former Liberal Commonwealth government's Prime Ministerial Task Group on Emissions Trading. The new government has committed to establishing an operational emissions trading scheme by 2010, and the approach ultimately adopted may draw from one or both of the two former proposals. It is also likely to be influenced by the report of the Garnaut Climate Change Review, due to be released in draft form by June 30, 2008. The Garnaut Review is an independent study by Professor Ross Garnaut, commissioned on April 30, 2007, by the state and territory governments, together with the Australian Labor Party (then in federal opposition). The Commonwealth government's own green paper on the design of its emissions trading scheme has been scheduled for release in July 2008.

Under Australia's system of federal government, Australian laws and regulations are developed at both the regional level, by state and territory governments, and the national level, by the Commonwealth government. The Australian Constitution divides legislative responsibility between the Commonwealth government and the state and territory governments, providing the Commonwealth government with powers to legislate in respect of an exhaustive list of issues that were, at the time the Constitution was drafted, considered to affect Australia as a nation rather than only people in a specific state or territory. The power to enter into and implement treaties such as the Kyoto Protocol and the UNFCCC is a Commonwealth government power, whereas the power to regulate the energy and forestry sectors lies within the broad state and territory legislative mandate. In terms of emissions trading legislation, an overlap clearly exists between the ability of the Commonwealth government to legislate on this issue (such legislation being a means of implementing Australia's obligations under the UNFCCC) and the ability of state governments to regulate their own industry sectors.

If the Commonwealth government passes a law that is properly within its power to pass, and if that law is designed to "cover the field" of an area that is already the subject of state and territory legislation, then the Commonwealth law overrides the state law. Therefore, a trading scheme established by the Commonwealth government would take priority over one created under state and territory legislation.

It is likely that any national emissions trading scheme would draw on two elements of existing Australian legislation in developing a forestry offset framework. One is the assessment and permanence risk management procedures implemented by the first operational emissions trading scheme in the world to recognize forest sink credits, the New South Wales Greenhouse Gas Abatement Scheme (NSW Scheme) or, alternatively, the procedures used under the Commonwealth government's Greenhouse Friendly initiative. The other element is the legislation developed by various Australian states and territories allowing property rights in sequestered carbon ("carbon sequestration rights") to be "unbundled" from other property rights and registered separately on land title. This allows transactions in carbon commodities to be secured by indefeasible property rights, providing greater certainty for the market.

In the rest of this section we describe the two existing high-level policy proposals for a national emissions trading scheme in Australia. We also examine the manner in which the NSW Scheme has developed rules and procedures for accrediting forest sink projects and managing the long-term permanence risks arising from such projects, including a discussion of the manner in which various Australian states have developed property legislation to "unbundle" carbon sequestration rights.

National Emissions Trading Proposals

In September 2006 the Australian state and territory governments released a discussion paper outlining a proposed design for a national emissions trading scheme (NETS).[1] Despite its opposition to the Kyoto Protocol, the former Liberal Commonwealth government soon followed suit. In late May 2007 the Prime Ministerial Task Group on Emissions Trading, an industry working group established by the Commonwealth government, recommended the establishment of an Australian national emissions trading scheme by 2011, which would include the ability to generate and trade offsets from forest sink projects.[2]

In its emissions trading scheme discussion paper (the "Garnaut Paper"), released on March 20, 2008, the Garnaut Review sets out its current thoughts on emissions trading scheme design as a basis for public comment and discussion. The Commonwealth government stated at COP/MOP 3 in Bali that the findings of the Garnaut Review would be central to policy development for an emissions trading scheme in Australia. In addition, the government may pick up the work already undertaken by the Labor state and territory governments and the Prime Ministerial Task Group on Emissions Trading in designing its national emissions trading scheme. We therefore provide overviews of both proposals, as well as the Garnaut Paper.

STATE AND TERRITORY PROPOSAL. Under the NETS proposal, all states and territories would adopt a cap-and-trade mechanism consistent with other international frameworks for carbon trading. The NETS was proposed to be fully operational by 2010. Its goal is to reduce emissions in 2050 to roughly 40 percent of their 2000 level. It is unusual in Australia for states to work together to harmonize their legislative frameworks. It is even more unusual that an issue such as emissions trading, which was originally unpopular among some of the high-emitting, lucrative, and politically persuasive resources industry members in Australia, should obtain state consensus. That such consensus to address climate change through market measures can exist across Australian states (as well as among the international community) is evidence of the political importance of the issue and the recognized need for concerted action.

Forestry is a priority abatement activity for inclusion in the NETS. It is proposed that offsets be created in accordance with emerging approaches being developed for the Joint Implementation mechanism of the Kyoto Protocol. Offsets would need to meet permanence and measurement criteria, and baseline methodologies must be robust.

The NETS discussion paper states that additionality would be a requirement for eligible offsets but that "financial additionality" would not be the appropriate measurement. It was not specified whether individual sink projects would be required to make additionality submissions or whether, for example, all sink projects on land cleared since January 1, 1990, would be deemed to be additional (as is the case in the NSW Scheme).

The discussion paper anticipates that the rules governing sink projects under the NETS would be based on the NSW Scheme. However, there was some uncertainty over whether the NETS would implement the NSW Scheme's requirement that sequestration be maintained for 100 years. The NETS may adopt a different definition of "permanence."[3] Furthermore, the "make-good" provisions may be adjusted to allow the surrender of any permits or offset credits (not just from forestry projects) if the sequestration is not maintained in accordance with the scheme's rules.

FORMER COMMONWEALTH GOVERNMENT PROPOSAL. The second proposal is that of the Prime Ministerial Task Group on Emissions Trading, which also proposed a national emissions cap-and-trade scheme, to be operational by 2011 with the following features:

—Government would set a long-term aspirational goal for emission reductions covering all greenhouse gases and a series of short-term annual caps for certain sectors, initially until 2020.

—Coverage of the scheme would be likely to extend to facilities with emissions exceeding 25 kilotonnes of carbon dioxide equivalent (CO_2e) a year and would include stationary energy, fugitives (excluding open-cut coal mines), industrial processes, and transport.[4] Coverage would generally be of direct emissions for large facilities, but with upstream coverage of fuel suppliers (nonindustrial coal, gas and petroleum) for other energy emissions. Agricultural and land-use emissions and possibly the waste sector would initially be excluded, with the possibility of later incorporation.

—Recognition would be given to a wide range of offset credits from abatement activities both in Australia and internationally, including credits from forestry and agriculture. In particular, recognition would be given from 2008 to offset credits generated from abatement activities accredited under existing standards, such as the Commonwealth government's Greenhouse Friendly standard, or the Clean Development Mechanism of the Kyoto Protocol. Credits from carbon forest sinks could be created and "banked" ahead of the commencement of permit trading.

In terms of land use and land-use change projects, the task group has stated that "an integral part of Australia's international climate change strategy should be to develop the elements of a future Australian approach to international offsets. It would need to be informed by discussions with a range of international partners. A programmatic model . . . is preferable—one that would lead to greater investment certainty, drive transaction costs lower and promote a longer-term perspective. Simplified approaches to offset recognition could be considered, provided that they could demonstrate robust abatement outcomes. Inclusion of land use, forestry and wood products and the recognition of credits for carbon geosequestration should be priorities. This would help to promote more flexible models for greater global participation."[5]

The proposed inclusion of harvested wood products and geosequestration is unique, but it is a particular focus for Australia given its economic identity as a net wood and coal exporter. Avoided deforestation and forest stewardship are also mentioned regularly throughout the task group's report.

The task group's report is still a very high-level document and contains little information about the technical aspects of the eligibility of land use, forestry, harvested wood products, avoided deforestation, and forest stewardship to generate tradable offsets. However, the report does propose that forestry projects be able to create credits from 2008 to be recognized for trading at a later stage. Much work remains to be done to canvas a range of technically complex and politically contentious issues involving forest sinks, including managing permanence risk on a legislative basis, accounting for harvested wood products, and assessing "additionality" for offsets (and determining whether additionality is a prima facie requirement).

Garnaut Review. The Garnaut Paper seeks to guide the design of an emissions trading scheme in Australia, and its proposed key design features are

—Caps should be expressed as a trajectory of annual emissions targets over time, with four trajectories being specified upon establishment of the emissions trading scheme. The first up to 2012 should be based on Australia's Kyoto commitments and the other three will reflect increasing levels of ambition post-2012.

—Coverage of the scheme should initially extend to stationary energy, industrial processes, fugitives, transport, and waste. Agriculture and forestry should be included as soon as practicable.

—All permits should be auctioned at regular intervals, with the possible exception of providing free permits to trade-exposed, emissions-intensive industries in lieu of cash payments.

—Determining limits on international purchases as well as strategic and policy parameters for international linking should be a role for government. Direct links with the CDM should be limited.

—Unlimited banking should be allowed. Official lending of permits by the independent authority to the private sector should also be allowed, but may be subject to limits in terms of quantity and time.

—Domestic offsets should be accepted without limits, but will have a small role given the scheme's broad coverage.

—A financial penalty should apply for noncompliance, along with a "make-good" provision.

The Garnaut Paper recognizes that there is considerable potential for sequestering carbon through change in land management and agricultural practices. However, it suggests that full inclusion of agriculture and forestry in Australia's emissions trading scheme would require that issues regarding measurement and monitoring of greenhouse gases be resolved. For example, inclusion of forestry emissions in the scheme would require an assessment of carbon sequestered in long-lived timber products.[6]

The lessons learned through the experience of the Australian states and territories, which were "early movers" in terms of forestry offsets and carbon property rights, could be a valuable basis on which to develop appropriate measures for the proposed national emissions trading schemes.

A World First: Forest Sinks under the NSW Scheme

New South Wales, the most populous Australian state, already has a functioning emissions trading scheme, the Greenhouse Gas Abatement Scheme (NSW

Scheme), which places mandatory emission targets on electricity retailers and large energy users in the NSW electricity market.

The NSW Scheme was the first operational emissions trading scheme in the world to recognize forestry sink projects. Sink project developers and the scheme administrator (the Independent Pricing and Regulatory Tribunal of NSW) have therefore had to develop a range of strategies dealing with property legal aspects, contractual entitlements, and forest management approaches, in order to address the 100-year "permanence" requirement under the scheme.

Because the NSW Scheme is designed to be compatible with the Kyoto Protocol's accounting approach for forest sinks, the experience gained by the scheme administrator and participants could be invaluable for participants in other jurisdictions where it is possible to develop forest sink offsets.

NSW abatement certificates (the relevant "credits" under the NSW Scheme) can be generated from afforestation and reforestation on the basis of a tonne of CO_2e sequestered by a Kyoto Protocol–eligible forest (that is, a forest created by human-induced activity on land that was cleared of trees on December 31, 1989). As such, they are potentially compatible with the LULUCF accounting requirements for afforestation and reforestation projects under the Kyoto Protocol.

LEGISLATIVE BACKGROUND TO THE NSW SCHEME. The NSW state parliament passed legislation in December 2002 that amended the Electricity Supply Act 1995 (NSW) to require electricity retailers and certain other entities, including large electricity users such as aluminum smelters—collectively referred to as "benchmark participants"—to meet mandatory targets for reducing or offsetting the emission of greenhouse gases from the production of the electricity they supply or use. Penalties are imposed on benchmark participants that fail to meet targets in any given year. The act provides only a broad legal framework for the operation of the scheme. Most specific provisions relating to the creation of abatement certificates are set out in the Regulations and the Greenhouse Gas Benchmark Rules. The legislative structure of the scheme is depicted in figure 18-1.

The scheme commenced on January 1, 2003, in NSW, and following passage of complementary legislation through the Australian Capital Territory's Legislative Assembly, it commenced in the Australian Capital Territory on January 1, 2005. Enforceable annual emission reduction targets have been set for 2003 to 2012, and the NSW government has stated its commitment to extending the scheme until 2020 with automatic fifteen-year rollovers. The first state emissions target is 8.65 tonnes of CO_2e emissions per head of state population. This will progressively decline to the final benchmark of 7.27 tonnes of CO_2e per person in 2007 (5 percent below the equivalent NSW per capita emissions in 1990), as shown in figure 18-2. Benchmark participants' targets are based on their contributions to total state electricity retail sales.

Figure 18-1. *Legislative Framework of the New South Wales Scheme*

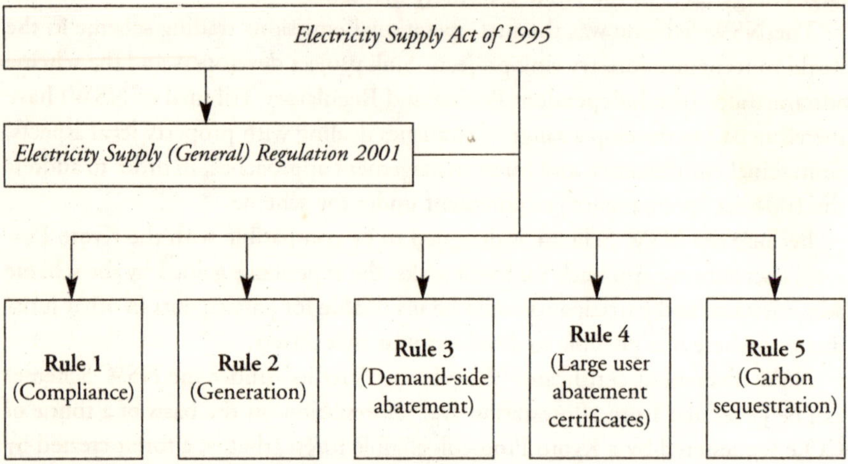

The scheme's Greenhouse Gas Benchmark (Carbon Sequestration) Rule 5 makes provision for abatement certificates to be created from forest sinks, which can subsequently be used by benchmark participants to meet their emission reduction obligations. The carbon sequestration rule is consistent with article 3.3 of the Kyoto Protocol and recognizes afforestation and reforestation.[7] Currently, avoided deforestation projects and forest management or revegetation projects (as described under article 3.4 of the Kyoto Protocol) are not included in the scope of the carbon sequestration rule, although there is some scope to introduce such projects at a later stage.[8]

The NSW government has indicated the NSW Scheme will remain in existence until 2020 or until commencement of a Commonwealth emissions trading scheme. A working group is currently considering options for the transitioning of abatement certificates into the national scheme so as not to disadvantage market participants that have made good faith investments under the NSW Scheme.

Eligibility Criteria. The eligibility criteria for an entity to create offsets from forest sinks under the NSW Scheme are summarized as follows:

—The entity must manage a "carbon sequestration pool" and obtain accreditation as an abatement certificate provider.

—The entity must exercise "control sufficient to enforce Carbon Sequestration Rights" over land in NSW or the Australian Capital Territory (or any other jurisdiction approved by ministerial declaration) that complies with article 3.3 of the Kyoto Protocol (that is, was cleared of forests on January 1, 1990).

—The entity must undertake afforestation or reforestation on the eligible land with the result that an "Eligible Forest" is created (minimum height, crown cover, and width restrictions apply).[9]

Figure 18-2. *New South Wales Benchmark, 2003–12*

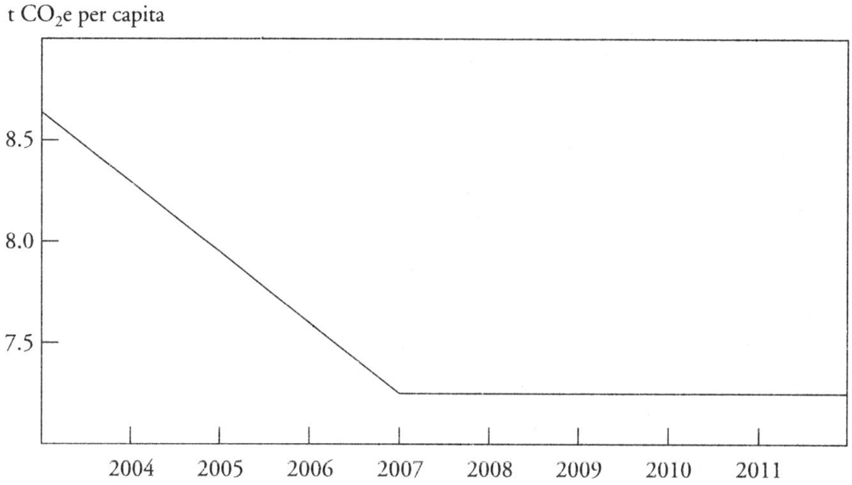

—The entity must be able to demonstrate to the scheme administrator that legal arrangements are in place so that it can maintain the abatement achieved by a carbon sequestration activity for 100 years from the date on which the resultant abatement certificates are registered. If sequestration is not maintained for 100 years, then the abatement certificate provider may be required to "make good" the sequestration shortfall by surrendering abatement certificates from other forest projects.

If all the NSW Scheme requirements are met, then an abatement certificate provider may claim abatement certificates for carbon sequestered after January 1, 2003. Details on each of these requirements are set out below.

CARBON SEQUESTRATION POOL. The NSW Scheme requires sequestration project developers to mitigate the risks arising from reliance on one plantation to generate credits by requiring that they manage a carbon sequestration pool on eligible land. A carbon sequestration pool is an aggregation of forests managed to provide carbon sequestration. The forests could be commercial plantations or native vegetation planted for biodiversity purposes. However, a single plot of land cannot be registered as a carbon sequestration pool. The purpose of the carbon sequestration pool requirement is to spread permanence risks (for example, risks of bush fires) across a range of forests in different locations, rather than accrediting all abatement to one forest that could be destroyed by a single event.

ABATEMENT CERTIFICATE PROVIDER. An abatement certificate provider is an entity accredited under the NSW Scheme to create abatement certificates on the basis of carbon sequestration. A range of entities have become abatement certificate providers, including state-owned forestry companies, private sector entrepreneurs, and nongovernmental organizations (NGOs). To become an abatement certificate provider, an entity must register for accreditation in respect of a specific carbon sequestration pool and must comply with the other eligibility requirements as described below.

CARBON SEQUESTRATION RIGHTS. A carbon sequestration right is a registrable property right to carbon sequestered on a particular piece of land. Under the Conveyancing Act 1916 (NSW) as amended by the Carbon Rights Legislation Amendment Act 1998 (NSW), the legal title to carbon sequestered by a forest on a piece of land is defined as a *profit a prendre* and a type of forestry right. It can be registered separately from the ownership of the land and also from the ownership of the trees on the land. Other Australian states have adopted similar legislation, drafted in slightly different terms.[10] The effect of designating a property-based carbon right as an independent form of proprietary interest is that benefits associated with carbon sequestration can be "unbundled" from the underlying land asset and are capable of being assigned separately from the land and the timber where the carbon is stored.

A carbon sequestration right is simply a registrable property right to any carbon sequestered on a certain piece of land. That a carbon sequestration right exists does not guarantee that any carbon is actually sequestered (for example, it is possible to own a carbon sequestration right over a car park). However, if there are plantations on the relevant land, it is possible to measure the carbon sequestered by those plantations to create carbon credits (or abatement certificates under the NSW Scheme).

Under the NSW legislation, the owner of the land has the prima facie right to register and assign a carbon sequestration right. If no carbon sequestration right is registered in respect of a particular piece of privately owned land, then the landowner is vested with the legal title to carbon sequestered on that land.[11] Each state's legislation treats the ownership and entitlements to carbon sequestration rights slightly differently, but in NSW the right is equated with a *profit a prendre*, a historical property law concept used to grant harvesters or hunters the right to enter land and take certain benefits from it. Although the analogy is imperfect (because a carbon sequestration right is a right to keep something *on* the land rather than remove it), the ability to register a carbon sequestration right on land title has a number of benefits. First, it puts anyone who wishes to deal in an interest in the land (purchasers, mortgagees, and so forth) on notice that the carbon sequestration right exists. Second, it "runs with" the land—that is, unless the holder of the carbon sequestration right terminates the right, future owners of the land are bound to recognize the holder's entitlement to sequestered carbon. Last, it provides assurance to purchasers and government regulators that they are dealing with the entity that is properly entitled to create and sell carbon arising from a plantation activity on a particular piece of land. This last point is particularly important in commercial forestry projects and community forestry projects in which a number of diverse entities may have some entitlement to the land and the trees.

In order to become an abatement certificate provider under the NSW Scheme, an entity must satisfy the scheme administrator that it exercises "control sufficient to enforce Carbon Sequestration Rights" over the eligible land. For this requirement to be met, carbon sequestration rights must be registered on the title of each parcel of land in the carbon sequestration pool.[12] Furthermore, if the entity applying to be the abatement certificate provider is not the owner of the carbon sequestration right, it must provide evidence that it has a legal entitlement to control the owner of the carbon sequestration right. Evidence of sufficient control could include, for example, copies of contracts with landowners.

ELIGIBLE CARBON SEQUESTRATION. Abatement certificates may be created only in respect of carbon sequestered on or after January 1, 2003, and not "brought to account or traded" for any purpose other than the creation of abatement certificates under the NSW Scheme. Abatement certificates are accredited for "net change" in carbon stock—the total amount in tonnes of CO_2 contained in a sequestration pool at a given time—over a certain period attributable to human-induced activity ("Carbon Sequestration Activity"), calculated in accordance with Australian Standard AS4978.1 2006, "Carbon Accounting for Greenhouse Sinks Part 1: Afforestation and Reforestation."

This standard allows the carbon stock to be estimated using a variety of methodologies. However, the forest manager must be able to demonstrate (on the basis of an uncertainty analysis) a 70 percent probability that the net increase in the area's carbon stock is greater than the number of abatement certificates it has created. This results in the creation of more certain or reliable abatement certificates and provides a direct financial incentive for participants to establish accurate systems for estimation. Forests must also be periodically monitored to verify the amount of carbon stored.

PERMANENCE. The NSW Scheme has no additionality criteria for forest sink abatement, but it does have strict permanence requirements. An abatement certificate provider must be able to demonstrate to the scheme administrator that it can maintain the abatement achieved by a carbon sequestration activity for 100 years from the date on which the resultant abatement certificates are registered. This is perhaps the most onerous aspect of the NSW Scheme relating to sink projects. If sequestration is not maintained for 100 years, then the abatement certificate provider may be required to "make good" the sequestration shortfall by surrendering abatement certificates from other forest projects.

In order to meet the permanence requirements, the approach to carbon sequestration must be consistent with the long-term maintenance of carbon stock. For example, a clear-fell forestry project with no plans to replant after clearing would be ineligible. Furthermore, for each eligible forest, a "Restriction on the Use of Land by a Prescribed Authority" ("restriction on use") must be registered for the benefit of the scheme administrator on the title of the respective parcel of eligible land. The requirements in the restriction on use are triggered if there is clear evidence that the carbon stock in the sequestration pool has fallen (or is likely to fall) below the minimum level. If this occurs, the landowner must take action (as required by the scheme administrator) to restore, replant, replace, reinstate, or re-establish any trees forming part of the eligible forest.

After an abatement certificate provider has created abatement certificates, the provider can reduce or sell its sequestration pool only by seeking the approval of the scheme administrator to assign its ongoing sequestration obligations to another accredited abatement certificate provider.

The permanence requirement is very difficult to satisfy, particularly given the long time periods involved, the generally accepted duration of contracts, and the variability of land-use over time. However, as discussed later, a number of carbon sink projects have nonetheless achieved registration under the NSW Scheme.

Risk Management. Relatively unpredictable hazards and risks such as fires, diseases, pests, and climate variability can significantly affect the amount of carbon stored in a sequestration pool. Consequently, to become accredited, an eligible abatement certificate provider must satisfy the scheme administrator that it has adequate risk management procedures in place with respect to hazards and risks that might affect the carbon stock. According to the scheme's guidelines, suitable risk management procedures must, at a minimum, identify hazards and risks, assess the hazards and risks for potential effect and likely occurrence, describe risk management strategies and practices to mitigate the hazards and risks, and include a review cycle to update the assessments and monitor the effectiveness of the arrangements.

Record-Keeping. Abatement certificate providers must also keep adequate records to satisfy the scheme administrator of their compliance with the NSW Scheme. To be eligible to accredit a particular carbon pool, abatement certificate providers' records must include the following:

—the location and area of each eligible forest site and the eligible land, including all spatial data necessary to uniquely identify the area of forest that will be used to create abatement certificates

—the carbon sequestration rights held in respect of the eligible land

—a description of the vegetation on each eligible forest site, including planting date, species, stocking rate, and productivity measures such as site index or other data that can be used to asses the growing capacity of the site

—the area of each eligible forest and the total gross area of the sequestration pool

—the sampling and measurement methodologies used to monitor the growth of the eligible forests and associated levels of uncertainty

—the models that have been applied—for example, site index models, tree growth models, wood density factors, biomass partitioning models, carbon proportion factors, and soil carbon models—and associated levels of uncertainty

The scheme administrator maintains a publicly accessible online registry containing details of abatement certificate providers and the ownership and status of abatement certificates at any point in time. Benchmark participants use the registry to surrender abatement certificates to meet their compliance obligations. However, trading of abatement certificates takes place outside the registry, which is not a trading platform.

CASE STUDY: MANAGEMENT THROUGH "POOLING." Forests NSW is a public trading enterprise within the NSW government Department of Primary Industries. In February 2005 it was fully accredited as an abatement certificate provider. This made Forests NSW the world's first independently audited and scheme-approved supplier of sink credits within an operational and mandatory greenhouse gas abatement scheme.

Forests NSW has adopted "carbon pooling" as an approach to manage its forest estate. This approach involves the grouping of individual sink projects into one centrally managed sequestration pool. Managers of carbon pools have responsibility for entering into agreements to acquire carbon sequestration rights from landowners and for on-selling abatement certificates to benchmark participants. Carbon pooling is generally more attractive to both sellers and purchasers of carbon rights because it represents an effective mechanism to spread risks across a portfolio of projects. It also has the potential to provide cost savings due to economies of scale.

Forests NSW's carbon pool currently consists of thirty-two individual forests located in northeastern NSW.[13]

The thirty-two properties making up the carbon pool cover a total area of 24,991 hectares, of which less than half is plantation. For environmental purposes and improved habitat provision, the remaining area consists of patches of native forest and riparian vegetation, which do not give rise to abatement certificates. Typically, plantation areas within Forests NSW's carbon pool contain several species of varying age classes. The 10,329 hectares of eligible plantation area are planted predominantly with four species that occur naturally in northeastern NSW: spotted gum, Dunn's white gum, blackbutt, and flooded gum. The oldest areas of plantation were planted in 1996, with the majority of forest planted between 1997 and 2000.[14]

Forests NSW uses a carbon accounting system based on field research undertaken over several years. For example, in order to verify that land meets the definition of reforestation, aerial photographs from the late 1980s are used. To meet the other requirements of the NSW Scheme, Forests NSW has created a range of procedures and models covering issues such as tree growth rates, rainfall data, soil type, mortality rates, risk management, harvesting, internal auditing, and tree species. These variables are modeled using uncertainty analyses to estimate the number of abatement certificates that can be created.

The first batch of Forests NSW abatement certificates, representing the sequestration of more than 130,000 tonnes of CO_2, was purchased by Energy Australia, a liable electricity retailer under the NSW Scheme, for more than A$1 million in 2005. Forests NSW mitigates the permanence risk by selling abatement certificates from a carbon pool made up of diverse vegetation types in diverse geographical locations. It is highly unlikely that a force majeure event would affect all the plantations in the pool, so it is unlikely (provided a "buffer" of carbon was retained and not sold) that a single event would place Forests NSW in breach of the NSW Scheme permanence requirements. The pooling approach is in fact mandated under the NSW Scheme to ensure that participants can adequately manage permanence risk.

Case Study: Sink Projects with Added Benefits. CO2 Group Limited is a boutique firm specializing in the provision of environmental services. Its wholly owned subsidiary, CO2 Australia Limited, aims to be the leading forestry sequestration company in Australia.[15] CO2 Australia is accredited as an abatement certificate provider under the NSW Scheme and establishes commercial-scale, permanent mallee eucalypt plantings in partnership with individual farmers in high priority revegetation agricultural regions of NSW. Mallee eucalypts are native to Australia, are drought resistant, and are thereby suitable for planting on marginalized land that might not otherwise support commercial forests. The eucalypts are generally planted in strips on unused portions of agricultural land (for example, between fields and along riparian corridors). CO2 Australia makes an up-front payment to landowners commensurate with the market value of a lease on the areas of their land on which the trees will be planted. In addition, CO2 Australia pays for the establishment and maintenance of the plantings and in return receives the carbon sequestration rights over the land. It is also granted access to the property to manage the plantation and to maintain the trees in situ for at least 100 years.

Typically, the mallee eucalypts are planted in belts and occupy 10 to 15 percent of participating properties. The plantations are not harvested and do not displace rural agriculture because they are integrated into cereal cropping systems. Along with greenhouse gas abatement, the plantations deliver multiple benefits such as dry land salinity mitigation, erosion management, improved soil quality, and regional employment opportunities. Mallee eucalypts have a root zone that can extend 30 meters below ground level, allowing them to survive extended droughts and to thrive in areas of low rainfall. Because of the high oil content in the leaves, mallee eucalypts are generally unpalatable to stock and insects. Indeed, in previous decades many Australian farmers expended significant effort in clearing their land of stubborn and robust mallee eucalypts in order to establish agricultural crops. It is interesting that improved environmental knowledge has resulted in farm management's undergoing a full circle—mallee eucalypts are now being used to manage the environmental difficulties caused by intensive agricultural practices and to provide income through carbon trading to the same farmers who cleared them.

In 2005, Carbon Banc, the trading arm of CO2 Group, entered into a deal to supply abatement certificates to Country Energy, another electricity retailer under the NSW Scheme, from 2007 until 2013. Up to 30,000 hectares of plantations were required to meet the volume of certificates under the supply agreement. The plantations were established on properties in the region for which Country Energy is the major energy retailer and provide the aforementioned benefits to landholders and the environment. This transaction was the first Kyoto-compliant carbon sink deal under an operating emissions trading system anywhere in the world.[16]

New Zealand

The New Zealand government has taken a proactive role in developing an international climate change regulatory regime. In December 2002, after broad public consultation, New Zealand became the 102nd country to ratify the Kyoto Protocol. The prime minister, Helen Clark, actively supported the ratification, stating that "the Kyoto Protocol is the international community's response to climate change and New Zealand is playing its part."[17] The government has also stated its awareness that tackling climate change requires a continuous reduction in greenhouse gas emissions throughout the twenty-first century and that it intends to seek continued and differential emission reduction targets beyond 2012. Furthermore, it has expressed a desire to use its status as a party to the Kyoto Protocol as a basis to persuade developing countries in the Asia-Pacific region to take on emission reduction obligations in future commitment periods.

New Zealand was one of the few countries to secure a "100 percent target" under the protocol, meaning that it is not required to reduce emissions below 1990 levels. However, after a revision of its greenhouse gas emission projections in May 2005, recent figures indicated that New Zealand may fall short of meeting its Kyoto Protocol target by 45.5 Mt CO_2e.[18] Accordingly, the government has initiated several changes to its climate change policy.

New Zealand has taken an approach different from Australia's to managing its carbon sink assets. Initially the New Zealand government adopted a policy reserving all rights and obligations with respect to emission reductions generated by forest sink activities. Recently, however, it has encouraged further private sector involvement in sink activities through the Permanent Forest Sink Initiative (PFSI) and a proposed emissions trading scheme (the NZ ETS). In the rest of this section we describe New Zealand's evolving approach to managing its forest sinks as part of its climate change strategy.

Initial Policy: Government Retention of Credits

Forests and their related sink credits form a key part of New Zealand's strategy to meet its commitment under the Kyoto Protocol. It is the government's view that all benefits, liabilities, and obligations under the protocol prima facie vest in the Crown. The government originally decided to limit or not devolve many of these benefits and liabilities. Indeed, the government has so far retained ownership of all sink credits allocated to New Zealand in the first Kyoto commitment period. As a result, sink credits generated under the protocol are not yet a property right of forest owners. The value of all carbon credits generated by post-1990 forest plantings have been retained, owned, and managed by the government. The government has so far also assumed all liabilities under the protocol in respect of these forests, and forest owners will not face deforestation or harvesting liabilities at any stage. For non-Kyoto forests (planted before 1990), the government has stated its intention to maintain a cap of 10 percent on the deforestation liability it will accept. This equates to 21 million tonnes of CO_2 emissions and is expected to cover any actual deforestation occurring over the first commitment period.

The government's initial policy was driven by its desire to manage New Zealand's overall response to climate change and to avoid potential major distortions to land values and uses created by the protocol's differential treatment of pre- and post-1990 forests.

"Carbon Farming" under the Permanent Forest Sink Initiative

On August 31, 2006, the New Zealand government announced that it would proceed with the PFSI in order to facilitate permanent (nonharvest) commercial forest sinks. Importantly, the government will still retain the forest sink credits and associated deforestation liabilities that New Zealand receives for commercial plantation forests established since 1990 (although the proposed NZ ETS would reverse this position, as discussed later). Under the PFSI, forest owners who establish new permanent forest sinks will receive fully tradable, Kyoto-compliant emission units in proportion to the carbon sequestered.[19] This is intended to allow farmers to make better economic use of their land, particularly when that land is isolated and erosion prone and therefore unsuitable for agriculture or clear-fell plantations.

For eligibility under the PFSI, land must conform to article 3.3 of the Kyoto Protocol. That is, it must not have been covered in forest at December 31, 1989. Kyoto-compliant exotic forests established after October 17, 2002, and Kyoto-compliant indigenous forests established from January 1, 1990, will also be eligible. To qualify for emission units, forests must be "direct human induced . . . through planting, seeding and/or the human-induced promotion of natural seed sources." Thus some form of active management will be required in establishing the forest. Emission units will be transferred to landowners only after the amount of CO_2 stored in the forest has been verified.

Initially the PFSI was intended to prohibit harvesting for at least thirty-five years. This prohibition was recently removed, perhaps as a result of significant pressure from commercial forestry interest groups.[20] Timber may now be harvested on a continuous canopy basis, which aims to ensure that the ground is always covered by a canopy of forest. This will allow the harvesting of roughly 20 percent of the trees in a forest at any one time. Clear-felling of forests will not be permitted and will incur penalty payments, although these harvesting restrictions will be removed after ninety-nine years. The New Zealand approach therefore reflects the carbon pooling approach and the permanence requirements in the NSW Scheme.

Rights and obligations under the PFSI will be formalized in a contract between landowners and the government. They will be registered against land titles and will run with the land. This means that the obligations will remain in perpetuity unless the government agrees to vary the contract with the landowner. Landowners will be required to meet all costs of entering and administering the PFSI, including ongoing monitoring and verification expenses. Where, for any reason, the amount of carbon stored in a forest decreases, the landowner will be required to "replace" emission units for the CO_2 released back into the atmosphere. Landowners who deliberately breach harvesting restrictions will be required to make a penalty payment.

The Climate Change Response Amendment Bill received assent in November 2006, making the changes to the Climate Change Response Act 2002 that were necessary for implementation of the PFSI. The Forests Act 1949 was also amended to establish regulation-making powers for execution of the PFSI. General PFSI regulations as well as rules for carbon accounting, fees, and levies are currently under development.

New Zealand's Emissions Trading Scheme

On September 20, 2007, the New Zealand government announced a proposed emissions trading scheme. The NZ ETS will link to the international market and cover the following sectors: forestry from January 1, 2008; liquid fossil fuels from

January 1, 2009; stationary energy and energy-intensive industry sectors from January 1, 2010; and agriculture, waste, and all other emissions from January 1, 2013.[21]

In a reversal of its initial policy, the government will devolve credits associated with forests established since 1990. The credits will be in the form of NZ Units (NZUs). Owners of eligible forests will be able to choose whether or not to take part in the NZ ETS. If they do opt to join the scheme, both NZUs and liabilities will be devolved. This means that if there is any decline in the carbon stock (for example, from harvesting or fire), the forest owner will be required to make good the loss by surrendering NZUs (or other acceptable credits). If the carbon stock increases, the owner will earn NZUs.

Owners of pre-1990 forests will not earn credits, but they will be liable for deforestation occurring after January 1, 2008. The government will assist affected owners by issuing them free NZUs. Furthermore, no liability will be incurred if pre-1990 forests are harvested and then replanted.[22]

The PFSI will remain in force as a complementary measure to the NZ ETS. In December 2007, the Climate Change (Emissions Trading and Renewable Preference) Bill was introduced to parliament as the first legislative step in establishing the NZ ETS.

Notes

1. National Emissions Trading Taskforce, "Possible Design for a National Greenhouse Gas Emissions Trading Scheme," August 2006 (www.emissionstrading.net.au/key_documents/discussion_paper).

2. Prime Ministerial Task Group on Emissions Trading, Commonwealth of Australia, "Report of the Task Group on Emissions Trading," 2007 (www.pmc.gov.au/publications/emissions/docs/emissions_trading_report.pdf).

3. For example, the Commonwealth government's Greenhouse Friendly program, under the voluntary Greenhouse Challenge Plus industry program, adopts a seventy-year permanence requirement.

4. "Facility" means (1) a geographically defined site or building, including all structures and all mobile equipment operating within site boundaries, or (2) a fleet of vehicles operating on public roads, or a fleet of aircraft, locomotives, or vessels, whether or not based at a single site.

5. Prime Ministerial Task Group on Emissions Trading, "Report of the Task Group on Emissions Trading."

6. Garnaut Climate Change Review, *Emissions Trading Scheme Discussion Paper*, www.garnautreview.org.au/CA25734E0016A131/WebObj/GarnautClimateChangeReviewTermsofReference2007/$File/Garnaut%20Climate%20Change%20Review%20Terms%20of %20Reference%202007.pdf (March 20, 2008).

7. Independent Pricing and Regulatory Tribunal of NSW, "Intro to GGAS" (www.greenhousegas.nsw.gov.au/documents/Intro-GGAS.pdf).

8. It is noted on page 2 of Greenhouse Gas Benchmark (Carbon Sequestration) Rule 5 that "it is proposed that Kyoto Protocol Article 3.4 sinks will be incorporated in this Rule in future, once accounting rules for such sinks have been developed."

9. The scheme also requires an eligible forest to be a minimum of 0.2 hectare, have at least 20 percent crown cover at maturity, and consist of trees with the potential to reach a height of 2 meters at maturity.

10. The other state legislation recognizing carbon sequestration rights (in various forms) is the Forestry and Land Title Act 2001 (Qld); the Forestry Rights (Amendment) Act 2001 (Vic); the Carbon Rights Act 2003 (WA); the Forest Property Act 2000 (SA); and the Forestry Rights Registration Act 1990 (Tas).

11. Crown land is subject to separate property legislation.

12. This is done through the NSW Department of Lands under section 87A of the Conveyancing Act, 1919.

13. Forests NSW, "An Overview of Forests NSW Carbon Pool" (Sydney, 2005).

14. Ibid.

15. See www.co2australia.com.au.

16. Some sinks transactions may have been undertaken for "prompt start" CDM projects, but in early 2005 the Kyoto Protocol had not yet entered into force.

17. "PM Signs Kyoto Protocol Ratification Document," December 10, 2002 (www.beehive.govt.nz/ViewDocument.aspx?DocumentID=15689).

18. "Updated Emissions Balance Reinforces Need for Action," September 20, 2007 (www.beehive.govt.nz/ViewDocument.aspx?DocumentID=15689).

19. PFSI forests will generate removal units (RMUs) for the Crown, which will then be devolved to landowners. If the landowner wishes to receive Assigned Amount Units (AAUs), the Crown may, if circumstances permit, transfer AAUs instead.

20. See, for example, Kyoto Forest Association, "Submission to the Commerce Select Committee on the Climate Change Response Amendment Bill 2005" (www.kfoa.co.nz/submission.htm).

21. "Launch of emissions trading scheme," September 20, 2007 (www.beehive.govt.nz/ViewDocument.aspx?DocumentID=30691).

22. "Forestry in a New Zealand Emissions Trading Scheme," September 20, 2007 (www.maf.govt.nz/climatechange/background-reports-and-analysis/forestry-in-nz-emissions-trading-scheme/forestry-in-NZ-emissions-trading-scheme.pdf).

CASE STUDY

The West Coast Development Trust, a New Zealand Example

SEAN WEAVER

New Zealand's West Coast Development Trust shows that when the right economic incentives are in place, forest conservation can compete economically as well as politically with deforestation drivers such as commercial logging. It also shows how funds might be managed in project-based initiatives through the establishment of a trust managed by local community leaders along with donors, local government, business interests, and NGOs.

In the late 1990s a major public conservation campaign, the West Coast Forests Campaign, was organized by national and local environmental NGOs to protect the biodiversity of 130,000 hectares of indigenous forests on the west coast of the South Island of New Zealand.[1] However, the majority of the local population of approximately 35,000, spread across several small towns and rural areas, supported the continued logging of the forests because of the economic benefits it generated for the community. These benefits included direct employment in logging and processing timber as well as support for industries such as transport and road engineering in a region where economic opportunities were relatively scarce. The government-owned timber company also sponsored local community initiatives.

These government-owned indigenous forests were being logged through a combination of clear cutting, high-intensity selective logging, and sustainable forest management. By the late 1990s, the west coast of the South Island was one of the last areas in the country where unsustainable logging of indigenous, old-growth, lowland rain forest was taking place. On private land, logging operations could be conducted only as ecologically based, sustainable forest management under government license. The government-owned forests on the west coast were exempt from this policy. Given the income logging provided to the local community, the conservation campaign could succeed only if it found a way for equivalent, alternative economic benefits to accrue to the community without logging the forests.

The conservation NGOs supported my proposal to protect the forests in the national interest in exchange for economic development assistance from the government to the local community. The original proposal was designed as an asset swap in which a set of government-owned plantations would be transferred to local ownership. In the end the government decided to offer the community the equivalent value of the plantations, NZ$100 million (U.S.$69 million), which the community was able to negotiate upward to NZ$135 million. In return for this funding, the forests became permanently protected areas managed by the New Zealand Department of Conservation.

Rather than being given to the community in cash to replace the lost income from logging, the majority of the money was placed in a locally controlled trust fund—the West Coast Development Trust. The trust is managed by the community as investment capital for local development initiatives. Community members are elected to the trust, which invests the funds and distributes the annual interest by means of a credit line to which local businesses can apply for business development finance. Loans are awarded on terms better than those offered by local banks.[2]

The West Coast Development Trust increased access to finance by the local community, creating an environment conducive to alternative economic development not based on deforestation and forest degradation. The development package also helped to resolve the conflict between conservation and local development for west coast communities, enabling the local culture to move into a new phase of (clean) development and associated attitudes.

Following the decision to protect these forests, the government decided to end not only this logging but all logging of indigenous forests on government-owned land. The demand for indigenous New Zealand timber continues to be met by private indigenous forest sources. Because the deal operated outside the carbon market, emission reduction offset credits were not generated, so the benefits to the atmosphere and the climate are absolute and additional.

The New Zealand example offers lessons for addressing the relationship between, on the one hand, communities in developing countries that seek to use their natural forest endowments for basic development needs and, on the other, communities or agencies in developed countries that seek to protect their forests in the interest of global climate change mitigation.

Natural forests still exist in many developing countries because the deforestation development path—which has generally been the norm since the Industrial

Revolution began—is yet to arrive. Significant economic incentives to deforest often exist as drivers of basic development. But these situations can be turned around if the right incentives are established to replace the deforestation drivers, as happened in New Zealand.

The New Zealand example can be thought of as a "direct barter" transaction between a local community and a government, in which development needs were met in return for agreeing to protect forests. The principle behind direct barter simply recognizes that a national forest asset amounts to an actual or a potential means of production for the host community or nation.[3] If an equivalently valued means of production is offered in exchange for the protection of this resource, then the economic costs to the host community or nation are potentially zero. The economic benefits to the host community or nation could include a more rapid rollout of strategic development systems and infrastructure (potentially in other sectors such as export manufacturing) that would be difficult to achieve without such a deal.

Direct barter transactions need not be only financial transactions. They could encompass anything the two parties are willing to trade in exchange for forest protection—trade deals, technology transfers, migrant labor schemes, education and training, cash, or any combination of these. Such transactions are also outside the carbon market, so any emission reductions or avoided emissions are absolute and additional.

Notes

1. The South Island's west coast is a stretch of land several hundred kilometers long wedged between the Tasman Sea and the Southern Alps. It is New Zealand's most heavily forested and sparsely populated region.

2. See West Coast Development Trust 2007, "West Coast Investment Opportunities" (www.immigration.govt.nz/migrant/stream/invest/investors/LinkAdministration/Resources Links/RegionalInvestmentOpportunities/WestCoastInvestmentOpportunities.htm).

3. For a more detailed description of direct barter as a potential means of avoiding deforestation, see the direct barter section in the 2007 Vanuatu Submission to the UNFCCC SBSTA working group on REDD, prepared by the author (http://search.un fccc.int/query. html?col=fccc&qt=REDD).

19

Using Forests and Farms to Combat Climate Change: How Emerging Policies in the United States Promote Land Conservation and Restoration

CATHLEEN KELLY, SARAH WOODHOUSE MURDOCK, JENNIFER MCKNIGHT, AND REBECCA SKEELE

Forests, grasslands, and agricultural lands across the United States play an integral role in sequestering carbon dioxide emissions. Estimates for 2005 show that this carbon "sink" absorbs 780 million metric tons of carbon dioxide equivalent each year, equal to approximately 11 percent of 2005 U.S. greenhouse gas (GHG) emissions. However, development and other land conversion activities have caused this carbon sink to decrease by more than 14 percent since 1990.[1] Including land-use offset opportunities in U.S. state, regional, and federal market-based GHG emission reduction programs could be an effective way to create incentives to protect and rehabilitate these valuable carbon sinks. In addition, including offsets can increase the flexibility and cost-effectiveness of these programs, thereby lowering program compliance costs and facilitating quicker and potentially greater emission reductions.

In this chapter we provide an overview of existing and emerging policies at the state, regional, and federal levels in the United States that create incentives for land management, conservation, and restoration activities that will reduce greenhouse gas emissions and sequester carbon. Although most of these policies are still under development or in the early stages of implementation, they offer evidence of the direction in which the policy debate is headed in the United States. We evaluate the design of these regulatory offset programs and explore the ways in which they promote the benefits and address the challenges associated with land conservation and restoration. We also review several voluntary climate change programs that have helped set precedents for the inclusion of land-use offsets in mandatory climate change programs and could potentially be linked to such programs in the future.

The Importance of U.S. Land Conservation, Management, and Restoration for Carbon Mitigation

Forests, grasslands, and agricultural lands represent about 79 percent of all land cover in the United States.[2] Implementing a variety of management activities on this land could greatly increase its capacity to absorb carbon dioxide from the atmosphere. These activities include afforestation and reforestation, sustainable forestry practices, the implementation of conservation tillage on existing cropland, increased harvest rotations, and selective thinning. If employed effectively, such land management practices would serve to maintain and possibly expand the United States' carbon sink capacity over the next 50 to 100 years.[3]

A 2006 study published by the Pew Center on Climate Change analyzed the potential for increased carbon sequestration through such activities.[4] The results suggested that if a carbon payment of up to U.S.$12.50 per metric ton of CO_2 were available, then 47 million hectares of marginal agriculture land in the United States would be converted to forest, and an additional 270 million metric tons of carbon dioxide would be sequestered annually for the next 100 years.[5] This would offset nearly 20 percent of current U.S. carbon dioxide emissions generated from the combustion of fossil fuels. This scenario represents nearly one-third of current cultivated cropland and therefore admittedly is overambitious. But at the same carbon payment, according to the Pew study, even a less drastic change in land management, such as the modification of current agricultural practices, could result in the annual sequestration of up to 70 million metric tons of carbon dioxide. This study demonstrated the important role that land use could play in the effort to stabilize GHG concentrations in the atmosphere.

Potential Benefits of Including Land-Based Offsets in Climate Legislation

Including offset opportunities in market-based climate change programs would increase the programs' flexibility and lower compliance costs. Land management and conservation offsets can increase the likelihood that market-based GHG programs will successfully decrease emissions over time in a cost-effective manner while generating additional environmental, social, and economic benefits. Specifically, improved land management and conservation offsets are beneficial for the following reasons.

First, they produce real, measurable, and verifiable emission reductions and increased carbon sequestration. Reliable, proven, and accurate methods for measuring, monitoring, and verifying the carbon sequestration of land management and conservation activities are already in widespread use. Methods for measuring and monitoring terrestrial carbon pools, based on commonly accepted principles of forest inventory, are well established and tested. Third-party verification by an accredited institution ensures the veracity of offsets being claimed. To further enhance credibility, tonnes can be discounted to account for any scientific uncertainties regarding offset measurements. In addition, viable methods and measures exist for ensuring the permanence of any emission reductions and sequestration achieved. Examples are permanent conservation easements and securing insurance for carbon offset credits.

Second, offset credits increase the flexibility and lower the costs of emission reduction programs. Crediting offsets increases the supply of cost-effective allowances available for purchase by the regulated entities and thus lowers overall compliance costs in a market-based program. By expanding the allowance market to include low-cost emission reductions from outside sources, offsets allow covered entities to take on tighter emission limits that are economically feasible without drastically increasing compliance costs.

Third, including offsets helps to protect an emissions trading market against price volatility and thus lessens the need for price control instruments such as safety valves and cost caps.[6] Once triggered, a safety valve or cost cap allows regulated companies to purchase additional permits issued by the government at predetermined prices. Once in effect, a safety valve allows emissions from regulated entities to exceed the emissions cap set by legislation or regulation and therefore undermines the environmental goals of the program.

Offsets would lower the overall compliance costs of the program, thereby reducing the need to include a safety valve or cost cap. If a cost-control instrument is included in an emissions trading program, then offsets help to delay the triggering of a safety valve or cost cap, thus increasing the environmental gains relative to not including offsets.

By also offering the option to purchase emission reduction credits in the marketplace, offsets can improve the liquidity of an emissions trading program, thereby reducing the price volatility of the market and thus the need for cost-control measures, as well as improving the ability of the regulated entities to meet their emissions caps and effectively plan for the future. Additionally, offsets allow covered entities to take on tighter emission limits that are economically feasible. Thus an appropriately stringent cap can be used to protect against the possibility of offset credits flooding the market.

The fourth advantage is that in addition to their climate benefits, land conservation, management, and restoration projects often generate other environmental, social, and economic benefits. These include watershed protection, resulting in reduced soil erosion and improved water quality; biodiversity preservation; open space preservation and sprawl abatement; and generation of project employment opportunities and local tax revenues.

U.S. Federal Climate Legislation

The following is a summary of some of the key climate change U.S. federal legislative proposals introduced in Congress in 2006 and 2007. Each piece of legislation sets a different reduction timetable. Most of the proposals include offsets as one of many ways in which regulated entities may achieve the outlined emission reduction goals. The allowed use of offsets differs from bill to bill in terms of which types of offsets are allowed, how many offsets may be used by the regulated entity to meet its cap, and whether offsets from outside the United States are allowed.

In an effort to generate broad support for climate change legislation in general, inclusion of provisions that allow the use of agriculture offsets that benefit the farm sector are considered a potential method for gaining political support from farmers (an important lobby group) for the enactment of climate change cap-and-trade legislation.

It is unclear why some legislation allows the use of internationally generated offsets while some does not. Some policymakers consider that offsets generated in the United States provide economic benefits to the country and should remain in the United States. Others consider the inclusion of international offsets a way to facilitate the linking of a U.S. program to international schemes such as the Kyoto Protocol.

Lieberman-Warner America's Climate Security Act of 2007

The Lieberman-Warner America's Climate Security Act, introduced by Senators Joe Lieberman (I-Conn.) and John Warner (R-Va.) in October 2007, would establish a cap-and-trade program in the United States to reduce greenhouse gas emissions between 2012 and 2050. It is estimated that the bill would reduce emissions by approximately 18–25 percent by 2020 and approximately 62–66 percent by 2050.[7] The bill, introduced in the Global Warming Subcommittee of the Environment and Public Works Committee, was voted favorably out of that subcommittee in November 2007 and out of the full committee in December.

The act allows covered entities to satisfy up to 15 percent of their allowance obligation by submitting offset allowances from agricultural, forestry, and other offset-producing projects in the United States. In addition, 5 percent of the Emission Allowance Account for each year is dedicated to support activities that reduce greenhouse gas emissions from the agriculture and forestry sectors of the economy or increase greenhouse gas sequestration from those sectors. Furthermore, 2.5 percent of the Emission Allowance Account for each year will be distributed for activities in countries outside the United States that reduce greenhouse gas emissions from deforestation and forest degradation or increase sequestration of carbon by restoring forests and degraded lands.

The Climate Stewardship and Innovation Act of 2007

The Lieberman-McCain Climate Stewardship and Innovation Act (CSIA) was introduced in the U.S. Senate in January 2007. This bill would regulate greenhouse gas emissions from major energy, industrial, and transportation sources. The legislation caps GHG emissions from the covered entities at 2004 levels by 2012 and then lowers the cap until it reaches one-third of year 2004 levels by 2050. It uses tradable allowances in a flexible, market-based approach. It also includes incentives for the development of climate-friendly energy technologies such as nuclear power. In June 2005, an earlier version of the bill was defeated in the U.S. Senate by a 60-38 vote.

Under the CSIA, 30 percent of an entity's total allowance submission requirement could be satisfied through offset projects, including international offsets. Offsets could be obtained through projects such as international allowance trading and forest carbon projects, including agricultural sequestration, geological sequestration, reforestation, and forest preservation. The bill includes provisions preventing the use or introduction of non-native species. If an entity chooses to meet more than 15 percent of its allowance submissions through offsets, it must meet at least 1.5 percent of its total allowance submissions through increased agricultural sequestration.

The Electric Utility Cap-and-Trade Act of 2007

In January 2007 Senator Dianne Feinstein (D-Calif.) introduced legislation that would reduce GHG emissions from the electricity sector to 6 percent below 2001 levels by 2020 and 50 percent below 2001 levels by 2050. The bill would establish a mandatory cap-and-trade system. Feinstein's bill allows covered entities to earn unlimited credits from offset programs such as agricultural projects, land conservation projects, reforestation, and avoided deforestation. A covered entity may satisfy up to 5 percent of the total yearly allowance through the use of forest management offsets. This restriction does not apply to other types of offsets such as grazing management or afforestation. Up to 25 percent of an entity's obligation could be met through international trading. The legislation would apply strict standards to ensure that credits are awarded for real, permanent emission reductions through sequestration efforts.

U.S. Federal Government Program 1605(b) for Voluntary Reporting of Greenhouse Gas Emission Reductions

The U.S. federal government has established a program, sponsored by the Department of Energy, for the voluntary reporting of greenhouse gas emission reductions and sequestration. Section 1605(b) of the Energy Policy Act of 1992 (EPACT) provides guidelines for the voluntary reporting of emissions, reductions, and sequestration of the main greenhouse gases.

The goal of section 1605(b) is to create a record that can be used "by the reporting entity to demonstrate achieved reductions in greenhouse gases." Private, public, and federal entities may report under the guidelines.

Section 1605(b) includes standards for assessing carbon sequestered in forests and soils. Sequestered carbon can be estimated by the use of look-up tables, models, or field analysis. The guidelines call for the measurement of live trees, standing dead trees, understory vegetation, down deadwood, and forest floor and soil organic carbon.

Section 1605(b) allows an entity to register for credits ex-ante (before they actually occur) from land restoration and preservation activities if the entity restores native habitat on the land and establishes an administrative restriction, such as a permanent conservation easement, that ensures that no human-caused carbon emissions will occur in the future. The entity can register up to 50 percent of the carbon stock increase expected over the next fifty years.

Source: U.S. Department of Energy, Technical Guidelines, Voluntary Reporting of Greenhouse Gases (1605[b]) Program, Section 2.4.3.4, "Reductions from Land Restoration and Preservation," March 2006, p. 274.

The Clean Air Planning Act

Senator Tom Carper (D-Del.) introduced the most recent version of his Clean Air Planning Act (CAPA) in the Senate in May 2006. Also referred to as the "Four Pollutant" bill, this legislation amends the existing Clean Air Act and sets mandatory emission limits for the four major air pollutants emitted by the electric generating sector—sulfur dioxide, nitrogen oxides, mercury, and carbon dioxide. Specifically, by 2015 CAPA would cut sulfur dioxide emissions by 82 percent, nitrogen oxide emissions by 67 percent, mercury emissions by 90 percent, and carbon dioxide emissions to 2001 levels. Covered entities would be permitted to meet this new CO_2 emission limit by reducing their own emissions or by buying CO_2 credits on the open market.

CAPA allows covered entities to use offset allowances from greenhouse gas reduction or sequestration projects. There is no explicit affirmation that land conservation or reforestation projects would be eligible for credit. However, the bill does mandate that the allowed offset projects "not cause or contribute to adverse effects on human health or the environment."

U.S. Regional and State Climate Programs

In the absence of a federal climate change program to reduce U.S. greenhouse gas emissions, several states have stepped forward to implement programs at the state and regional levels to reduce emissions. Here we outline the most important regional and state programs.[8]

The Northeast Regional Greenhouse Gas Initiative

The Northeast Regional Greenhouse Gas Initiative (RGGI) is the most developed regional climate change effort in the United States.[9] Scheduled to go into effect in 2009, this multistate program is designed to limit CO_2 emissions from power plants. Currently, Connecticut, New York, New Jersey, Vermont, Maine, New Hampshire, Massachusetts, Maryland, Rhode Island, and Delaware have committed to enact or adopt the draft regulation.

If ultimately carried out, the plan would freeze power plant emissions of CO_2 for the region at current levels until 2015 and then reduce them by 10 percent by 2019. RGGI would also allow trading of carbon "allowances" to achieve compliance.

Public agency staff from the environmental, energy, and public utility departments in each of the participating states have worked together to devise the "Model Rule." The Model Rule is a regulation that each state must adopt as drafted in order to ensure that all the states' regulations remain compatible and consistent. A few provisions remain up to each state to decide, such as the size of the cap assigned to each regulated entity and the percentage above 25 percent of allowances that will be auctioned and how that revenue will be spent.

The RGGI Model Rule allows covered sources to get credit for a limited portion of their emissions by buying allowances for GHG emission mitigation activities. Offset credits may be awarded so long as they "represent carbon dioxide equivalent emission reductions or carbon sequestration that are real, additional, verifiable, enforceable, and permanent." The following is a summary of the key provisions related to land-based offsets.

Percentage of emissions covered by offsets: Each regulated entity covered by the RGGI program may use offsets equal to only 3.3 percent of its CO_2 emissions for a given control period, equivalent to approximately half a source's emission reduction requirement.[10] The ability to use a greater quantity of offsets is built into the program in order to provide greater flexibility for compliance if certain allowance price triggers are reached (that is, it has a safety valve). If the average allowance price reaches U.S.$7 per ton of CO_2 over a twelve-month period, then regulated entities would be able to use offsets equal to 5 percent of their carbon dioxide emissions. If the allowance price reached U.S.$10 per ton over a twelve-month period, then entities would be able to use offsets equal to 10 percent of their emissions, and offsets from international trading programs would be allowed.

Geographic coverage and environmental co-benefits: Afforestation is currently the only land-use activity that is eligible for offset credit under the Model Rule, although other project types may be added in the future. To qualify, a project must be carried out on land that has been unforested (10 percent forest coverage or less) for ten years or more. Afforestation projects must "promote the restoration of native forests" and be "managed in accordance with widely accepted environmentally sustainable forestry practices." Commercial timber harvesting may occur on project sites so long as certification is obtained through a RGGI state-approved sustainable forestry organization.

The Model Rule allows for the use of offsets from projects located in any state or other U.S. jurisdiction, so long as that non-RGGI state has established a cap-and-trade program or commits to certain administrative responsibilities to ensure the credibility of offset allowances from that state.

Additionality: The Model Rule specifies that carbon sequestration is to be determined using a base-year approach in which the amount of carbon sequestered is measured as a net increase in carbon relative to the base-year measurement. The base-year measurement is a static snapshot of the sequestered carbon at that time.

Permanence and guarantees: The Model Rule requires that the project sponsor place the project site under a legally binding, permanent conservation easement. The easement would require the land to be maintained in a forested state in perpetuity and the carbon density within the offset project boundary to be maintained at levels at or above the levels achieved by the end of a carbon dioxide offset-crediting period. The conservation easement must also require that the land be managed in accordance with environmentally sustainable forestry practices. If afforestation offsets are lost because of wildfires or other damaging events, the user of the credits is required to replace those credits in order to remain in compliance.

To account for potential leakage, the Model Rule also requires that a 10 percent deduction be applied to the total amount of carbon created through a sequestration project. However, the 10 percent deduction is not required if the project sponsor retains long-term insurance that guarantees the replacement of any lost sequestered carbon for which CO_2 offset allowances were issued.

Measurement and monitoring: The carbon sequestration and emission reductions from the afforestation projects must be monitored and measured at least every five years. Each individual carbon pool must be directly measured using a measurement protocol and sample size that achieves 95 percent confidence that the reported value is within 10 percent of the true mean for the combined carbon pool measurement. Direct measurement procedures shall be consistent with current forestry good practice and the guidance outlined in the U.S. Department of Energy's 1605(b) guidelines.[11]

Applications for offset projects are required to include a verification report signed by a RGGI-accredited third-party verifier. These third-party verifiers must meet a set of minimum qualifications and comply with rules of conduct outlined in the Model Rule.

Enforcement: If an entity regulated under the RGGI program does not have sufficient allowances to cover its emissions during a compliance period, then it must provide offsets worth three tons for every ton of excess emissions. Each RGGI state has the authority to level additional penalties and fines if emission violations occur.

Oregon: The Climate Trust

In 1997 the Oregon state governor signed into law Oregon House Bill 3283, the first-ever legislative carbon dioxide emissions regulation passed in the United States.[12] Referred to as the Oregon Standard, this bill authorizes the state's Energy Facility Siting Council to set CO_2 emission efficiency standards for all newly constructed energy generating facilities. The covered facilities may comply with these reduction requirements through the management of their own offset or cogeneration projects or through a "monetary path" in which they provide funding for offset projects through a qualified third party.

The Oregon-based Climate Trust is recognized as a "qualified organization" under the statute and has been implementing offset projects since 1997. The Climate Trust is a nonprofit organization that invests funds in projects that will reduce GHG emissions or sequester carbon in forests. The scope of the trust's offset portfolio includes energy efficiency, renewable energy, biological sequestration, cogeneration, material substitution, and transportation efficiency projects. These projects have so far amounted to more than U.S.$4.9 million in investments and more than 1.9 million metric tons of carbon dioxide emission offsets.

Before investing in a project, the Climate Trust reviews the project proposal to ensure that all the organization's requirements are met. The criteria related to land-based offsets include the following:

Geographic coverage and environmental co-benefits. Although the Climate Trust is willing to invest in projects all over the world, preference is given to projects located in Oregon. International projects must involve a strong U.S. partner as well as a partner in the host country. Host country approval and support is strongly preferred. There is no explicit requirement that a project include environmental co-benefits, but environmental as well as health and socioeconomic factors are taken into consideration in the selection process, and preference is given to offset projects that include additional environmental, health, or socio-economic benefits.

Additionality. Projects are considered only in situations where the carbon benefits would not otherwise accrue in the absence of offset project funding. The Climate Trust "retires" offsets once they are acquired, so they are ineligible for any future trading. Offset credits that have already been awarded under another organization or trading scheme are ineligible.

Permanence and guarantees. The Climate Trust reviews each project proposal to ensure that proponents offer guarantees for achieving the specified carbon and ecological benefits. As a form of guarantee, the trust may consider a pay-for-performance approach under which it pays a fixed amount per ton of carbon dioxide over a specified period of time. It typically pays for the forward credits that will be generated in the future; forestation projects in the trust's current portfolio have lengths ranging from 50 to 100 years.

Measurement and monitoring. The Climate Trust does not explicitly require any specific method of carbon measurement or monitoring. But CO_2 benefits must be measured using state-of-the-art methods and protocols, and the Climate Trust prefers review and verification by a qualified third party.

The Climate Trust's carbon forestry program focuses on reforestation and protection of critical habitat. The trust partners with local and international organizations to fund forest conservation projects all over the world. Land-use projects may include forest preservation, reforestation, afforestation, and forest management efforts. To date such projects have succeeded in restoring and protecting more than 1,600 hectares of forest habitat, resulting in offsets of more than 600,000 metric tons of carbon dioxide over the next 50 to 100 years.

California

California has been a nationwide leader in the effort to reduce greenhouse gas emissions. Given California's status as the world's twelfth largest GHG emitter, its leadership on this issue is particularly important.[13]

THE CALIFORNIA CLIMATE ACTION REGISTRY. Established in 2001 by California legislation, the California Climate Action Registry (CCAR) is a nonprofit public-private partnership that serves as a voluntary greenhouse gas registry to protect, encourage, and promote early actions to reduce GHG emissions.[14] The expectation is that early actions reported under CCAR would be eligible for crediting under any future California regulations.

The registry has developed and published two reporting protocols. The General Reporting Protocol (GRP) contains instructions for the way entities should calculate and report their emissions. The Certification Protocol (CP) provides instruction to the registry's approved certifiers on how to conduct a standardized third-party assessment of the reported data. The registry encourages all sectors to report their greenhouse gas emissions. However, the GRP and CP provide general reporting guidance for the most common nonbiological emissions, which include those from stationary combustion, mobile combustion, and indirect electricity-heat-steam consumption.

CALIFORNIA FOREST PROTOCOLS. In 2002 California state legislation further expanded the Climate Action Registry's legislative mandate to include reporting guidance for forest landowners and industry entities as well as forest projects. These so-called California Forest Protocols are detailed standards instructing on how to account for, measure and monitor, and verify emission reductions and sequestration from forest conservation, management, and restoration in California. The protocols provide comprehensive guidance to timber companies and other forest entities in the United States in accounting for and reporting their biological carbon stocks and emissions.

Geographic coverage and environmental co-benefits: Project eligibility under the Forest Protocols extends only to forestry projects located in the state of California. Project management must follow a "natural forest management" regime that promotes a diverse native forest composed of trees of different ages and species. Projects are eligible only if they do not exacerbate other local environmental problems such as problems involving water quality and biodiversity.

Additionality: All projects must exceed applicable land laws and regulations. Forest management practices must exceed the requirements outlined in the California Forest Practice Rules, Option C. Forest conservation practices must not be required by law, and project activity must exceed baseline activity. The baseline can reflect either a site-specific immediate threat or countywide conversion trends. Reforestation projects must not be required by law and must occur on land that has not been forested or has been out of forest cover for at least ten years.

Permanence and guarantees: Projects are required to be secured with a perpetual conservation easement in order to guarantee the permanence of the climate and environmental benefits. Conservation easements are legally binding documents that dictate how the land must be managed over time and are attached to the land title.

Measurement and monitoring: The California Forest Protocols require an annual report of GHG inventories and complete GHG data transparency for all forestry projects. Each annual report must include a declaration and description of any on-site activity-shifting leakage that has occurred. Each annual report must be reviewed by a third-party certifier approved by the California Energy Commission and the California Climate Action Registry. In addition, third-party certification of carbon benefits is required in years one and five, six and ten, and so forth. Project developers can enter into a contract with an approved certifier for up to six years. This forced rotation of certifiers is intended to increase the accuracy of carbon quantification and reporting.

THE CALIFORNIA GLOBAL WARMING SOLUTIONS ACT. The California Global Warming Solutions Act of 2006 (formerly Assembly Bill 32) was signed into law by California governor Arnold Schwarzenegger in September 2006. The statewide legislation caps California's GHG emissions at 1990 levels by 2020, a reduction of 25 percent from estimated 2020 business-as-usual levels. The legislation is significant because it is "the first enforceable state-wide program in the United States to cap all greenhouse gas emissions from major industries that includes penalties for non-compliance."[15] In addition to allowing for the creation of a market-based trading system to achieve emission reductions, the act includes a safety-valve feature that allows the governor to "suspend the emissions caps for up to one year in the case of an emergency or significant economic harm."[16] Furthermore, "where appropriate and to the maximum extent feasible," the new legislation will "incorporate the standards and protocols developed by the California Climate Action Registry."[17]

It has not yet been determined how the act will address offsets. According to the time line set out in the act, a potential market-based cap-and-trade system will not be implemented until January 1, 2012. The legislation instructs the California Air Resources Board to consider the standards and protocols of the California Climate Action Registry when developing regulations to implement the Global Warming Solutions Act. Therefore, it is likely that the regulations established under the new law will reflect the aforementioned forest protocols as described in the registry.

Lessons Learned

Significant progress has been made in the United States toward establishing state, regional, and even federal programs related to greenhouse gas emission reductions from forestry projects. Clearly the California Registry, the Climate Trust, the Department of Energy's 1605(b) program, and the Northeast Regional Greenhouse Gas Initiative will all serve as models for federal programs that emerge. As Congress debates the legislative contents of programs aimed at abating GHG emissions, inclusion of offsets continues to play a central role in the discussions.

Land conservation, management, and restoration have a significant role to play in reducing emissions. Because of the noteworthy global contribution of deforestation as a source of GHG emissions, it is important that land-use activities be considered when entities seek to reduce those emissions. Given the abundance of forests, grasslands, and agriculture land in the United States, an important opportunity exists to implement programs and policies that promote carbon sequestration and emission reduction through activities carried out on such land. These activities could also result in many additional environmental benefits, such as habitat protection, land conservation, and water quality protection.

Offsets included in market-based programs play an important role in mitigating price volatility and allowing for deeper emissions cuts. As the role of offsets continues to be considered and debated, it is critical that program design ensure that the quality, accuracy, and enforcement of offsets produced is high, such that a metric ton of carbon emissions reduced from offset projects equals a metric ton of emissions reduced from regulated sources. Without such fungibility (based on robust scrutiny and precision), confidence in offsets will diminish, and the future use of this important vehicle will likely be compromised.

Yet the prospect for the inclusion of offsets in federal policy seems good. Support for the use of offsets, particularly offsets generated from land conservation, restoration, and management, is strong among entities such as power companies that are likely targets of carbon cap legislation. Many such companies view these projects as viable, credible, and affordable means for reducing their climate effects. They view investments in carbon forestry projects as a means for achieving carbon emission reductions now, while they begin switching to cleaner technologies.

Others view land conservation, restoration, and management offsets as a means for expanding the base of support for mandatory climate change emission reduction legislation by gaining the support of those in the U.S. agriculture sector. With potential payments to farmers and foresters for projects that sequester carbon or reduce greenhouse gas emissions, the farmers and foresters stand to gain financially if these provisions are included in climate change legislation. The so-called farm lobby has long enjoyed a powerful voice in Washington, D.C., and is viewed as a potentially influential force for demonstrating support for climate change legislation.

No doubt the discussion and debate surrounding land-based carbon offsets will continue as Congress steps up its effort to draft legislative proposals aimed at mitigating climate change.

Notes

1. Energy Information Agency, "Emissions of Greenhouse Gases in the U.S. for 2005" (www.eia.doe.gov.oiaf/1605/ggrpt/pdf).

2. K. Richards, R. Sampson, and S. Brown, *Agricultural and Forestlands: U.S. Carbon Policy Strategies* (Arlington, Va.: Pew Center on Global Climate Change, 2006).

3. M. Manion, "Forest Carbon Sequestration," *Catalyst* (Union of Concerned Scientists) 3, no. 2 (2004): 18–19.

4. Richards, Sampson, and Brown, *Agricultural and Forestlands.*

5. A carbon payment is defined as the estimated marginal cost per metric ton of carbon, calculated using net present value, over 100 years.

6. Safety-valve provisions are being debated and considered as U.S. legislation is developed. Mainly they are seen as a means to control the cost of a cap-and-trade program for

the regulated entities and its overall effect on the national economy. A variation of a safety valve was included in the Northeast Regional Greenhouse Gas Initiative (RGGI) Program Model Rule, discussed later. In RGGI, instead of the safety valve's acting as a price cap, it acts as a trigger to allow expanded use of offsets by the regulated sources as a means to meet their emission caps. The use of a safety valve in this manner was viewed as a way to provide greater flexibility and mitigation of cost risk to regulated entities in meeting their caps without resulting in an increase of emissions above the overall emission cap set by the RGGI Program.

7. Natural Resources Defense Council, *NRDC Legislative Fact Sheet*, Lieberman-Warner Climate Security Act (Washington, D.C., December 2007).

8. This chapter was drafted prior to the establishment of the Western Climate Initiative and the Midwest Regional Greenhouse Gas Reduction Accord.

9. See www.rggi.org.

10. Pew Center on Global Climate Change, "Q & A: Regional Greenhouse Gas Initiative" (www.pewclimate.org/what_s_being_done/in_the_states/rggi/rggi.cfm, accessed December 20, 2006).

11. Department of Energy, Technical Guidelines, Voluntary Reporting of Greenhouse Gases (1605[b]) Program, Chapter 1, Emissions Inventories; Part 1 Appendix: Forestry; Section 3: Measurement Protocols for Forest Carbon Sequestration (March 2006).

12. See www.climatetrust.org.

13. California Office of the Governor, "Gov. Schwarzenegger Signs Landmark Legislation to Reduce Greenhouse Gas Emissions," September 27, 2006 (http://gov.ca.gov/index.php?/press-release/4111/).

14. See www.climateregistry.org.

15. Pew Center on Global Climate Change, "California Global Warming Act," July 1, 2007 (www.pewclimate.org/what_s_being_done/in_the_states/ab32/index.cfm).

16. Ibid.

17. "California Climate Action Registry Forest Protocols Overview, Climate Registry," June 2004 (www.climateregistry.org/docs/PROTOCOLS/Forestry/04.06.14_Final_Forest_Protocols_Board_Overview.pdf).

CASE STUDY

The Van Eck Forest Management Project in California

MICHELLE PASSERO, RACHAEL KATZ, AND LAURIE WAYBURN

The Pacific Forest Trust submitted California's first greenhouse gas (GHG) emission reduction project with the California Climate Action Registry (CCAR) on behalf of the Van Eck Forest Foundation in July 2006.[1]

Some 45 percent of California was naturally forested, yet over 40 percent of this was lost to conversion by 1990. Nonetheless, the state is home to highly productive forests with significant carbon stocks and the potential to store carbon for the long term. Indeed, the single most productive forest type in the world, the coast redwood (*Sequoia sempervirens)* temperate rain forest, can store more carbon than any other. California is also a major timber-producing state and one that has encouraged sustainable forest and natural forest management.

Recognizing that forest change and loss is the second largest source of CO_2 emissions globally and that California is also losing its forest base at a rapidly increasing rate, new legislation was signed in 2002 to link reductions in net CO_2 emissions with increased natural forest conservation and sustainable management. The legislation (SB 812) was established with a set of distinct principles to ensure this linkage. The forestry principles are centered on key GHG project requirements, including additionality, permanence, native habitat conservation, and natural forest management. Under California's program, additionality requires forest projects to provide emission reductions above what would occur under "business-as-usual" practices, which are defined as the regulatory requirements for land use and forestry. To secure the permanence or long-term duration of emission reductions, forest projects must be secured with a perpetual conservation easement.[2] Further, project activities must restore or promote native forest types and manage forests to maintain naturally occurring conditions, such as complex habitats and a broad range of ages and species across the landscape (California is home to some of the most diverse conifer and oak forests in the world).

The Pacific Forest Trust manages the Van Eck forests and holds a perpetual conservation easement on them that secures the forest land base. The Van Eck project is expected to achieve more than 500,000 tons of net CO_2 emission removals through the sustainable management of more than 2,100 acres of forest land over a 100-year period. These climate benefits are inventoried and then certified using the accounting practices of CCAR's Forest Protocols.

Our project, which started in 2004, is based in Humboldt County, along the northern coast of California in the coastal redwood zone. It produces significant CO_2 emission removals through changes in forest management that increase overall carbon stocks on the landscape and in wood products over time. Changing silviculture from business-as-usual is essential to the net carbon stock gains achieved in this project. Overall, one-third of what is grown is harvested annually, allowing the inventory of timber in the forest, and therefore live carbon pools, to increase over time. This also yields greater harvested wood products. Generally, small group selection and variable retention harvesting is practiced. Riparian buffers on all streams supporting aquatic life have been extended significantly above those required by law. Natural regeneration of redwoods (stump sprouting) and Douglas firs is augmented with planting from local seed stock as necessary to ensure full restocking after harvest. Large standing and lying dead trees are also managed and retained on site, increasing the dead pool of carbon as well.

Emission removals are secured with a perpetual conservation easement that protects the project area from conversion to nonforest uses and sets performance goals and guidance for the growth and retention of carbon stocks on the project site. The forests in this region are typically managed to hold less than one-tenth of the carbon stocks they can naturally hold. Thus, a key requirement of the conservation easement is to restore more of the inventory of timber, and therefore carbon, to the site, restoring carbon levels to natural capacity. To ensure this, roughly 1 million board feet of timber per year are harvested while 3 million feet are grown.

To prepare our project for registration, our foresters established an inventory of the carbon stock on the total acreage of the project site according to CCAR requirements. Then, using computer simulation models, we estimated the carbon stocks of the project on the basis of California Forest Practice Rules, the business-as-usual scenario pursuant to the registry's Forest Protocols. Projected emission removals were estimated on the basis of the difference between the baseline scenario carbon stocks and the increase in carbon stocks that would occur from the project activity. The CCAR protocols require regular direct sampling and annual reporting of the carbon stocks to confirm emission reductions and any emissions we produce. Currently, Van Eck's carbon inventory and projections have been completed, recorded with the registry, and verified by an approved third party.

In addition to GHG emission reductions, the Pacific Forest Trust's Van Eck forest project will provide many other public benefits. The older, more complex native forest, which supports many threatened, endangered, and rare species of fish, wildlife, and plants, is being restored. Biodiversity and wildlife habitat are being protected and enhanced. The project is also yielding more timber sustainably over time than would otherwise occur, keeping the land in timber production and maintaining the local forest economy.

The Pacific Forest Trust's project with the Van Eck forests demonstrates how the forest sector can be broadly incorporated in the fight against global warming. Private, managed forests constitute more than 60 percent of U.S. forests overall. As such, their conservation and sustainable management is critical to both reducing forest emissions and increasing net stocks of carbon (that is, removing CO_2 from the atmosphere). This is the first project of many to come in California, and we are working to replicate the approach in other states throughout the United States. The work done in California and other states will be key in shaping the ongoing national debate over what can and should be done at the federal level to reduce the United States' effect on the global climate.

Notes

1. The Pacific Forest Trust is a nonprofit organization that works to preserve, enhance, and restore America's private, productive forests in order to sustain their many public benefits.

2. A perpetual easement is a permanent legal contract, entered into voluntarily, that is affixed to the land title and limits or removes development rights.

20

Carving a Niche for Forests in the Voluntary Carbon Markets

KATHERINE HAMILTON, RICARDO BAYON, AND AMANDA HAWN

The voluntary offsetting of greenhouse gas (GHG) emissions started as early as 1989, when Applied Energy Services Corporation (AES), an American electricity company, decided to invest in an agroforestry project in Guatemala.[1] With this investment AES wished to offset the GHG emissions of its new cogeneration plant in Uncasville, Connecticut, by paying farmers in Guatemala to plant 52 million pine and eucalyptus trees on their land. AES, like other companies since, invested in the project to reduce its "carbon footprint" for reasons beyond legislation and helped to kick-start the voluntary carbon markets. More than fifteen years later, voluntary carbon markets have begun to grow rapidly and diversify significantly. However, forestry projects remain the most widely used source of offsets in the marketplace.[2] In this chapter we briefly introduce the voluntary carbon markets, highlight their potential advantages, explore why the voluntary side of the carbon markets is fertile ground for forestry-based offsets, and outline the role of standards in the marketplace.

A Snapshot of the Markets Today

By definition the voluntary carbon markets consist of carbon offset trades that are not required by regulation. These trades include purchases in the rapidly growing retail offset markets, purchases of credits by organizations directly from project developers for retirement or resale, and donations to GHG reduction projects by corporations, which then receive credits. The voluntary markets also include transactions under the Chicago Climate Exchange (CCX), a voluntary but legally binding U.S.-based cap-and-trade system.[3] In this chapter we focus on the voluntary carbon markets outside the CCX system, which we refer to as the "over-the-counter (OTC) voluntary carbon markets" or the "carbon offset markets."

Sellers in the carbon offset markets include retailers selling offsets online, conservation organizations hoping to harness the power of carbon finance, and developers of potential Joint Implementation (JI) or Clean Development Mechanism (CDM) projects with credits that, for a variety of reasons, do not qualify for the regulated markets. Buyers range from multinational corporations and nonprofit organizations to individuals. Buyers' motivations include the wish to manage climate change effects, interest in innovative philanthropy, desire for public relations benefits, the need to prepare for (or deter) regulations, and plans to resell credits at a profit.

Like the regulated carbon markets, the voluntary markets are new, quickly evolving, and complex. Unlike the regulated markets, the voluntary markets lack consistently used standards and widely available, impartial information. Hence the voluntary carbon markets remain a caveat emptor marketplace: emerging organizations offer offsets from a variety of sources and project types, the supply chain for carbon offsets is complex, the markets are fragmented, and the product is highly abstract. In an effort to increase the user friendliness of the marketplace, standards and certification programs are developing in the voluntary markets to help clarify the definition of a valid credit.

A recent marketwide study by the organizations Ecosystem Marketplace and New Carbon Finance recorded 13.4 million tonnes of offset credits transacted in the OTC markets in 2006. Combining this figure with publicly available data on the Chicago Climate Exchange shows that the voluntary carbon markets transacted at least U.S.$93 million worth of credits in 2006.[4] Although this number is significant, the markets remain a fraction the size of the regulated carbon markets. The World Bank's "State and Trends of the Carbon Market 2007" valued the regulated markets at more than U.S.$30 billion in 2006.[5]

Despite the comparatively small size of the voluntary carbon markets, some investors believe they are poised for explosive growth, and many companies see potentially lucrative business opportunities associated with the creation of carbon-neutral products for retail consumption.[6] The World Bank's 2007 report describes the voluntary carbon markets as "expected to expand exponentially in the coming years with growing popular interest in mitigating climate change."[7] A report by the consulting group ICF echoes this high growth prognosis, predicting that the voluntary carbon markets could have a volume of 400 megatonnes of carbon dioxide equivalent (Mt CO_2e) per year by 2010.[8] Proponents of the markets equate this growth with enabling stakeholders to further leverage offsets as one strategy in the fight to mitigate greenhouse gas emissions. For example, Michael Molitor, cofounder of the world's first voluntary carbon offset fund, noted that it is impossible to "fight Godzilla with Underdog. You need a monster that's at least as big."[9]

The Ultimate "Flexible Mechanism"

Amid such optimistic predictions, voluntary markets are rich not only in opportunity but also pitfalls. In response to a lack of broadly accepted certification or registration procedures, market critics describe situations in which purchased credits are simply "hot air." For example, recent news articles have highlighted projects that would have happened regardless of carbon financing or that failed to accomplish their advertised sustainable development benefits.[10]

Many of the challenges facing the voluntary markets do not apply to the regulated markets, which were born out of the Kyoto Protocol's flexibility mechanisms: international emissions trading, the Clean Development Mechanism (CDM), and Joint Implementation (JI).[11] Unfortunately, the regulated markets' regulatory requirements have created high cost and time barriers that exclude many project developers from entering the markets.

These barriers point to one of several unique, positive attributes of the voluntary side of the carbon markets. Relative to the CDM or JI, the voluntary markets could be considered the ultimate flexible mechanism. Although this flexibility is connected with potential market risks, including lack of additionality or double counting, it is also the source of numerous market strengths, such as potentially reduced transaction costs, lower barriers to entry, promotion of innovation, and incentives for credits rich in environmental and social co-benefits beyond emission reductions. Innovation, flexibility, and lower transaction costs can benefit buyers as well as suppliers. When an organization purchases carbon offsets to meet a public relations or branding need, creativity, speed, cost-effectiveness, and the ability to support specific types of projects are often clear and valuable benefits.

This combination of strengths is particularly true for what in the parlance of the Kyoto Protocol negotiations is referred to as "land use, land-use change, and forestry" (LULUCF) projects. Proponents of LULUCF projects suffer from the high costs associated with the CDM and the slow acceptance of LULUCF credits in the regulated markets. As of mid-2007, seven different afforestation-reforestation methodologies had been accepted by the CDM Executive Board.[12] Yet only one LULUCF project, in comparison with more than 700 non-LULUCF projects, had actually been registered under the CDM and was generating Certified Emission Reductions. An additional barrier for LULUCF credits is that the European Union's

Emissions Trading Scheme, the largest potential market for carbon offsets, excludes LULUCF credits of any kind.[13]

In addition to the challenges to getting LULUCF projects internationally approved, many of them face financing hurdles. The significant start-up funds required to try to register a project under the CDM add to the investment costs and are particularly challenging for many project proponents in the developing world. According to one estimate, the cost of getting a carbon offset project approved by the CDM Executive Board under the Kyoto Protocol ranges from U.S.$50,000 to $250,000. For a typical small-scale CDM project, the United Nations Development Program (UNDP) calculates that the project's total up-front costs account for 14 to 22 percent of the net present value of its revenue from carbon credits.[14]

Facing CDM hurdles, a variety of LULUCF project developers have found their niche in the voluntary markets. Martha Ruiz Corzo, founder and director of the Sierra Gorda Biosphere Reserve, is one example of a vocal proponent of using the voluntary markets. This reserve, a biodiversity "hot spot," is home to fourteen ecosystems, to jaguars, brown bears, and exotic orchids, and to a population of about 100,000 human residents, whom Ruiz describes as living in "extreme poverty." After successfully convincing the Mexican government to create the Sierra Gorda Biosphere Reserve in the Mexican state of Querétaro, Ruiz recognized a need for additional financing and began looking to opportunities in the markets for ecosystem services.[15]

Seeking means of financing reforestation and creating sustainable livelihoods for the local population, Ruiz was optimistic about using the CDM to help protect and reforest the area. But after battling for seven years to certify a carbon sequestration project in Sierra Gorda under the CDM, Ruiz explained, "I finally realized that CDM rules are too high and too expensive to create a good deal for the people of Sierra Gorda."[16] Frustrated, she began exploring the voluntary carbon markets and immediately sold 5,230 emission reduction credits—"solid, verifiable credits," she said—to the UN Foundation. The carbon project, developed with Sierra Gorda's partner organization, Bosque Sustentable, with assistance from Woodrising Consulting, Inc., and funding from the Global Environment Facility, reforests land previously converted to agricultural and livestock use in the reserve. As a form of self-insurance to address permanency concerns, Bosque Sustentable retains 20 percent of the credits generated by the projects.[17]

At the same time that organizations such as Bosque Sustentable are using the voluntary carbon markets to help finance reforestation and afforestation projects, which have accepted methodologies under the CDM, other organizations are using the markets to incubate methodologies for activities that are not approved under the CDM. For example, conservation organizations such as the Nature Conservancy and Conservation International are working toward obtaining carbon financing for forest protection projects, often referred to as "avoided deforestation projects" or "reduced emissions from deforestation and degradation (REDD)," a concept not currently approved to produce carbon credits within the CDM process.

The Noel Kempff Climate Action Project is one example of the way conservation organizations are linking REDD with carbon finance. In 1997 the Nature Conservancy, the Bolivian government, and the Bolivian Fundación Amigos de la Naturaleza joined forces to protect 832,000 threatened hectares of tropical forest bordering the Noel Kempff Mercado National Park in northeastern Bolivia. According to the Nature Conservancy, the project was "expected to reduce, avoid and mitigate up to 17.8 million tonnes of carbon dioxide in the atmosphere over 30 years by avoiding logging and agricultural conversion of the land."[18] With financing not only from the Nature Conservancy but also from American Electric Power Company, BP America, and PacifiCorp, the preserved land became part of the national park and a pilot project for measuring carbon sequestration via avoided deforestation. The government of Bolivia has rights to 49 percent of the offset credits and is required to invest any profits from sales back into park management. In turn for their investment in the project, the investing companies have rights to 51 percent of the carbon credits generated.[19]

Buying into Charismatic Carbon

In contrast to the described constraints of developing LULUCF CDM projects, LULUCF projects may not only face lower financing and bureaucratic hurdles in the voluntary markets but also be valued more highly for providing "charismatic carbon." Ruiz describes the credits she is selling as "not just a carbon credit, but . . . a green jewel" and believes the biodiversity and sustainable development co-benefits of her projects make them particularly viable in the voluntary markets.[20] Many buyers seem to agree. After facilitating the 1989 Applied Energy Services offsetting program in Guatemala, Paul Faeth, of the International Institute for Environment and Development (IIED), noted, "The carbon fixation is nice, but the social benefits are even greater. There will be fuel wood, better soil, higher crop yields, jobs. If I never do anything else in my career, I feel I've helped do a little something good here."[21]

Without a regulatory driver, demand in voluntary offset markets is shaped by consumer choice variables that barely exist in the regulated markets. The demand curve for offset purchases has as much in common with the markets for fair trade or organic cotton as it does with the regulated carbon markets. Although CDM projects may have a range of co-benefits, the majority of credits are simply sold as regulatory commodities used to comply with a particular regulation. Voluntary offset markets, on the other hand, are characterized by strong links to philanthropic and public relations motives, and the demand for strong sustainable development benefits associated with offset credits is an important driver in the markets.

Factors contributing to a credit's appeal include sustainable development and biodiversity co-benefits and measurable land saved. In the realm of offset projects these co-benefits are a competitive advantage for LULUCF projects.[22] As well as being charismatic, forestry-based credits have the benefit of being critical, well-known actors in the carbon cycle. Hence, for a business seeking to communicate the benefits of offsets, a forestry project is a simple choice for communicating the concept. "Trees [are] one area of carbon sequestration that everyone understands, even little kids understand . . . People get it."[23]

Numerous buyers also recognize the critical role LULUCF projects can play in some of the poorest areas of the world. As noted by the World Bank's BioCarbon Fund, which was set up to invest in forestry and land-use change projects, LULUCF activities "are of critical importance to many economies in transition, and developing countries and their agricultural and forestry sectors can benefit from carbon finance flows. Many developing and primarily agrarian countries have small energy and industrial sectors, hence opportunities for emission reductions are few. Even in countries where other opportunities exist, LULUCF activities can foster sustainable rural development and directly affect the lives of the poorest people."[24]

A recent Ecosystem Marketplace and New Carbon Finance report supports this statement quantitatively. The organizations recorded more than 500,000 offset credits created through African projects and sold into the voluntary carbon markets in 2006. More than half these credits originated in LULUCF projects.[25]

From Hot Air to Transparency: The Role of Carbon Verification Standards

As LULUCF carves a niche in the current carbon offset markets, standards and registries are the tools shaping the future of these markets. Whether they are interested in the markets as a tool for conservation or in new tradable commodities, experts agree that the voluntary carbon markets are at a critical juncture in proving their viability. At present several related and unrelated efforts are under way to make the voluntary carbon markets more transparent by creating registries and standardizing the credits being sold.

In the following sections we outline the major independent standards and certification programs in the voluntary markets (table 20-1) and then compare several elements of the systems. It is important to note that numerous offset retailers and

Table 20-1. *Major Certification Programs and Standards Available or Soon to Be Available for the Voluntary Carbon Offset Markets*

Program	*Description*	*Focus on environmental and social benefits?*	*Reporting/ registration*	*Includes LULUCF methodology?*	*Geographical reach*	*Start date*
WBCSD/WRI Protocol	GHG accounting guidelines for projects and organizations	No	Does not include registry	Protocol created for LULUCF	International	2001
Voluntary Carbon Standard	Carbon verification standard	No	Use Bank of New York; other registry to be determined	Yes; methodologies to be determined	International	2007
CCB Standards	Certification program for offset projects	Yes	Projects on website	Only LULUCF (including avoided deforestation)	International	First projects certified 2007
Plan Vivo	Methodology and certification for offset projects and carbon credits	Yes	No	Community-based agroforestry	International	2000
Gold Standard	Certification for offset projects and carbon credits	Yes	VER registry in development	Energy projects	International	First project validated 2006; first credits verified 2007
CCAR	Registry protocol	No	Reporting protocols used as standards	Yes, first protocol	Forestry, California; livestock, U.S.; registry, international	First protocol 2005

ISO 14064	Certification program for emissions-reporting offset projects, carbon credits	No	No	Yes	International	Methodology released 2006
Social Carbon	Certification for offset projects and carbon credits	Yes	Creating its own registry system	Yes (including avoided deforestation)	South America and Portugal	First methodology applied 2002
ECIS Standard	Certification for offset projects and carbon credits	No	To be determined	Follow CDM or JI methodology	International	To be determined
VER+	Certification program for offset projects, carbon credits, and carbon-neutral products	No	TÜV SÜD BlueRegistry	Includes a JI or CDM methodology	International	2007
Green-E	Meta-certification program for offset sellers	No	Registry incorporated	Accepts other standards that include LULUCF	Aimed at North America; international possibilities	2007
Greenhouse Friendly	Certification program for offset sellers and carbon-neutral products	No	No	Yes	Australia	2001
CCX	Internal system for CCX offset projects and CCX carbon credits	No	Registry incorporated with trading platform	Yes (including avoided deforestation)	International	2003

Source: K. Hamilton and others, "State of the Voluntary Carbon Markets 2007: Picking Up Steam," Ecosystem Marketplace and New Carbon Finance, Washington and London, 2007.

the Chicago Climate Exchange have created their own standards to be used internally. We define a certification program as not only abiding by specific standards but also using a logo or brand to label a product or project that has been verified according to these standards. The systems we summarize have a range of purposes and characteristics but can be divided into two broad categories. The first set—the Voluntary Carbon Standard, the Climate, Community, and Biodiversity Standards, Plan Vivo, and the Gold Standard—is focused on standards for offset projects. The second set, including all the remaining systems we describe, is focused on certifying offset sellers, products, or services. The WBCSD/WRI GHG Protocol and Australian Greenhouse Friendly program are used to verify both offset projects and greenhouse-neutral products and services.

The WBCSD/WRI GHG Protocol for Project Accounting

The "Protocol for Project Accounting" created by the World Business Council for Sustainable Development (WBCSD) and the World Resources Institute (WRI)—known by the staggering acronym WBCSD/WRI GHG Protocol—is a set of guidelines that can be used by project developers and has also been incorporated into numerous standards.[26] This protocol was created along with a corporate accounting and reporting standard for greenhouse gases. Neither the GHG Protocol nor the corporate accounting standard is a certification system or verification standard itself. The GHG Protocol "aims at harmonizing GHG accounting and reporting standards internationally to ensure that different trading schemes and other climate related initiatives adopt consistent approaches to GHG accounting."[27] The GHG Protocol includes standards for corporate accounting and reporting as well as project accounting. In addition to the general GHG Protocol, a set of sector-specific guidelines for LULUCF was recently created.

The Voluntary Carbon Standard

The Climate Group, the International Emissions Trading Association (IETA), and the World Economic Forum have developed the Voluntary Carbon Standard (VCS), a carbon verification standard launched at the end of 2007. The VCS aims to provide a credible but simple set of criteria that will lend integrity to the voluntary carbon markets and underpin the credible actions that already exist. Its goal is to become the "quality threshold" in the markets.[28] It is meant to give actors in the emerging voluntary offset markets confidence in the integrity of their investments. The VCS covers agriculture, forestry, and other land-use projects, referred to as AFOLU within the standard. The VCS uses a buffer withholding approach to address nonpermanence risks associated with LULUCF projects, resulting in fully fungible (and nontemporary) Voluntary Carbon Units generated by LULUCF and other project types.

The Climate, Community, and Biodiversity Standards

The Climate, Community, and Biodiversity (CCB) Standards focus exclusively on land-use-based carbon mitigation projects.[29] Development of the CCB Standards was spearheaded by the Climate, Community, and Biodiversity Alliance (CCBA), a partnership between a group of nongovernmental organizations, corporations, and research institutions including Conservation International, CARE, the Nature Conservancy, Pelangi Indonesia, BP, Intel, Weyerhaeuser, CATIE, CIFOR, and ICRAF. The CCB Standards are designed to be used in either the voluntary or the regulatory markets to ensure, in addition to carbon benefits, that there is a net community and biodiversity benefit to LULUCF projects. A unique feature of the CCB Standards is that project developers can use them for designing and certifying "multiple-benefit" forestry projects that support local communities and protect biodiversity. The CCB Standards are project design standards, so they can be applied early in the project development cycle, even before carbon benefits have been generated on the ground.

In February 2007 the first two projects were certified with the CCB Standards. The first project resulted from a partnership between Conservation International and the Nature Conservancy in Yunnan province, China.[30] The second is a partnership between the organizations Futuro Forestal and CO2OL-USA to reforest degraded land on the Pacific coast of Panama, in Chiriqui and Veraguas provinces.[31] Several dozen projects are currently using the CCB Standards, mostly in the developing world.

Plan Vivo

Plan Vivo is a scheme focusing exclusively on LULUCF projects. It describes itself as "a system for promoting sustainable livelihoods in rural communities, through the creation of verifiable carbon credits."[32] The system was created by the Edinburgh Center for Carbon Management (ECCM) and is now managed by the nonprofit organization BioClimate Research and Development (BRandD). Whereas the CCB Standards include forest restoration and forest conservation projects, Plan Vivo is applicable only to community-based agroforestry. The Plan Vivo system is unique in that it provides a complete set of processes and internal requirements, from project design through operation and certification. Plan Vivo currently has three fully operational projects, in Mexico, Uganda, and Mozambique, which are producing carbon for the sale of Plan Vivo carbon offsets.

The Gold Standard

The Gold Standard seeks to define the high-end market for carbon credits but is focused exclusively on renewable energy and energy efficiency projects and explicitly excludes land-based offsets. The standard is based on an initiative of the World Wildlife Fund and was developed with other NGOs, businesses, and governmental organizations in response to the concern that the majority of CDM projects do not have significant sustainable development aspects, despite the approval of sustainable development benefits by host country governments. Although the standard was originally created to supplement CDM projects, it now also certifies voluntary offset projects. The standard is in the midst of creating registry procedures for Verified Emission Reductions (VERs), or units for voluntary carbon credits, to avoid double counting (and double selling) of voluntary carbon credits.[33]

The California Climate Action Registry Protocols

The California Climate Action Registry (CCAR) was established by California statute as a nonprofit voluntary registry for greenhouse gas emissions. Although CCAR has developed a general protocol and additional industry-specific protocols that give guidance on how to inventory GHG emissions for accounting in the registry (that is, what to measure, how to measure, the backup data required, and certification requirements), the registry has also developed project protocols that allow for the quantification and certification of GHG emission reductions. It is these protocols that essentially serve as a "verifiable" quasi-standard for voluntary carbon offsets. Already some U.S. companies such as Pacific Gas and Electric have announced that they intend to buy voluntary carbon offsets that meet the CCAR emission reduction protocols. CCAR currently has approved reduction protocols for livestock activities and forest carbon sequestration. The CCAR forestry protocols rely on legally binding conservation easements to mitigate carbon impermanence risk, which makes the protocols difficult to apply in developing countries.

ISO 14064

The standard known as ISO 14064/65 is part of the International Organization for Standardization (ISO) family of standards. The standard currently includes four components:

—Organization reporting: guiding organizations' quantification and reporting of GHG emissions (ISO 14064 Part 1)

—Project reporting: guiding project proponents' quantification, monitoring, and reporting of GHG emission reductions (ISO 14064 Part 2)

—Validation and verification: guiding the validation and verification of GHG assertions from organizations or projects (ISO 14064 Part 3)

—Accreditation of validation and verification bodies: guiding the accreditation or recognition of competent GHG validation or verification bodies

The ISO standards were created not to support a particular GHG program but to be "regime neutral," such that they could be used as the basis for any program. ISO does not certify or register GHG emissions or credits but provides accreditation, validation-verification, quantification, and reporting architecture.[34]

Social Carbon

The Social Carbon methodology and certification program was created by the Brazilian NGO Ecologica. The methodology is based on a sustainable livelihoods approach focused on improving "project effectiveness by using an integrated approach which values local communities, cares for people's potential and resources, and takes account of existing power relations and political context."[35] Although it was originally created for Kyoto Protocol carbon projects, the program's methodology is now also used for voluntary market projects. The Social Carbon methodology has been used in hydrology, fuel switching, and forestry projects in Latin America and Portugal since 2000. Recently the program launched a connected certification program to verify project use of the methodologies and credits resulting from these projects.

European Carbon Investor Services' Voluntary Carbon Offset Standard

In June 2007 a group of more than ten banks and financial institutions, organized under the name European Carbon Investor Services (ECIS) and including ABN Amro, Barclays Capital, Citigroup, Credit Suisse, Deutsche Bank, and Morgan Stanley, announced that it was creating a standard for carbon credits in the voluntary markets.[36] Imtiaz Ahmad, of Morgan Stanley, vice president of ECIS, described the standard as "a robust benchmark with environmental integrity in the voluntary market." The voluntary offset standard is aimed at bringing "the voluntary market up to the level of the regulated and standardized procedures of the compliance market." The standard is broadly similar to the CDM and JI, but it applies methodologies to an "eligible geographical area beyond those countries that have ratified the Kyoto Protocol" and is focused largely on the United States' and Australia's precompliance markets. Notably, it excludes carbon credits arising from the destruction of industrial gases such as HFC-23.

The VER+ Standard

In May 2007, project verifier TÜV SÜD announced its VER+ Standard, which will certify carbon neutrality as well as certify credits from voluntary carbon offset projects. The standard will be based on CDM and JI methodology. Martin Schröder of TÜV SÜD described the standard as "streamlined" with Kyoto. In tandem with VER+, TÜV SÜD announced BlueRegistry, which aims to be a platform for managing verified emission reductions from a variety of other standards, including the CCX and the Voluntary Carbon Standard, as well as green certificates.

The Green-E Greenhouse Gas Product Standard

The Green-E Greenhouse Gas Product Standard is not a project- or credit-based standard but instead was developed primarily to provide certification services to retail providers retiring carbon credits to sell as carbon offsets to customers. This standard is aimed primarily at North American retail providers and sellers of carbon offsets. The standard, launched in early 2008, will rely on other accepted project-based standards, such as the Gold Standard and the Voluntary Carbon Standard.[37]

The Greenhouse Friendly Initiative

The Greenhouse Friendly Initiative is the Australian government's voluntary carbon offset scheme for encouraging GHG emission reductions at several levels, including "providing businesses and consumers with the opportunity to sell and purchase greenhouse neutral products and services." The initiative provides two services: carbon neutrality certification and "Greenhouse Friendly Abatement Provider" (offset project) approval.[38]

Greenhouse Friendly "carbon-neutral" accreditation requires the preparation of a full, independently verified life-cycle assessment, an emissions monitoring plan, and annual reports. To be approved as a Greenhouse Friendly carbon-neutral product or service, offsets used must also be approved as Greenhouse Friendly carbon offsets.

The program allows forest sink projects that comply with strict eligibility criteria to create and sell GHG abatements nationally.[39]

Synthesis

These certification programs can be compared at numerous levels. They can be categorized according to where they fit in the supply chain and according to the process included. Standards can also be compared according to whether and how they emphasize sustainable development co-benefits and whether they include LULUCF projects or accept offsets from these projects. Along with standards, registries are also evolving in the markets and are critical accounting tools for protecting against double counting.

Conclusion

As literal, figurative, and political storms pound the concept of climate change into the public consciousness, companies, governments, and concerned citizens are reaching for a portfolio of solutions to this complex problem. Voluntary carbon markets cannot be a substitute for regulatory cap-and-trade systems, and carbon sequestration projects must be considered in addition to directly reducing the production of fossil fuel emissions. However, even under the most aggressive reduction targets it is highly unlikely that all GHG emission-producing entities will be regulated. The voluntary side of the carbon markets provides a means other than direct emission reductions for these entities to manage their climate effects. At the same time, the contributions of land-use change to climate change cannot be ignored, and interest is growing in projects that reduce emissions from deforestation, which currently represents 20 to 25 percent of global carbon emissions.[40] Both LULUCF projects and the voluntary carbon markets remain key tools in the portfolio of solutions for mitigating climate change. Moreover, the strengths of these two arenas can be complementary. Taken together, they can lead to greater innovation, broader participation, and deeper emission cuts than would otherwise be possible.

Notes

1. "Burning Coal in Connecticut, Planting Trees in Guatemala," Donella Meadows Archive (www.sustainabilityinstitute.org/dhm_archive/index.php?display_article=vn264aesed, accessed February 25, 2007).
2. K. Hamilton and others, "State of the Voluntary Carbon Markets 2007: Picking Up Steam," Ecosystem Marketplace and New Carbon Finance, Washington and London, 2007.
3. See www.chicagoclimatex.com/content.jsf?id=821.
4. Hamilton and others, "State of the Voluntary Carbon Markets 2007."
5. K. Capoor and P. Ambrosi, "State and Trends of the Carbon Market 2007" (Washington: World Bank, 2007).
6. C. Walker, "The Voluntary Carbon Market's Difference Maker: Michael Molitor," Ecosystem Marketplace, Washington, April 28, 2006.
7. Capoor and Ambrosi, "State and Trends of the Carbon Market 2007," p. 26.
8. ICF Consulting, "Voluntary Carbon Offsets Market: Outlook 2007," 2007 (www.icfi.com/markets/energy/doc_files/carbon-offsets.pdf).

9. Walker, "Michael Molitor."

10. Mark Honigsbaum, "Is Carbon Off-Setting Part of the Solution (or Part of the Problem?)" *The Observer,* June 10, 2007; "Another Inconvenient Truth," *Business Week,* March 26, 2007; F. Harvey and S. Fidler, "Industry Caught in a Carbon Smokescreen," *Financial Times,* April 25, 2007.

11. The Kyoto Protocol, the major driver of regulated markets, requires industrialized countries to reduce their carbon emissions by an average of 5 percent below 1990 levels by 2012. The CDM was established under article 12, and JI under article 6.

12. See http://cdm.unfccc.int/methodologies/index.html (July 13, 2007).

13. Directive 2004/101/EC of the European Parliament and of the Council of 27 October 2004 amending Directive 2003/87/EC establishing a scheme for greenhouse gas emission allowance trading within the Community, in respect of the Kyoto Protocol's project mechanisms.

14. T. Krolik, "The Argentine Carbon Fund Helps Developers Dance the Dance," Ecosystem Marketplace, Washington, May 25, 2006.

15. D. Ross and M. Ruiz Corzo, "A Project Developer's Perspective on the Voluntary Carbon Market: Carbon Sequestration in the Sierra Gorda of Mexico," in *Voluntary Carbon Markets: A Business Guide to What They Are and How They Work,* edited by R. Bayon, A. Hawn, and K. Hamilton (London: Earthscan, 2006).

16. Quoted in R. Bayon, "Case Study: The Mexico Forest Fund," Ecosystem Marketplace, Washington.

17. Ross and Ruiz Corzo, "A Project Developer's Perspective."

18. See www.nature.org/initiatives/climatechange/work/art4253.html.

19. P. May and others, "Local Sustainable Development Effects of Forest Carbon Projects in Brazil and Bolivia: A View from the Field," International Institute for Environment and Development, 2004 (www.poptel.org.uk/iied/docs/eep/MESpercent20Series/MES5BrazilandBoliviacarbonreport.pdf).

20. K. Hamilton, "Gourmet Carbon vs. Commodity Carbon: An Exploration of the Voluntary Market for GHGs," Ecosystem Marketplace, Washington, January 29, 2007.

21. "Burning Coal in Connecticut, Planting Trees in Guatemala."

22. A 2006 survey of twenty-six offset retailers found that 56 percent of their offset portfolios originated in LULUCF projects. E. Harris, "The Voluntary Retail Carbon Market: A Review and Analysis of the Current Market and Outlook," master's thesis, Imperial College London, 2006.

23. Erin Kelley, manager of environmental affairs, Interface Corporation, personal communication, May 7, 2006. Erin Meezan of Interface explained that the company chose forestry credits from major tree-planting projects by the nonprofit American Forests in California and the American South (two regions where Interface has plants) and Northeast for offsetting in-house emissions, with the goal of "employee education and engagement in our sustainability mission." With this goal in mind, forestry projects became particularly appealing.

24. World Bank, Carbon Finance Unit (http://carbonfinance.org/Router.cfm?Page=BioCFandFID=9708andItemID=9708andft=About, accessed June, 11, 2007).

25. Hamilton and others, "State of the Voluntary Carbon Markets 2007."

26. See www.ghgprotocol.org/templates/GHG5/layout.asp?MenuID=849).

27. World Business Council for Sustainable Development and World Resources Institute, "GHG Protocol Initiative: For Project Accounting" (www.ghgprotocol.org/).

28. IETA, Carbon Group, and World Economic Forum, "The Voluntary Carbon Standard Verification Protocol and Criteria: Version 1 for Consultation," March, 27 2006; M. Kebner, "The Voluntary Carbon Standard," paper presented at the GreenT Forum, "Raising the Bar for Voluntary Environmental Credit Markets," New York, May 2–3, 2006.

29. See www.climate-standards.org/standards/index.html.

30. See www.climate-standards.org/.

31. See www.climate-standards.org/news/news_feb2007.html (accessed 13 July 2007).

32. See www.planvivo.org.

33. The Gold Standard website (www.cdmgoldstandard.org/how_does_it_work.php (accessed July 13 2007)).

34. C. K. Weng and K. Boehmer, "Launching of ISO 14064 for Greenhouse Gas Accounting and Verification," ISO Management Systems, 2006 (www.csa.ca/climatechange/downloads/pdf/ISO_Management_Systems_14064_Article.pdf).

35. See www.socialcarbon.com.

36. Fiona Harvey, "Banks Take Step toward Carbon Credit Regulation," *Financial Times Limited,* June 28, 2007.

37. See www.green-e.org/getcert_ghg_intro.shtml.

38. See www.greenhouse.gov.au/greenhousefriendly/publications/gf-guidelines.html.

39. See www.environment.gov.au/minister/env/2007/pubs/mr26mar07.pdf.

40. The Intergovernmental Panel on Climate Change (IPCC) reports that carbon dioxide emissions associated with land-use change are a significant contributor to climate change. IPCC, "Climate Change 2007: The Physical Science Basis. Summary for Policy Makers," 2007 (www.ipcc.ch/SPM2feb07.pdf).

CASE STUDY

Reflections on Community-Based Carbon Forestry in Mexico

RICHARD TIPPER

My involvement in community-based carbon management started one night in 1993 when I was awakened in Edinburgh by a phone call from someone at Mexico's Instituto Nacional de Ecología (INE) informing me that the institute had approved funding to investigate the feasibility of supporting agroforestry and forest restoration in indigenous areas of Chiapas using carbon service payments.

The idea of using carbon payments for afforestation and forest conservation was then quite new. Some grandiose schemes for massive afforestation programs to balance the global carbon cycle had been postulated, but little had been done to examine the practical issues of how carbon payments could be used as an effective financial mechanism from the farmer's and rural community's perspectives.

I had become interested in the idea of using ecosystem payments to encourage more sustainable land use as a result of working for peasant farmers' cooperatives at the receiving end of government and for World Bank forestry and agroforestry programs. These "top-down" packages never seemed to deliver what was needed, when it was needed, where it was needed. Our local extension team had yearned for more flexible access to finance and skills to enable them to implement things that the communities themselves had decided were necessary and feasible.

This initial study, undertaken by scientists at Mexico's Colegio de la Frontera Sur, John Grace and me at the University of Edinburgh, and technicians from a regional credit union in Chiapas, sought to provide answers to some fairly basic questions. Would farmers want to undertake activities on the basis of a carbon service agreement? What types of forestry activities did farmers and communities wish to pursue, and what carbon benefits could be achieved? What level of payment would be needed and at what intervals? How could carbon services from many individual farmers or groups be aggregated in a way that could be taken to "buyers" of the service?

The results of the study were published in a government report and a scientific paper that was presented at interminable workshops on what was then the theoretical concept of "joint implementation."[1] But "the proof of the pudding was in the eating"—literally. I was invited to lunch by Max Mosley, president of the Federation Internationale de l'Automobile, the organization responsible for running Formula 1 motor sport. By the time coffee was served I had an offer to take back to the farmers in Mexico.

The project started as an internal budget line within the Unión de Credito Pajal, with about thirty farmers from six Tzeltal and Tojolobal communities participating. We soon realized that if this were going to be done properly, we needed a more formal system to ensure that we could keep track of successive sales of carbon, ensure proper consultation, planning, and implementation at each location, and provide adequate support to the farmers. The UK Department for International Development's Forestry Research Programme (now terminated) obliged by supporting the development of what is now known as the Plan Vivo system.

The Plan Vivo system offers planning guidelines for farmers and technical support teams; technical specifications providing straightforward, measurable indicators of carbon uptake for various types of forestry systems; standard carbon service agreements between farmers and a central trust fund; a governance structure and rules for the central trust fund; procedures for monitoring progress in making payments; and databases for keeping track of carbon and money.

Much of the time since 1997 has been spent refining and developing this system to meet the requirements of the expanding voluntary carbon market. Institutionally, the project developed through the formation of an independent trust fund, Fondo Bioclimatico, and later a process for issuing Plan Vivo certificates from a central Plan Vivo agency run by a United Kingdom–based NGO. From 2001 the project became financially independent of development assistance, and it is now a carbon-offset-financed enterprise.

In 2005 the project encompassed some 900 individual farmers plus several large groups in 43 communities representing 8 ethnic groups from different parts of Chiapas and Oaxaca. Through sales of offsets of more than 250,000 tonnes of CO_2, it has established around 1,000 hectares of new agroforestry and forestry systems and brought 4,000 hectares of communal forest into conservation management.

Over this time there have been various flirtations with the formal CDM system. The attractions of a large and steady demand for carbon are considerable, but the complications and uncertainties of the process, along with the complexity of implementing temporary carbon accounting at the field level, provide no clear route to value for projects involving communities and farmers.

Some key lessons have been learned over the past ten years. First, carbon-based finance for community agroforestry, forest restoration, and conservation *can* work; indeed, we believe this approach is likely to be more effective in achieving sustainable outcomes than aid-based support. Aid-based support may still be important in the early stages of a project—technical capacity-building and carbon-modeling—but should be phased out after a few years.

Second, successful implementation requires the building of institutional confidence and competence over a number of years. Most farmers and communities need to see a project working for some time before they have the confidence to join the process. It is generally necessary to start with a few highly motivated persons and build gradually.

Third, organizations must approach this market with a long-term view. Unlike aid-based funding initiatives, which are based on two- to five-year funding cycles, the carbon market is essentially a sales-based model and should be viewed in the same way as setting up a shop or factory. The carbon market demands that instead of managing a donor-beneficiary relationship, a carbon project must manage a customer-supplier relationship.

Fourth, getting a key investor or purchaser willing to commit to a basic-level carbon offset purchase for several years is of great importance in providing stability and confidence.

Fifth, the main constraints on establishing successful projects are the scale and reliability of the market for carbon offsets; the institutional stability and capability of organizations operating at the grassroots level; access to funding for feasibility and preparation work; and finding the right sort of operational staff (the Plan Vivo approach requires a different dynamic from that found in many public sector or large NGO forest programs).

Sixth, the main constraints on expanding established projects are the scale and reliability of the market for carbon offsets; the availability of good-quality indigenous tree seedlings; and access to good-quality banking and escrow facilities.

Note

1. G. Montoya and others, *Desarrollo forestal sustentable y captura de carbono en las zonas tzeltal y tojolobal del estado de Chiapas* (México, D.F.: Instituto Nacional de Ecología, 1995); B. H. de Jong and others, "Community Forest Management and Carbon Sequestration: A Feasibility Study from Chiapas, Mexico," *Interciencia* 20, no. 6 (1995): 409–16.

21

Developing Forestry Carbon Projects for the Voluntary Carbon Market: A Practical Analysis

MARISA MEIZLISH AND DAVID BRAND

Regulated carbon markets create value through governments' legislation of emission reduction requirements and definition of market rules. Such regulatory activity creates standardized units of trade and rules regulating, among other things, the creation of offsets that can be sold into the market. In addition to the regulated carbon market, a vibrant voluntary market in carbon offsets has developed. This market includes trade in a myriad of types of offsets, or "carbon credits," with varying qualities and prices.[1] Whereas the role of forestry carbon in markets established directly under the Kyoto Protocol is limited, significant opportunity exists in the voluntary market because of a simple market driver: People like trees.[2] In addition to helping mitigate climate change, forestry projects can deliver multiple other benefits, such as biodiversity habitat and water quality preservation. This advantage distinguishes forestry credits from credits arising from energy and industrial emission reduction projects—for example, projects that improve energy efficiency or reduce industrial gas production—which apply technology-based solutions.

Our objective in this chapter is to explain how forestry projects can capture the imagination of buyers in the voluntary market to deliver credible, marketable carbon products that meet a growing demand for carbon offsets. We consider the development of voluntary markets and the role of forestry projects—specifically reforestation, avoided deforestation, and managed forests—and then discuss the way changes in ownership of commercial forests have enabled the development of business models that include forestry carbon in addition to timber production. We then look at the management systems required to develop credible forestry carbon products and provide a checklist of elements for product development. Finally, we identify some potential buyers of voluntary carbon and present strategies for the future development of the voluntary market.

Carbon Markets Take Shape

Long before there was a Kyoto Protocol or a European Union Emissions Trading Scheme (EU ETS), carbon transactions were taking place. Pioneering deals in the late 1980s and early 1990s (see chapter 20) paved the way for retail and voluntary markets to develop. By the late 1990s, companies whose entire business focused on creating and trading carbon credits were born, and the concept of voluntarily offsetting greenhouse gas emissions was being adopted.[3]

The overall carbon market shuddered for two or three years after the United States withdrew from the Kyoto Protocol in March 2001, but shortly afterward the voluntary markets started to diversify. For example, the Chicago Climate Exchange (CCX) was established in 2003 as the first trading platform for voluntary carbon credits. Retail carbon companies, whose business is to market emission reduction products to companies and consumers who wish to offset their carbon footprints, proliferated. At the time of writing there are nearly ninety such companies worldwide.[4] On the demand side, the concept of businesses offsetting some or all of their emissions began to form part of corporate environmental and sustainability practices. It is now beginning to be used to differentiate brands and products, such as "carbon neutral" products, airline travel, and car loans.[5] Some investment funds that focus specifically on acquiring carbon credits are now also allocating portions of their portfolios to acquiring Verified Emission Reductions (VERs), the units for voluntary carbon credits.[6]

Even with these developments, the voluntary market remains difficult to categorize, and its boundaries difficult to define. The entry into force of the Kyoto Protocol in February 2005 raised the profile of carbon trading not only in the regulated but also in the voluntary carbon market. Estimations are that the market is growing by 100 percent or more per year and could be trading 400 million tonnes of carbon dioxide equivalent (t CO_2e) by 2010. Prices for VERs in 2006 averaged U.S.$10, up from $7 in 2005.[7]

Generating Carbon Credits from Forestry Projects

The emergence of carbon markets can be good news for the forestry sector, insofar as it creates incentives for the conservation and sustainable use of forests. Not only can forests provide renewable energy fuels and the lowest embodied energy building material, but trees also sequester carbon dioxide from the atmosphere as they grow. Combined with the additional environmental and social benefits of forests, the growth of renewable energy and carbon markets provides a new playing field for forestry investors. There are now emerging market values for a range of ecosystem services including water purification, nutrient retention, and biodiversity conservation, not just the value of timber.[8]

With these additional values being recognized, forestry investors and owners can consider the financial benefits of selling VERs in investment decisionmaking. In general, projects that can generate forestry VERs fall into three categories: afforestation-reforestation, avoided deforestation, and forestry management.

Afforestation and reforestation (AR) projects are those that plant trees on land currently under nonforest use. Applying the rules for Clean Development Mechanism (CDM) projects in an analogous manner, AR projects must take place on land that was unforested before 1990. This requirement is to remove any incentive to clear-cut forests in order to apply for carbon financing by reforesting the land. Under the Voluntary Carbon Standard (VCS), a new standard for accrediting voluntary carbon projects, AR projects must occur on land that was converted to nonforested use at least ten years before the project start date.

AR can refer to permanent plantings, which are often designed to deliver additional benefits such as salinity reduction and biodiversity habitat. It can also refer to commercial forestry, in which the primary focus is timber production. The incentives are designed to encourage forest management in a way that maximizes the economic benefits of both timber and carbon. This is accomplished through management systems that balance timber harvesting with the requirements of maintaining carbon stocks associated with credits sold in the long term. Consequently, if managed properly, commercial forestry plantations on deforested land can yield carbon benefits equivalent to those from other forestry projects.

Conservation, or avoided deforestation, projects prevent forests from being converted to nonforest land uses. Carbon credits are created by accounting for the existing volume of carbon sequestered in the forest. In order to claim the carbon associated with emission reductions, avoided deforestation projects need to establish that an immediate conversion threat exists and demonstrate that carbon revenues are essential to conserving the forest. These project types are not currently eligible under the Kyoto Protocol's CDM but are popular in voluntary markets. Their popularity might be due to the additional biodiversity and social equity benefits that are often associated with conservation projects—for example, protecting orangutan habitat by selling carbon credits associated with a conserved forested landscape.

The third type of project relies on changing forestry management practices in a way that increases the sequestration rate or maintains a higher level of carbon stock over time. Changed practices can include reduced harvesting regimes, longer tree rotations before harvesting, and restocking forests that have not naturally recovered after depletion events such as fires and disease. These projects aim to manage forests for conservation and timber values. Forest management projects are not eligible in the CDM and have so far supplied few VERs to the carbon market. In California, however, the success of such projects has been demonstrated, because of the specific inclusion of "conservation based forest management projects" in the voluntary California Climate Action Registry. For example, California governor Arnold Schwarzenegger recently purchased offsets from redwood forests managed in this way.[9] Improved forest management is also included in the VCS.

Commercial Forestry Investment

Concurrent with the growing interest in the carbon value of forests is a recent upsurge of private investment in the forestry sector. The vast amount of capital invested in timber production is likely to play an important part in enhancing the role of carbon sequestration in both voluntary and regulatory markets, as investors look to maximize returns. It is also likely that forestry projects focused on conservation or noncommercial uses will incorporate sustainable harvesting in the future, because the value of standing forests needs to compete with economic drivers that are rapidly converting forests to agribusiness plantations.

These economic drivers are most clearly evidenced in the growth of oil palm plantations in Indonesia and Malaysia where old-growth forests are being converted to agribusiness.[10] To help address this trend, the government of Australia and the World Bank both launched funds in 2007 focused on developing a system to finance projects that reduce emissions from deforestation and degradation (REDD).[11] REDD was one of the key issues discussed at the United Nations Framework Convention on Climate Change (UNFCCC) meeting in Bali in 2007, where negotiators acknowledged the role of forest conservation and called for countries to implement demonstration activities.[12] Market signals such as these indicate that in the future, forestry investment will encompass a wider range of stakeholders as well as find mechanisms to value nontimber assets, making the forestry asset class more attractive to investors.

Traditionally, however, governments, the forest products industry, and real estate investors were the owners of and investors in forests. Until the mid-1980s, capital markets did not view forests as an asset class of interest. Since the late 1980s, institutional investors have emerged as dedicated forestry investors driven by two key factors.[13] First, forest product companies began selling off their forestry assets to institutional and private equity investors in order finance growth in the processing side of their businesses. The era of the leveraged buyout in North America put even more pressure on timber companies to improve their balance sheets, further accelerating this trend. Second, institutional investors, who maintain large asset portfolios worth hundreds of millions of dollars or more, became interested in broadening portfolio allocations beyond traditional investments to encompass private equity and alternative assets, such as forests. With readily available capital from these large investors and willing timberland sellers, the transition in forestry ownership progressed steadily.

These trends, which began in the United States, are now being seen in other countries, and investment in forestry assets has increased 20 percent a year for the past twenty years.[14] One 2005 study estimated that more than U.S.$40 billion was invested in forestry assets, mostly in the United States but also in South America, Oceania, South Africa, and Europe. It estimated that the total investable and leasable forest land in the world was 870 million hectares, worth U.S.$480 billion.[15]

Investors in such assets are financially motivated: they will take the next step of moving into the carbon market if they determine that it makes commercial sense to do so.

This potential for pursuing carbon sales has yet to become mainstream at large commercial scales, but as carbon prices increase and innovative financial structuring and investment strategies are brought to the market, this may change. Currently, forest conservation and reforestation projects in developing countries can produce commercial returns at comparatively low prices for carbon (for example, U.S.$1–$5 per tonne of CO_2e), whereas in Europe, North America, and Australia, carbon prices need to be between U.S.$10 and $50 per tonne of CO_2e to drive significant investment.[16] Nevertheless, as carbon prices rise, investors will look favorably at pure conservation projects, including revegetation and avoided deforestation projects, as well as changes in forestry management practices. The risks associated with such projects will become more commercially viable with higher carbon prices.

Elements of Forestry Carbon Project Design

The challenge for investors is to develop carbon products that are acceptable to the marketplace. Before discussing the design and successful marketing of forestry carbon projects, it is important to note the challenges that forestry offsets continue to face from a number of influential nongovernmental organizations. Arguments tend to focus on the accuracy of carbon sequestration methodologies, the ensuring of permanence, and the establishing of additionality. For example, an NGO position paper on emissions trading in Australia recommended that carbon offsets not be included in a national emissions trading scheme because setting verifiable baselines and accounting for additionality were too difficult.[17]

At a minimum there is a need to demonstrate that the sequestered carbon stock can be measured or estimated with acceptable accuracy, and there is a need to demonstrate a system to ensure that carbon is stored in the forest for a significant period of time. The challenge for project developers and managers is often to communicate the complex modeling systems, simulations of risk factors, and carbon pooling vehicles that are able to address most of the concerns that have been raised. Critics of forestry credits have often relied on simplistic arguments such as "Carbon measurements are unreliable in trees" and "What if the forest burns down?" to discredit forestry-based carbon credits.[18] Therefore, accounting, carbon retention, and risk management are paramount to demonstrating the quality of the offset product. Carbon products with the greatest market acceptance have sophisticated and documented management systems that include the accurate estimation and verification of carbon sequestration, evidence that requirements for retention and permanence are met, and a process for registering credits and retiring them once they are sold.

Regarding the estimation of carbon sequestration, carbon accounting systems and financial models are used to determine the potential volume and value of carbon credits that can be generated from a land-based asset. This requires quantifying sequestration volumes on the basis of area statistics, growth rates, and models to convert from biomass to carbon dioxide equivalent units. In the case of conservation projects, sophisticated carbon accounting models use this information to determine the total volume of carbon emissions that is avoided by forgoing the deforestation activity (for example, conversion to palm oil), which represents the volume of carbon credits available to the marketplace. The most credible voluntary products sell carbon ex-post, meaning that the volume of sequestration or avoided emissions in 2007 will equate to carbon credits available for sale in 2008. Alternatively, products can sell carbon today that will be sequestered in the future—for example, over the life of a tree. These ex-ante products have a higher risk factor because sequestration has not yet occurred.

With regard to carbon verification, the most credible VERs will be verified to a recognized standard by a third-party service provider knowledgeable in forestry carbon. The authors of chapter 20 in this volume identified various voluntary carbon standards and discussed some of the challenges related to forestry carbon. What is important to reiterate is that different standards may carry different costs of compliance. Determining which standard to pursue can be determined by considering the potential carbon buyer. Certain buyers may require more stringent standards, according to the requirements or preferences of their stakeholders. Because transactions are voluntary, one strategy is for project developers to negotiate appropriate verification procedures directly with potential buyers.

Another strategy is to take a broader approach and gain accreditation to a standard considered by the market to be of high quality. This could increase upfront costs, but it can also increase the potential number of buyers who will con-

The Greenhouse Friendly Program

Greenhouse Friendly (www.greenhouse.gov.au/greenhousefriendly) is a voluntary program under the Australian government's Greenhouse Challenge Plus that allows businesses to market greenhouse-neutral products and services. Abatement projects that generate permanent, verifiable greenhouse gas emission reductions, avoidances, or sequestration are eligible to participate in the program. Eligible forestry projects can be Kyoto-compliant forests managed for environmental or natural resource purposes or harvested forests.

A primary determinant for project accreditation is that the activity be beyond a "business-as-usual" baseline. This additionality requirement is defined as investment or behavior change that would not normally be undertaken as part of a company's established operating practices, under existing internal investment requirements, or under federal, state, or territory regulations. Forest projects may be eligible when the business case for the project is that it would be commercially viable only when the carbon value of the project is factored in. Of twenty-seven abatement projects accredited under the Greenhouse Friendly program at the time of writing, four are forestry projects.

sider the project and the potential price paid for the offsets. For example, New Forests developed a carbon offering on behalf of its clients in accordance with the standards of the New South Wales Greenhouse Gas Abatement Scheme.[19] Although the credits could not be registered as part of the scheme because they were located in reforestation projects outside New South Wales, they were verified against the NSW rules and used similar accounting, risk management, and permanence criteria.

Also in Australia, the federal government has created a national voluntary carbon standard under the Greenhouse Friendly program. Projects certified against the Greenhouse Friendly standard need to demonstrate that implementation depends on carbon finance and are thus "financially additional." Projects that are financially viable without carbon revenues or that have received government grants for implementation are ineligible, on the argument that they would have occurred anyway and therefore provide no true carbon benefit to the planet. The ineligibility of commercial forestry projects and conservation and reforestation projects that have received government grants or funding has created a dearth of voluntary forestry-based carbon credits under this scheme.

The question of financial additionality in forestry credits can be complex. It could be argued that financial additionality is subjective and potentially based on different costs of capital to forestry investors. There could be circumstances in which two forest owners, side by side, planting the same trees on identical land, are treated differently because of different commercial conditions. Often there is a requirement for evidence such as board papers that "carbon value was part of the consideration."[20] The argument against financial additionality is that the atmosphere cannot differentiate between a tonne of carbon removed through a commercial forestry project and a tonne removed through a subcommercial forestry project. Both projects provide the positive externality of carbon sequestration.

Credible forestry carbon products must have measures in place to demonstrate convincingly that carbon dioxide has been removed from the atmosphere permanently, or at least for many decades. That is, they must show that carbon stocks will be managed on a long-term basis and not reemitted into the atmosphere a few years after the carbon offsets are sold.

Market standards for permanence vary. The California Climate Action Registry requires a permanent land title easement over forests generating carbon credits, and New Zealand's Permanent Forest Sink Initiative requires a 99-year covenant on the land title.[21] The NSW Greenhouse Gas Abatement Scheme requires legal arrangements guaranteeing a 100-year retention liability, and the Greenhouse Friendly program requires 70 years. The Chicago Climate Exchange has no fixed mandatory permanence period. Under the CDM, forestry credits are only temporary, with storage requirements ranging from 5 to 60 years. In voluntary markets there is usually an expectation that the project developer will address permanence by clearly defining the terms or period of retention.

Permanence creates a long-term liability that can be challenging to manage. One strategy that has been effective is carbon pooling. Carbon pools enable project developers to plan and manage multiple forest stands in a way that optimizes the carbon asset and maintains permanence obligations. By staggering harvests, pooling carbon from multiple projects, and using other management strategies, permanence can be managed in a cost-effective and credible way.

Figure 21-1 shows this concept by modeling an aggregate carbon profile for several reforestation projects. The carbon available for sale is limited by the volume associated with the minimum carbon stock level, such that the volume of carbon in the pool of forests never falls below the volume sold to the marketplace. The total available carbon for sale over time is 2 million tonnes of carbon credits, which is reached in 2040. The pool sells carbon in each year in which there is net positive sequestration, creating "vintages" that represent the carbon sequestration that occurred that year. For example, a 2007 vintage equals the difference between carbon stock levels at the end of 2006 and the end of 2007. Credits are sold until the minimum stock level of 2 million tonnes is reached, which may occur before or after 2040.

Additional costs to be considered are those arising from the marketing of carbon credits and the transaction process. Marketing requires identifying, contacting, and negotiating with buyers, which can often take months or even years. The physical process of creating a product for sale requires extensive documentation, which can include a product specification document that defines the product

Figure 21-1. *Hypothetical Carbon Profile for a Plantation Estate, Showing Carbon Stock Levels over Time, 2006–60*

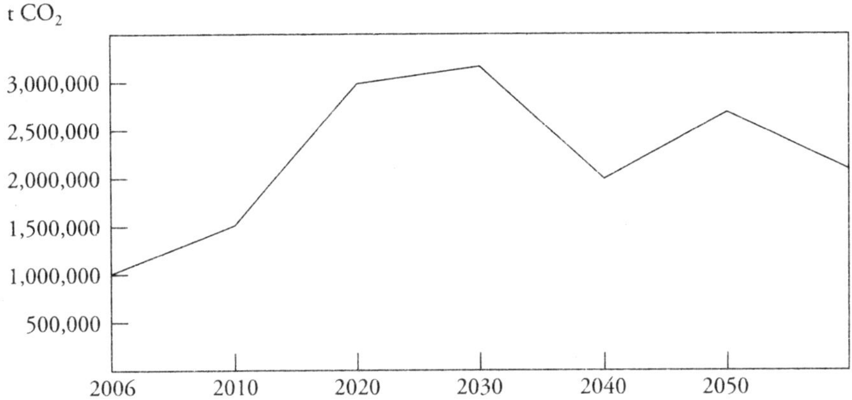

characteristics, legal guarantees for permanence or ownership of carbon rights, and third-party verification reports, including monitoring reports over time. Although these processes can be managed in-house, the decision to do so instead of outsourcing them will depend on the potential value of the carbon asset and any competing long-term management commitments.

In summary, a checklist for developing forestry carbon credits includes the following:

—*Eligibility.* AR projects in the voluntary carbon markets using a 1990 baseline (woody vegetation planted after January 1, 1990, on land cleared before then) are likely to be widely acceptable. The VCS requires that trees be planted on land cleared at least ten years before project implementation, which may be easier for some projects to demonstrate than others. Conservation and forest management projects are increasingly being included in regulatory rules and standards and are gaining acceptance in voluntary markets.

—*Carbon verification.* Project developers need to provide a transparent and independently verified method of quantifying sequestration volumes on the basis of area statistics, growth rates, and models to scale up from biomass to carbon. Verification should be undertaken against a published and reputable standard.

—*Permanence.* The project developer should demonstrate a commitment to managing the liability associated with the long-term storage of carbon dioxide, often defined as a period of time such as 100 years. This may include the use of land title restrictions, easements, or pooling vehicles.

—*Additionality.* Stringent definitions of additionality may attract a wider group of interested buyers but may prove difficult for project developers to demonstrate. At a minimum, projects need to be climate additional—in other words, emission reductions are achieved relative to a baseline scenario. Regulatory additionality may also be considered—that is, the project is not being undertaken to comply with existing legal or regulatory requirements. Financial additionality demonstrates that the project is not commercially viable without revenue from carbon credits.

—*Accounting.* Some retail forestry credits are based on ex-ante accounting (credits are sold before sequestration has occurred). Other products are based on ex-post accounting (credits are sold year-on-year after sequestration has occurred and been verified), which is compatible with the Kyoto Protocol approach.

—*Geographic location.* Some buyers of credits may specify that offset projects occur in the same state or region in which emissions are created.

—*Additional benefits.* Most buyers appreciate any additional social or environmental benefits that can be documented as part of the pedigree of the product. This can be a competitive differentiator of forest-based projects over other types of carbon projects, such as energy efficiency.

Products can exhibit any combination of these characteristics, and in general, the price per tonne should reflect the rigor with which the product addresses these issues. Although price often plays a prominent role in buying decisions, other characteristics related to forest carbon, such as the credibility of the vendor, the sustainability of the management of the underlying forest (for example, Forest Stewardship Council Certification for forestry-management-based projects), arrangements on permanence, and additional benefits, are often equally important. Products that are fully documented with supporting materials, independent verification, and a clear story line on the pedigree tend to be more acceptable to knowledgeable buyers.

Voluntary Carbon Buyers

The primary buyers of voluntary carbon are large businesses, often in sectors with significant carbon exposure, such as oil and energy, or with international brand recognition to maintain, such as banks and clothing companies. They may issue tenders, as HSBC bank did to purchase 170,000 tonnes of CO_2e in 2005, or negotiate private transactions with project developers, as BP did to offset 500,000 tonnes of CO_2e with its BP Global Choice product in Australia.[22]

Carbon retailers are another buyer group. About ninety retailers worldwide maintain portfolios of carbon offsets, which they usually sell to small to medium-size businesses and individuals primarily to offset emissions for individual events or activities.[23] For example, New Forests bought 140 tonnes of CO_2e in offsets from a carbon retailer to neutralize the emissions from its first year in business. Owners of forestry projects can potentially sell directly to retailers rather than to the ultimate buyers, which can reduce transaction costs as well as the price per tonne. In general, retailers tend to favor forestry-based carbon with multiple additional environmental benefits, such as enhancing biodiversity and providing local employment, because these are marketable "stories" that can be effectively communicated to consumers.

Governments are also increasingly purchasing carbon credits to offset activities or annual emissions. For example, the state government of Victoria in Australia conducted a competitive tender process in 2006 to purchase credits for offsetting the government's emissions.[24]

Outlook for the Future

In 2004, few observers would have predicted how quickly and dynamically the carbon market would grow in two short years. Although growth seems inevitable, there are likely to be surprises, adjustments, and unexpected developments as the market evolves, particularly as the post-2012 Kyoto world is determined. For forestry operators, a few trends will define future involvement in the voluntary market.

One of these is consolidation. The voluntary market encompasses a plethora of product types, business models, verification standards, and transaction structures. As has been seen in other new and fast-growing markets, such as the dot-com boom and bust of the late 1990s, only the most robust will have long-term staying power past the initial phase of excitement and discovery. This initial phase may already be ending as the EU ETS corrects its initial shortcomings and moves into a second reporting year and as some voluntary offset characteristics, such as ex-ante accounting, are being questioned by the market. As the market consolidates, businesses that can aggregate projects or credits will be able to afford sophisticated verification and marketing, producing the highest quality products for the least cost per tonne.

A second trend is businesses development. Many forestry projects looking to generate carbon credits begin as conservation efforts. Often these are not-for-profit or based on philanthropic funding. Carbon has moved on from this approach to become a profitable, attractive investment proposition requiring sophisticated analysis and forecasting. In other words, forestry carbon is a business now, applying mainstream business fundamentals to environmental values that have not traditionally been priced. For-profit businesses with robust business plans and knowledgeable personnel will drive the market forward. This has been seen in the U.S. conservation banking industry, in which for-profit businesses have been more financially successful than not-for-profit banks established for similar purposes.[25]

A third trend has to do with permanence. For the past decade, company boards have been asked to consider longer-term business strategies that better align with the goals of sustainable development. They are pushing their horizons out to five or perhaps ten years. The 70- to 100-year permanence requirements for forestry carbon credits will always be a mental roadblock, because even the longest-term investors, such as pension funds, do not make decisions on these time scales.

The forestry sector needs to develop approaches that meet permanence requirements while retaining options and "out" clauses, which are important to attract investment. Carbon pooling is one potential approach, whereby multiple landowners contribute to a single carbon pool that is managed to maximize carbon revenue and maintain permanence requirements. Within carbon pooling, the development of more sophisticated payment structures will allow greater flexibility and ultimately attract more landowners into participating. One such approach is that of "tonne years," in which landowners are paid a small amount annually for each year they retain carbon rather than being paid up front and carrying a long-term liability.

The future will not look like the past in relation to greenhouse gas emissions. A cost will be associated with this form of pollution, to which businesses and individuals will respond. Landowners and particularly forest owners have an advantage in this carbon-constrained world because of the sequestration capabilities of their primary asset. Understanding how carbon markets operate and assessing carbon assets can build competitive advantages by uncovering new value streams. The voluntary markets provide the most immediate route to commercializing carbon assets, and early movers are already positioning themselves to become reliable suppliers of rigorous offset products.

Notes

1. A carbon credit in this chapter is equal to one tonne of carbon dioxide equivalent.

2. A major energy retailer in Australia surveyed its customer base in 2005, offering a range of carbon offset options. Results, which were shared with the organization New Forests, indicated that forestry was solidly the preferred choice. According to a New Forests study in early 2006, a majority of the forty carbon retailers worldwide offered forestry-based carbon offsets, and about half offered only these products. A recent report by the Australian Institute also noted that forestry projects were the most popular type of carbon offsets. C. Downie, "Carbon Offsets: Saviour or Cop-out?" Research Paper 48 (Australian Institute, 2007) (www.tai.org.au/documents/downloads/WP107.pdf).

3. For example, the CarbonNeutral Company (formerly Future Forests) began in the mid-1990s and is now one of the leading companies in developing and sourcing voluntary carbon programs for businesses and individuals. See www.carbonneutral.com/. The Face Foundation was established in 1990 to help abate emissions by planting and protecting forests. See www.stichtingface.nl/. In early 2000, TEPCO, a Japanese company and the world's largest privately owned power utility, entered into an agreement with New South Wales State Forests in Australia to supply carbon offsets through tree plantations.

4. See www.chicagoclimatex.com/; K. Hamilton and others, "State of the Voluntary Carbon Market 2007: Picking Up Steam," Ecosystem Marketplace and New Carbon Finance, Washington and London, 2007.

5. Examples are GE's Ecomagination product line (http://ge.ecomagination.com/site/index.html), Virgin Blue's Carbon Offset Initiative (www.virginblue.com.au/carbon

offset/), and Bendigo Bank's Generation Green product range (www.bendigobank.com.au/public/generationgreen/index.asp).

6. Examples are Climate Change Capital (www.climatechangecapital.com/index.asp) and the World Bank's BioCarbon Fund (http://carbonfinance.org/Router.cfm?Page=BioCF).

7. "Upset about Offsets," *The Economist,* U.S. Edition, Business Section, August 5, 2006; K. Capoor and P. Ambrosi, "State and Trends of the Carbon Market 2007" (Washington: World Bank, 2007); IETA and World Bank,"State and Trends of the Carbon Market 2006: Update," presentation, Nairobi, November 13, 2006 (www.ieta.org/ieta/www/pages/getfile.php?docID=1955).

8. Ecosystem Marketplace is an online news and information portal developed in response to these emerging markets. For market data, analysis, and reports, see www.ecosystemmarketplace.com.

9. See www.climateregistry.org/Default.aspx?refreshed=true; Pacific Forest Trust, "Governor Schwarzenegger to Reduce Carbon Footprint Using Emissions Offsets from Forest Conservation Project," press release, May 12, 2007.

10. World Wildlife Fund Germany, "Rainforest for Biodiesel: Ecological Effects of Using Palm Oil as a Source of Energy," 2007 (http://www.wupperinst.org/uploads/tx_wibeitrag/wwf_palmoil_study_en.pdf); R. Swick, "Will the Global Supply of Nutrients Continue to Meet the Demands of the Feed Industry?" paper presented at the World Nutrition Forum, 2006 (www.worldnutritionforum.info/cms/biomin_wnf.nsf/($Jobs)/$C9EB88C4267941EAC12573AF00492515?OpenDocument).

11. These funds are the government of Australia's Global Initiative on Forests and Climate (www.greenhouse.gov.au/international/forests/about.html) and the World Bank's Carbon Partnership Facility (http://carbonfinance.org/Router.cfm?Page=FCPFandFID=34267andItemID=34267andft=About).

12. UNFCCC, Decision/CP.13, "Reducing Emissions from Deforestation in Developing Countries: Approaches to Stimulate Action" (http://unfccc.int/files/meetings/cop_13/application/pdf/cp_redd.pdf).

13. C. Binkley and J. Earhart, "A Global Emerging Markets Forestry Investment Strategy," International Forestry Investment Advisors and Global Environment Fund, 2005.

14. C. Binkley, "Rethinking Investments in Forestry," presentation to the Katoomba Group, Switzerland, October 29–30, 2003.

15. International Woodland Company, "Global Forestland Investment Study," 2005 (www.iwc.dk/publications/Globalpercent20Forestlandpercent20Investmentpercent20study.pdf).

16. This analysis is based on commercial models developed for New Forests' clients, who are investing in these regions. Forestry projects that are not commercially attractive become economically viable at these higher carbon prices.

17. Climate Action Network Australia, "Position Paper on Emissions Trading," 2007 (www.cana.net.au/Policies_positions/ETPositionPaperMarch07.pdf).

18. For example, "Flaws of the Concept" (www.sinkswatch.org); Downie, "Carbon Offsets."

19. See www.greenhousegas.nsw.gov.au/.

20. See the Clean Development Mechanism's guidelines for establishing baselines in carbon sequestration projects at http://cdm.unfccc.int/Panels/ar.

21. See www.climateregistry.org/PROTOCOLS/FP/ and www.maf.govt.nz/forestry/pfsi/.

22. See www.hsbc.com/1/2/newsroom/news/news-archive-2005/hsbc-goes-carbon-neutral-three-months-early and www.bp.com/subsection.do?categoryId=9012553and contentId=7024333.

23. Hamilton and others, "State of the Voluntary Carbon Market 2007."

24. Office of the Premier, Government of Victoria, Australia, "Government Steps Up Response," 2007 (www.dtf.vic.gov.au/domino/Web_Notes/newmedia.nsf/bc348d5912436a9cca256cfc0082d800/b2bbd9e769c3ad71ca25725e000507f2!OpenDocument).

25. J. Fox and A. Nin-Murcia, "Status of Species Conservation Banking in the United States," *Conservation Biology* 19, no. 4 (2005): 996–1007.

CASE STUDY

Carbon Sequestration in the Sierra Gorda of Mexico

DAVID PATRICK ROSS

Bosque Sustentable, A.C., a nongovernmental organization working in the Sierra Gorda Biosphere Reserve and surrounding areas in eastern central Mexico, signed a contract with the United Nations Foundation in March 2006 for the sale of 5,230 emission removal units (t CO_2e). The contract was the culmination of years of hard work, and our experience with the international carbon market during this time highlights the difficulties and opportunities for organizations interested in developing carbon sequestration projects in rural, poor areas.

The Sierra Gorda Biosphere Reserve is located in the Sierra Madre Oriental in the northern extreme of the state of Querétaro. Situated in a transition zone between the Nearctic and Neotropical biogeographical regions, the reserve constitutes the most ecosystem-diverse natural protected area in Mexico and is home to all six species of Mexican felines—the jaguar, puma, bobcat, margay, ocelot, and jaguarundi.

Despite its natural wealth, the reserve is an area of poverty. Approximately 90,780 inhabitants live in 629 communities throughout the reserve, and all five municipalities are ranked as highly or very highly marginalized. More than 70 percent of the economically active population in Pinal de Amoles, the site of carbon sequestration for the UN Foundation, earns less than U.S.$8 a day.

Bosque Sustentable, founded in 2002, works in close coordination with the Sierra Gorda Biosphere Reserve and its civil society partner organization, Grupo

Ecológico Sierra Gorda. From 1998 to 2004 the organizations of the Sierra Gorda focused their carbon efforts on looking for opportunities to enter the carbon market created by the Kyoto Protocol.

The barriers we encountered when trying to enter the Clean Development Mechanism (CDM) are common to many poor areas in Mexico and throughout Latin America. At the most basic level they include a lack of capital for developing projects and a lack of forest management skills among local landholders. Even when local capacity has been developed, high costs for verification and certification of carbon sequestration benefits mean that more carbon money goes to international consultants than to local people planting and protecting trees.

Another important barrier is the pattern of landownership in the Sierra Gorda, which lacks large, compact properties. Bosque Sustentable works with small landholders whose average plantation size is 1 hectare. This means that for a project of 400 hectares—small by international standards—Bosque Sustentable must work with approximately 400 different landholders scattered throughout the mountains. These properties lack telephone service and are accessible only by hours of driving on rough, unpaved roads, dramatically increasing the per-unit costs of carbon sequestration. Indeed, many properties originally envisioned as part of the project have had to be left out because of a combination of their remote location and insufficient growth rates. In addition, the majority of landholders do not hold title to the property in their own name. In most cases title is in the name of a deceased relative, and although possession is not in dispute, legal costs and exorbitant notary fees prevent the landholders from updating the titles.

Not surprisingly, impoverished farmers often require payments in the early years of the plantations. Although government programs support tree planting, mortality is high, and the payments are increasingly insufficient to attract participation. Carbon payments provide an additional incentive for participation, as well as small payments to landholders who otherwise simply cannot afford the investment of time and resources to establish and ensure the survival of a plantation. Although some buyers in the CDM market will make up-front investments, the additional risk involved usually entails a lower purchase price.

The unresolved issue of additionality was also difficult for us to navigate. With the support of an international consultant with CDM experience, Bosque Sustentable argues that although its project includes land located within a federal Natural Protected Area, the land is private and under no legal requirement for reforestation; therefore the CDM requirement for project additionality can be met. Other consultants and certain nongovernmental organizations, however, continue to argue that reforestation within Natural Protected Areas should not be considered additional for CDM purposes.

For these reasons and others, Bosque Sustentable and its partner organizations decided to give up on their effort to enter the CDM market. "For years we heard that the Clean Development Mechanism was a tool for sustainable development," explained Martha Isabel Ruiz Corzo, director of the Sierra Gorda Biosphere Reserve. "The reality is that the CDM is light years away from the needs of areas of poverty."

Now Bosque Sustentable is focusing on the voluntary carbon market. Its program is targeted to organizations, businesses, and individuals who want not only to contribute to the fight against global warming but also to fight poverty and conserve biodiversity.

This Sierra Gorda carbon sequestration project, developed with the assistance of Woodrising Consulting, Inc., sequesters carbon by reforesting land previously converted to agricultural and livestock uses in the Sierra Gorda Biosphere Reserve and its area of influence in the state of San Luis Potosí. Participants include private and communal landowners as well as landholders when there is no dispute regarding possession (as indicated by a record of possession obtained from the local municipal authority).

As currently structured, the project includes payments for carbon already captured as well as for the sequestration of carbon thirty years into the future. Emission reduction credits are calculated every five years following verification. Bosque Sustentable retains 20 percent of the emission reductions as a form of self-insurance.

The sale to the UN Foundation was the first of several carbon transactions that have now been achieved in the Sierra Gorda. Our experience shows that the voluntary carbon market has the potential to play an important role in sustainable development efforts around the world. To achieve this potential, however, the development of rigid, Kyoto-like standards for the voluntary market must be avoided. Instead, flexible but reliable criteria should be used to meet the needs of areas of poverty.

Contributors

FRÉDÉRIC ACHARD
European Commission Joint Research Centre, Italy

RICARDO BAYON
EKO Asset Management Partners, United States

DAVID NEIL BIRD
Joanneum Research, Austria

DAVID BRAND
New Forests, Australia

SANDRA BROWN
Winrock International, United States

EDMUNDO CLARO
Recursos e Investigación para el Desarrollo Sustentable, Chile

STÉPHANE COUTURE
Institut National de la Recherche Agronomique, France

PAUL DETTMANN
Greenhouse Balanced, Australia

MICHAEL DUTSCHKE
BioCarbon Consult, Germany

JOHANNES EBELING
EcoSecurities, United Kingdom

MANUEL ESTRADA PORRUA
National Institute of Ecology of the Ministry of Environment and Natural Resources, Mexico

HUGH D. EVA
European Commission Joint Research Centre, Italy

Sandro Federici
Environment and Territory Protection Agency, Italy

Jan Fehse
EcoSecurities, United Kingdom

David Freestone
World Bank

Andrea García-Guerrero
Ministry of Environment, Colombia

Karen Gould
Baker & McKenzie, Australia

Giacomo Grassi
European Commission Joint Research Centre, Italy

Sandra Greiner
Climate Focus, The Netherlands

Katherine Hamilton
Ecosystem Marketplace, United States

Amanda Hawn
New Forests, United States

Dominique Hervé
Faculty of Law, Diego Portales University, Chile

Toby Janson-Smith
Conservation International, United States

Martina Jung
Ecofys, Germany

Rachael Katz
Pacific Forest Trust, United States

Cathleen Kelly
German Marshall Fund, United States

Eric Knight
Baker & McKenzie, Australia

Franck Lecocq
AgroParisTech, and Institut National de la Recherche Agronomique, France

Bruno Locatelli
CIRAD (Centre de coopération internationale en recherche agronomique pour le développement), CIFOR (Center for International Forestry Research), Indonesia

Jennifer McKnight
Nature Conservancy, United States

Marisa Meizlish
New Forests, United States

Monique Miller
Baker & McKenzie, Australia

Danilo Mollicone
University of Alcalà, Spain

Paulo Moutinho
Instituto de Pesquisa Ambiental da Amazônia, Brazil, and Woods Hole Research Center, United States

Sarah Woodhouse Murdock
Nature Conservancy, United States

Edward Nir
Forest Research Institute, Papua New Guinea

ROBERT O'SULLIVAN
Climate Focus, United States

SONAL PANDYA
Conservation International, United States

MICHELLE PASSERO
Pacific Forest Trust, United States

TIMOTHY PEARSON
Winrock International, United States

LUCIO PEDRONI
Centro Agronómico Tropical de Investigación y Enseñanza, Costa Rica

LAURA LEDWITH PENNYPACKER
Conservation International, United States

ROSIMEIRY PORTELA
Conservation International, United States

JEANNICQ RANDRIANARISOA
Conservation International, United States

TONY RINAUDO
World Vision, Australia

CARMENZA ROBLEDO
Swiss Foundation for Development and International Cooperation, Switzerland

DAVID PATRICK ROSS
Sierra Gorda Biosphere Reserve, Mexico

ZENIA SALINAS
Centro Agronómico Tropical de Investigación y Enseñanza, Costa Rica

BERNHARD SCHLAMADINGER
TerraCarbon LLC, Austria

SEBASTIAN M. SCHOLZ
World Bank

ERNST-DETLEF SCHULZE
Max Planck Institute for Biogeochemistry, Germany

STEPHAN SCHWARTZMAN
Environmental Defense Fund, United States

JÖRG SEIFERT-GRANZIN
Fundación Amigos de la Naturaleza, Bolivia

SCOTT SETTELMYER
TerraCarbon LLC, United States

REBECCA SKEELE
Nature Conservancy, United States

HANS-JÜRGEN STIBIG
European Commission Joint Research Centre, Italy

CHARLOTTE STRECK
Climate Focus, The Netherlands

RICHARD TARASOFSKY
Trade Law Bureau, Canada

RICHARD TIPPER
Edinburgh Centre for Carbon Management, United Kingdom

Patricia Tobón
Swiss Foundation for Development and International Cooperation, Switzerland

Assefa Tofu
World Vision, Ethiopia

Eveline Trines
Treenes Consult, The Netherlands

Ben Vitale
Conservation International, United States

Sarah Walker
Winrock International, United States

Laurie Wayburn
Pacific Forest Trust, United States

Sean Weaver
Victoria University and Carbon Partnership, New Zealand

Kelly J. Wendland
Conservation International and University of Wisconsin–Madison, United States

Martijn Wilder
Baker & McKenzie, Australia

Index